Agricultural Commercialization, Economic Development, and Nutrition

**Other Books Published in Cooperation with the
International Food Policy Research Institute**

Agricultural Change and Rural Poverty: Variations on a Theme by Dharm Narain
Edited by John W. Mellor and Gunvant M. Desai

Crop Insurance for Agricultural Development: Issues and Experience
Edited by Peter B. R. Hazell, Carlos Pomareda, and Alberto Valdés

Accelerating Food Production in Sub-Saharan Africa
Edited by John W. Mellor, Christopher L. Delgado, and Malcolm J. Blackie

Agricultural Price Policy for Developing Countries
Edited by John W. Mellor and Raisuddin Ahmed

Food Subsidies in Developing Countries: Costs, Benefits, and Policy Options
Edited by Per Pinstrup-Andersen

Variability in Grain Yields: Implications for Agricultural Research and Policy in Developing Countries
Edited by Jock R. Anderson and Peter B. R. Hazell

Seasonal Variability in Third World Agriculture: The Consequences for Food Security
Edited by David E. Sahn

The Green Revolution Reconsidered: The Impact of High-Yielding Rice Varieties in South India
By Peter B. R. Hazell and C. Ramasamy

The Political Economy of Food and Nutrition Policies
Edited by Per Pinstrup-Andersen

Agricultural Commercialization, Economic Development, and Nutrition

EDITED BY JOACHIM VON BRAUN AND EILEEN KENNEDY

Published for the International Food Policy Research Institute

The Johns Hopkins University Press
Baltimore and London

© 1994 International Food Policy Research Institute
All rights reserved
Printed in the United States of America on acid-free paper

The Johns Hopkins University Press
2715 North Charles Street
Baltimore, Maryland 21218-4319
The Johns Hopkins Press Ltd., London

A catalog record for this book is available from the British Library.

Library of Congress Cataloging-in-Publication Data

Agricultural commercialization, economic development, and nutrition / Joachim von Braun, Eileen Kennedy, editors.
    p. cm.
    Includes bibliographical references.
    ISBN 0-8018-4759-1 (hc : acid-free paper)
    1. Produce trade—Developing countries—Case studies. 2. Food supply—Developing countries—Case studies. 3. Cash crops—Developing countries—Case studies. 4. Food crops—Developing countries—Case studies. 5. Poor—Developing countries—Nutrition—Case studies. 6. Developing countries—Economic conditions—Case studies. I. Von Braun, Joachim, 1950- . II. Kennedy, Eileen T., 1947- . III. International Food Policy Research Institute.
HD9018.D44A363 1994   93-30256
338.1′91724—dc20   CIP

Contents

List of Tables and Figures xi

Acknowledgments xix

PART I Introduction

1 Introduction and Overview 3
   JOACHIM VON BRAUN

PART II Conceptual Framework

2 Conceptual Framework 11
   JOACHIM VON BRAUN, HOWARTH BOUIS, AND EILEEN KENNEDY

PART III Household-Level Effects: A Synthesis

3 Production, Employment, and Income Effects of Commercialization of Agriculture 37
   JOACHIM VON BRAUN

4 Consumption Effects of Commercialization of Agriculture 65
   HOWARTH BOUIS

5 Health and Nutrition Effects of Commercialization of Agriculture 79
   EILEEN KENNEDY

PART IV The Driving Forces of Commercialization

6 Commercialization of Agriculture and Food Security: Development Strategy and Trade Policy Issues 103
   NURUL ISLAM

7 China's Experience with Market Reform for Commercialization of Agriculture in Poor Areas 119
TONG ZHONG, SCOTT ROZELLE, BRUCE STONE, JIANG DEHUA, CHEN JIYUAN, AND XU ZHIKANG

8 Investment in Rural Infrastructure: Critical for Commercialization in Bangladesh 141
RAISUDDIN AHMED

9 Agricultural Processing Enterprises: Development Potentials and Links to the Smallholder 153
JOHN C. ABBOTT

10 Contract Farming and Commercialization of Agriculture in Developing Countries 166
DAVID GLOVER

11 Income and Employment Generation from Agricultural Processing and Marketing at the Village Level: A Study in Upland Java, Indonesia 176
TOSHIHIKO KAWAGOE

PART V The Diverse Experience

12 Nontraditional Vegetable Crops and Food Security among Smallholder Farmers in Guatemala 189
JOACHIM VON BRAUN AND MAARTEN D.C. IMMINK

13 The Nutrition Effects of Sugarcane Cropping in a Southern Philippine Province 204
HOWARTH BOUIS AND LAWRENCE J. HADDAD

14 The Effect of a Short-Lived Plantation on Income, Consumption, and Nutrition: An Example from Papua New Guinea 218
JOHN R. McCOMB, M. P. FINLAYSON, J. BRIAN HARDAKER, AND PETER F. HEYWOOD

15 Why Should It Matter What Commodity Is the Source of Agricultural Profits? Dairy Development in India 239
HAROLD ALDERMAN

16 Effects of Sugarcane Production in Southwestern Kenya on Income and Nutrition 252
EILEEN KENNEDY

17 Commercialization of Rice and Nutrition: A Case from West Kenya 264
RUDO NIEMEIJER AND JAN HOORWEG

18 The Triple Role of Potatoes as a Source of Cash, Food, and Employment: Effects on Nutritional Improvement in Rwanda   276
JÜRGEN BLANKEN, JOACHIM VON BRAUN, AND HARTWIG DE HAEN

19 Maize in Zambia: Effects of Technological Change on Food Consumption and Nutrition   295
SHUBH K. KUMAR AND CATHERINE SIANDWAZI

20 Tobacco Cultivation, Food Production, and Nutrition among Smallholders in Malawi   309
PAULINE E. PETERS AND M. GUILLERMO HERRERA

21 Smallholder Tree Crops in Sierra Leone: Impacts on Food Consumption and Nutrition   328
FRIEDERIKE BELLIN

22 Nutritional Effects of Commercialization of a Woman's Crop: Irrigated Rice in The Gambia   343
JOACHIM VON BRAUN, KEN B. JOHM, AND DETLEV PUETZ

PART VI Conclusions

23 Conclusions for Agricultural Commercialization Policy   365
JOACHIM VON BRAUN AND EILEEN KENNEDY

References and Bibliography   377

Contributors   397

Index   401

Tables and Figures

Tables

2.1 Patterns of time allocation effects under different exogenous changes   19
2.2 Assumed production function per plot   31
2.3 Assumed production function of domestic chores   31
2.4 Farming and household chore equilibrium (Stage 1)   32
2.5 Allocation of time for improved profitability (Stage 2)   32
2.6 Labor specialization and income maximization (Stage 3)   33
3.1 Overview for case study settings of commercialization features and production, employment, and income effects   39
3.2 Regression analysis of effects of income on nutrition of children for six study areas undergoing commercialization   44
3.3 Effect on children's nutritional status (weight-for-age) of a 10 percent increase in income of the poor (at US$100 per capita)   46
3.4 Average farm size and food production in study areas   48
3.5 Change in cropping pattern with adoption of new cash crops and crops under new crop technology (averages for the middle tercile farm-size groups)   49
3.6 Degree of participation of scheme participants in commercialization schemes by farm-size terciles   50
3.7 Staple food production per capita with new cash crops and crops under new crop technology (averages for the middle tercile farm-size groups)   51
3.8 Yields of major cereal crops of farms participating and not participating in commercialization schemes   51
3.9 Net returns to land and family labor (gross margins) of new cash crops under new technology and of staple foods (subsistence crops), in 1984–85 U.S. dollars   53

xii *Tables and Figures*

3.10 Change in women's labor use when agriculture is more commercialized 55
3.11 Marketed surplus (gross) of staple foods when agriculture becomes more commercialized 56
4.1 Overview for case study settings of food expenditure and consumption effects 67
4.2 Expenditure elasticities and budget shares for selected nonfood items 70
4.3 Calories purchased per U.S. dollar 73
4.4 Selected determinants of household demand for calories 75
4.5 Sources of food acquisition (in value terms) 75
4.6 Household-level calorie adequacy ratios, by total expenditure tercile 76
5.1 Overview for case study settings of health and nutrition effects 80
5.2 Percentage of preschoolers in participant and nonparticipant households who were ill 82
5.3 Percentage of time that preschoolers are ill with any illness and with diarrhea in particular, participant and nonparticipant households 83
5.4 Age at which infants are totally weaned, participant and nonparticipant households 84
5.5 Prevalence of stunting, wasting, and malnutrition among preschoolers, participant and nonparticipant households 85
5.6 Selected coefficients for relationship between program participation, socioeconomic variables, and preschoolers' morbidity (all preschoolers) 86
5.7 Selected coefficients for preschooler growth model for height-for-age 88
5.8 Selected coefficients for preschooler growth model for weight-for-age 89
5.9 Selected coefficients for preschooler growth model for weight-for-height 90
5.10 Women's body mass index (BMI) and percent time ill, by income per capita tercile, participants and nonparticipants 93
5.11 Time ill, caloric intake, and nutritional status of preschoolers by gender of head of household and cash crop production, Kenya and Malawi 95
6.1 Distribution of total crop output between basic food crops and cash crops, developing country regions, 1961–63 and 1983–85 104
6.2 Recent trends in terms of trade of different commodity groups, 1948–86 and 1962–86 107

6.3 Changes in farmgate prices in selected developing countries   110
6.4 Indices of nominal rates of protection, 1969–71 = 100   111
6.5 Instability of world market prices, 1962–87   112
6.6 Distribution of crop scientists working on different crops in different regions, and importance of respective crop groups, 1980–85   116
6.7 Percentage change in yield value per hectare of different crop groups in developing-country regions, 1962–70 and 1970–84   117
7.1 Marketing rates for farm products, 1978, 1984, and 1987   135
7.2 Foodgrain production, procurement, and resale, by income strata, in poor counties, 1983 and 1987   137
8.1 Differences in the use of modern technology and price levels between infrastructurally developed and underdeveloped villages, Bangladesh, 1982   145
8.2 Marketing of paddy by landownership group in infrastructurally developed and underdeveloped villages, Bangladesh, 1982   147
8.3 The effect of infrastructure on employment, Bangladesh, 1982   148
8.4 Effect of infrastructural development on consumption pattern, Bangladesh, 1982   150
8.5 Participation in income-earning activities in infrastructurally developed and underdeveloped villages, Bangladesh, 1982
8.6 Percent of women practicing family planning in electrified and nonelectrified households and villages, Bangladesh, 1985   151
9.1 Case studies of agricultural processing enterprises   156
10.1 Network of case studies in East and Southern Africa, by crop   169
10.2 Contract farming linkages to food security and policy implications   174
11.1 Value added and employment indicators for selected crops per hectare of intercropped harvested area, Garut, Indonesia, 1986 (average of 18 sample farms)   181
11.2 Prices and marketing margins of selected agricultural commodities at various points in the marketing chain   183
11.3 Income and employment generation from production, processing, and marketing of selected upland crops, per hectare of harvested area   185
12.1 Production of subsistence maize and use for household consumption in farm households, by farm size, Guatemala, 1985   194

xiv *Tables and Figures*

12.2 Determinants of budget shares to food in total sample, cooperative nonmember and member households, Guatemala, 1985   197
12.3 Determinants of availability of calories to households in total sample and the lowest two quartiles of cooperative members and nonmembers, Guatemala, 1985   199
12.4 Multivariate analysis on effects of income and income source and composition on the level of nutritional status of children aged 6–120 months   202
13.1 Income, expenditures, and calorie availability and intake, by income and expenditure quintile and crop-tenancy group, Bukidnon, the Philippines, 1984–85   207
13.2 Regression results for preschooler calorie intakes as a function of household calorie intakes   210
13.3 Z-scores for height-for-age, weight-for-age, and weight-for-height for preschool children, by age and expenditure quintile, Bukidnon, the Philippines, 1984–85   212
13.4 Regression results for relationships between preschoolers' calorie intake and weight-for-height   214
14.1 Selected demographic features and crop areas of nonwage and wage households, Papua New Guinea, 1987–88   224
14.2 Household income and expenditure of nonwage and wage households, Papua New Guinea, 1987–88   225
14.3 Mean cash and subsistence income received by males and females over ten years of age, by source of income, Papua New Guinea, 1987–88   227
14.4 Expenditure shares of foods and nonfoods of nonwage and wage households, Papua New Guinea, 1987–88   228
14.5 Regression analysis of energy availability   229
14.6 Frequency of diarrhea in children of nonwage and wage households, by income tercile, Papua New Guinea, 1987–88   230
14.7 Means of three nutritional indices and percentages of children below cut-off points for each index for Karimui children, by nonwage and wage households, Papua New Guinea, August 1987, February 1988, and August 1988   231
14.8 Regression models for change in percentage of standard for weight-for-age, height-for-age, and weight-for-height per month of children in sample households   234
14.9 Regression model for attained levels of anthropometric standards of children   236
15.1 Changes in milk, calorie, and protein consumption attributed

Tables and Figures xv

	to dairy cooperatives in Karnataka, India, January 1983–April 1984 246
15.2	Difference in expenditures and earnings sources in cooperative and control villages, Karnataka, India, 1983–84 248
16.1	Mean annual income per capita per year, by source and activity group, study 1 compared to study 2, cohort group, South Nyanza, Kenya, 1984–85 and 1985–87 254
16.2	Characteristics of farming patterns for agricultural households, cohort group, South Nyanza, Kenya, 1984 and 1986 255
16.3	Comparison of household caloric intake for cohort and total sample, 1984–85 and 1985–86, South Nyanza, Kenya 256
16.4	Household consumption function, 1985–87 257
16.5	Total time ill and time ill with diarrhea of preschoolers and women, stratified by income quartiles, South Nyanza, Kenya, 1985–87 258
16.6	Z-scores for children, in 1984–85 study and 1985–86 study, South Nyanza, Kenya 259
16.7	Regressions of preschooler Z-scores for height-for-age, weight-for-age, and weight-for-height, 1985–87 study 261
17.1	Resource base of sample rice farmers in West Kenya, 1984 270
17.2	Estimates of annual income of four groups of farmers in West Kenya, 1984 271
17.3	Nutritional conditions among farmers in West Kenya, by degree of income diversification, 1984 272
18.1	Land use of households with and without additional Gishwati land, Rwanda, 1985–86 279
18.2	Average land and labor productivity of different cropping systems, Rwanda, 1985–86 281
18.3	Labor allocation in households with and without additional Gishwati land, by gender, Rwanda, 1985–86 283
18.4	Overall degree of subsistence orientation in agricultural production and share of marketed surplus by crop, Rwanda, 1985–86 285
18.5	Increase in farm size and calorie production in households with additional Gishwati land, by quartile of calorie production per consumer-equivalent, Rwanda, 1985–86 286
18.6	Estimation results for relationship between calorie consumption and subsistence orientation 288
18.7	Multivariate analysis of determinants of nutritional status of children, 6–72 months 290
19.1	Characteristics of farms, by size, Eastern Province, Zambia, 1986 298

19.2 Daily per capita consumption of nutrients, by high- and low-adoption areas in plateau region, Eastern Province, Zambia, annual sample average, 1986   300
19.3 Child malnutrition, Eastern Province, Zambia, February 1986   303
19.4 Child malnutrition, by adoption of hybrid maize, Eastern Province, Zambia, February 1986   305
20.1 Per capita income for different farm households, by type of income, Malawi, 1986–87   311
20.2 Subsistence ratios for main sample groups, Malawi, 1986–87   315
20.3 Total maize supply available to sample households, Malawi, 1986–87   317
20.4 Absolute value of per capita nonfood purchases, food purchases, and home consumption, and their share of total expenditure (including home consumption), Malawi, 1986–87   322
20.5 Correlations of child size (in terms of Z-scores) with household variables   323
20.6 Correlations of child size with household resource variables   324
21.1 Household resources of old plantation farmers, subsistence farmers, and new adopters, Sierra Leone, 1988–89   332
21.2 Labor allocation to food crop cultivation, tree crops, and other work, by farm household and gender, Sierra Leone, 1988–89   333
21.3 Annual expenditure and expenditure share of farm households, by expenditure tercile, Sierra Leone, 1988–89   338
21.4 Per capita calorie and protein intake and anthropometric indicators of children aged less than three years, by farm household type and expenditure tercile, Sierra Leone, 1988–89   339
21.5 Determinants of calorie consumption   340
21.6 Determinants of weight-for-age for children   341
22.1 Food energy consumption and nutritional status of children, by household type per capita income quartile and season, The Gambia, 1985–86   347
22.2 Yields, distribution of rice fields, labor allocation, and marketed surplus, by type of rice technology, The Gambia, 1984–85   349
22.3 Determinants of access of individual women to fields in the new rice scheme (PROBIT estimate)   351
22.4 Yield fluctuations in irrigated and upland crops, Jahally-Pacharr project area and neighboring villages, The Gambia, 1984–87   353
22.5 Marginal and average labor productivity of major crops in the wet season, The Gambia, 1985–87   354

22.6 Calorie consumption by quartile of cereal availability from own production and by quartile of total expenditure, The Gambia, wet season 1985–86   355
22.7 Calorie consumption model for the wet season, 1985   357
22.8 Determinants of children's nutritional status, 1984–85   361

Figures

2.1 Commercialization at the household level: Determinants and consequences for income, consumption, and nutrition   13
2.2 Allocation of household time between home goods production, farming for the market, wage earning, and leisure   18
2.3 Resource allocation to market versus subsistence production under risk   21
2.4 Commercialization of agriculture: Potential pathways through which women's and preschoolers' health and nutritional status may be influenced   30
3.1 Local terms of trade of sugarcane/maize in the Kenyan and Philippine study areas, 1974–86   57
3.2 Income and income sources of scheme participants and nonparticipants   60
7.1 Poor regions of China and their dominant staple food crops   130
7.2 Distribution of 664 poor counties by per capita foodgrain production, 1983 and 1987   132
7.3 Distribution of 664 poor counties by real per capita income, 1983 and 1987   138
11.1 Typical marketing and processing channels for farm products at the village level   180
14.1 Cardamom production and wages at the Karimui Spice Company, 1981–88   221
14.2 Total wages paid and daily wage rates at Karimui Spice Company, August 1987 to July 1988   221
19.1 Daily per capita consumption of calories, by high- and low-adoption areas in Plateau Region, Eastern Province, bimonthly sample average, 1986   302
21.1 Seasonal work time profiles by type of work and gender   334

Acknowledgments

The research contained in this volume represents the collaborative efforts of many individuals and institutions throughout the world. First and foremost, this volume would never have materialized had it not been for the time and effort of the hundreds of households that participated in the individual case studies. These unnamed households are the main reason why our understanding of the effects of commercialization of agriculture has been advanced.

Each study included in this synthesis benefited from the support of national institutions and the advice of individual researchers and policy advisers. We are particularly grateful for these inputs to Jorge Alarcón, Joel Asentista, Anne M. S. Coles, Joanne Csete, Tulio Garcia, Yvan de Jaeger, Priscilla Kariuki, Abbas Kasseba, Sambou Kinteh, Michael Latham, Antonio Ledesma, Azucena Limbo, Francis Madigan, Roberto Montalvan, Bernard Mutwewingabo, Frank Noy, Dismas Nyongesa, Festes Omoro, John Omuse, Ruth Oniang'o, Mireya Palmieri, Serge Rwamasirabo, Luke Wasonga, and Lourdes Wong.

Per Pinstrup-Andersen was instrumental in helping to conceptualize and initiate the studies. Other current or former IFPRI researchers, including Bruce Cogill, Elizabeth Jacinto, Ellen Payongayong, Deborah Rubin, Graciela Wiegand, and Yisehac Yohannes are thanked. Rajul Pandya-Lorch significantly contributed to the editing and streamlining of the entire volume. Jay Willis's word processing assistance is greatly appreciated.

Earlier drafts or portions of this manuscript benefited from reviews by Mohamed Arift, Jaque Arnault, Joan Atherton, Romulo Caballeros, Frances Davidson, Rodolfo Estradra, Rodolfo Florentino, Govert Gipbers, Jorge Gonsalez del Walle, Ken B. Johm, James Kiele, Michael Lipton, Richard Longhurst, Paul Lunven, Melanie Marlett, Reynaldo Martorell, James Otieno, Nancy Pielemeier, Carlos de Leon Prera, Ammar Siamwalla, Robert Tripp, and Chen Xiwen.

A number of donor agencies were generous in their support of this research, including the U.S. Agency for International Development, the Federal German Ministry for Economic Cooperation and Development, the International Fund for Agricultural Development, the Rockefeller Foundation, and the Australian Centre for International Agricultural Research.

While we have tried to be comprehensive in our acknowledgments, important contributors to the thoughts and findings included in this volume may have been inadvertently omitted in the above listing. Our special thanks are to them. Any errors or omissions are solely the responsibility of the authors.

Joachim von Braun
Eileen Kennedy

PART I

Introduction

1 Introduction and Overview

JOACHIM VON BRAUN

Why should there be a book about the commercialization of subsistence agriculture, economic development, and nutrition? There are two compelling reasons. First, concerns and suspicions about adverse effects on the poor of commercialization of subsistence agriculture persist and influence policy of developing countries and of donor agencies. Second, in rural areas of low-income countries, nutritional welfare is determined by many complex factors whose relationships to agricultural commercialization and economic development need to be traced in order to design optimal rural growth policies that benefit the poor. In view of the challenges of rapid urbanization and the chances of commercialization, the question is not *if* subsistence agriculture should be overcome, but *how*. Thus, the purpose of this book is to clarify concepts, add comprehensive factual information, and assist policy and program analysts in identifying potentials and risks of promoting commercialization of agriculture for poverty alleviation.

Specialization and the development of markets and trade that characterize commercialization are fundamental to economic growth.[1] The principal advantages of market-oriented policies and the powerful forces of trade for development are unquestionable. However, the risks of policy and market failures, deficiencies in knowledge and information of actors in production and markets at all levels, and household-level complexities and intrahousehold conflicts are real, too, and need to be recognized as determinants of inefficiencies and inequities. Therefore, it should not be taken for granted that the transformation of traditional agriculture progresses efficiently, not to mention equitably, even if the point of departure — subsistence agriculture — happens to be in a state of "poor but efficient" (Schultz 1983). Subsistence production for home

---

1. An overview on development of cash cropping and food crop growth in the context of economic development is given in von Braun and Kennedy (1986).

consumption is chosen by farmers because it is subjectively the best option, given all constraints. In a global sense, however, it is one of the largest enduring misallocations of human and natural resources, and, due to population pressure and natural resource constraints, it is becoming less and less viable. Today, about 440 million farmers in developing countries still practice subsistence production to a significant extent.[2] A large proportion of land resources in low-income countries is devoted to subsistence cropping (von Braun and Kennedy 1986), yet land constraints, ecological problems (especially in marginal areas), and rapid urbanization call for change.

One might expect that the process of commercialization, by raising incomes, actually improves welfare, food security, and nutritional status that could otherwise have been worse. But as commercialization takes place, how are higher average incomes distributed among various economic and social groups? Does higher household income necessarily lead to improved food consumption and better nutrition for all household members? Could a different approach to agricultural development—for instance, one of regional or village or household food self-sufficiency—more efficiently alleviate the particular problem of undernutrition and still meet the objectives of economic growth and higher incomes?

It is widely argued in a large body of literature that commercialization of agriculture mainly has negative effects on the welfare of the poor. The related widespread conclusion—commercialization of agriculture is bad for nutrition—probably emerged from a mix of historical, real, ideological, and methodological factors: (1) adverse effects that resulted from coerced cash crop production that constrained the capacity of smallholders to cope with risks under some colonial production schemes; (2) noted exploitation of smallholders under monopsonistic conditions of projects; (3) general suspicion that commercialization leads to adverse "capitalist" production and market relationships (whereas subsistence agriculture was idealized as providing food sufficiency); and (4) noncomprehensive studies and anecdotal evidence. A number of studies in the past have suggested that the introduction or expansion of cash crop production may adversely affect food security and nutritional status. A review of related research (von Braun and Kennedy 1986; Longhurst 1988) found that many of the studies from which these generalizations were extrapolated (for example, Lappé and Collins 1977) were conceptually flawed. Often, comparative studies disregarded potentially confounding factors but just compared nutrition with and without cash

---

2. This rough estimate (about 300 million in Asia, 110 million in Africa, and 30 million in Latin America and the Caribbean) is arrived at on the basis of subsistence crop production and total agricultural labor force estimates by country.

crops (Hernandez et al. 1974; Lambert 1979) or were based on very small, potentially biased samples (Gross and Underwood 1971).

It may be impossible to answer the above-mentioned questions definitively because there are so many possible policy variations within the competing paradigms of specialization and self-sufficiency, because of the considerable variation in economic and social conditions across countries and regions, and, finally, because inevitably there are winners and losers, at least relatively, in any process of change. However, it is possible to study the *process* of commercialization in specific contexts and to identify key factors that lead to either beneficial or detrimental outcomes in income, consumption, and nutrition. This will provide guidance for policy formulation in this area. In designing and implementing future projects and policies for commercialization, then, the goal would be for policymakers to find ways to enhance the beneficial outcomes, while minimizing the potentially harmful ones.

The objectives of the book are threefold:

1. identify the *driving forces* of increasing commercialization of subsistence agriculture and assess their *effects* on household real incomes, food consumption, expenditures on nonfood goods and services, and nutritional and health status in different settings;
2. describe the *process* by which increasing commercialization affects income, consumption, and nutritional status, in order to identify the most important elements of this process as well as to estimate how each of these elements is influenced by the transition in different settings;
3. evaluate *policies and programs* to cope with possible adverse impacts and to foster beneficial ones on income, consumption, and nutrition that arise during the commercialization transition.

It is not just short-term policies aimed at making the best of situations with deficient markets that are essential to agricultural commercialization, but also long-term policies for creating and developing agricultural markets where they are thin and particularly risky. Macroeconomic policies, including trade and exchange rate policies, play a fundamental role in commercialization of agriculture. The critical role of incentives is now well understood (Krueger, Schiff, and Valdés 1988; Mellor and Ahmed 1988). A synergistic relationship exists between agricultural commercialization and technological change (Binswanger and von Braun 1991). The potential lead role of technological change in food crops for agricultural and rural growth is also now well understood (Hazell and Ramasamy 1991; Mellor 1986). What is less well known, it seems, is the potential for agricultural commercialization to take a lead position in agricultural growth stimulation too. The speed at which commercialization of agriculture is occurring in some countries while in others there is stagnation or even a reversal to subsistence agriculture on a large scale, combined with the challenge of economic restructuring and shift from planned to market-oriented economies in many low-income countries, urgently demands enhanced knowledge to stimulate and supplement the process of agricultural commercialization, and, where needed, to guide it in the interests of the poor.

This book does not provide a blueprint on how to commercialize and diversify agriculture. Instead, it attempts to let real-world experience with agricultural commercialization and the analysis of experience speak for itself. Nevertheless, as will be shown, policy has a key role to play in shaping the successes and failures of commercialization from a growth and poverty perspective. The reader most interested in lessons for policies and programs may find the studies on the driving forces of commercialization (Part IV) as well as the case studies (Part V) of greater relevance.

To address the above issues, the International Food Policy Research Institute (IFPRI) in collaboration with other institutions conducted comprehensive microlevel studies in five countries—The Gambia, Guatemala, Kenya, the Philippines, and Rwanda—at carefully selected program or project sites where farm households had recently switched from semisubsistence staple food production with low levels of external inputs to production of more crops for sale in the market or to production with new inputs and technology. The study sites were selected especially for their recent experiences with the transition to commercialization, in order to capture potential adjustment problems and to identify appropriate corrective measures. Comparable studies undertaken by IFPRI and other research institutions in India, Kenya, Malawi, Papua New Guinea, Sierra Leone, and Zambia that to some extent followed a similar research protocol are also included in this synthesis. These case descriptions, based on original data collections, are telling examples of diverse experiences. They may prove useful to policy analysts and planners concerned with choices and design of commercialization policies and programs, as well as to advanced students in applied development economics, agriculture, and nutrition.

Commercialization of agriculture may evolve in the context of dual structures. For instance, large-scale plantation agriculture, with wage labor derived from subsistence-oriented smallholder farms, can coexist with smallholder farms. While the role of agribusinesses and large-scale farms in commercialization has been much debated (Dinham and Hines 1984), this book concentrates on smallholder commercialization, which is much more widespread. Yet, smallholder commercialization is frequently tied through contract farming with large business operations (see chapter 10), as, for instance, in the cases of the sugarcane outgrowers' schemes in Kenya and the Philippines, the tea schemes in Rwanda, and export vegetable production in Guatemala.

At first glance, this book appears to overemphasize household and microlevel features of the commercialization process. Yet, most of the contentious issues and valid concerns regarding poverty alleviation and nutrition effects are at those levels, which are also where the biggest gaps in knowledge have to be filled. Thus, the book begins with a conceptual chapter (Part II) and follows it with syntheses of the production, employment, income, consumption, nutrition, and health effects of commercialization (Part III). While providing a general overview, these syntheses in no way capture the unique features of the case studies in Part V, each of which traces crop-, country-, and organization-specific commercialization experiences through the production-nutrition chain.

As outlined in more detail in the conceptual framework (Part II), commercialization of agriculture is viewed as the result of a set of interacting driving forces composed of demographic change, technological change and market creation, infrastructure development, and macroeconomic and trade policy.

Part IV addresses these driving forces. First, the key roles of development strategy and trade policy for commercialization are discussed in a global context (chapter 6) and in the context of the largest commercialization experiment of the 1980s, China's domestic trade and agricultural market reform (chapter 7).

Second, the role of infrastructure for agricultural commercialization is analyzed for the case of Bangladesh (chapter 8). Third, it is clarified that the nature of the backward linkages from agricultural processing to farms is very much a function of project design and legal arrangements in the more integrated modernizing agriculture. There is a lot of scope for project policy and legal arrangements to foster success and prevent adverse effects of commercialization in the smallholder sector (chapters 9 and 10). However, in a rapid rural growth environment, such as Indonesia in the 1980s, the broad-based agricultural commercialization process—in contrast to a narrow, isolated "project approach" to commercialization—reveals impressive employment and welfare effects for the poor (chapter 11). This Indonesian case can be viewed as a more mesolevel analysis and complement to the China case. Both Indonesia and China represent cases where agricultural commercialization plays a critical role for growth, employment expansion, and social welfare, including dramatic improvements in rural nutrition. We point at these here in order to stress that the major improvement in nutrition observed in the fast-growing Southeast Asian economies in the 1970s and 1980s is closely related to the developmental effects of commercialization of agriculture—an insight that may not be gained fully from the case study material in Part V, which has a household and project focus.

The findings of this book lend ample support to an approach that aims to capture the employment and income gains from commercialization that are so beneficial for the poor, and to combine this approach with policy awareness to prevent market failures in factor and insurance markets as well as with complementary public action to provide the public goods for nutritional improvement, that is, in the health and sanitation area.

If this book has contributed to clarifying and correcting what has for a long time prevailed as conventional wisdom regarding commercialization-poverty-nutrition linkages and, more important, if it has contributed to assisting policy analysts and their policymaker clientele in their considerations for related policies, then it was worth the valuable time spent by so many rural households in building the empirical base of the studies assembled here.

PART II

Conceptual Framework

2 Conceptual Framework

JOACHIM VON BRAUN
HOWARTH BOUIS
EILEEN KENNEDY

Definitions

In this chapter, the basic theoretical relationships and definitional issues related to the commercialization of agriculture are described. Simply speaking, cash crops can be defined as crops for sale. The listing of typical agricultural processing enterprises in chapter 9 gives a rough overview of cash crops (also, see von Braun and Kennedy 1986). Yet, commercialization of agriculture as a process and a characteristic of agricultural change is more than whether or not a cash crop is present to a certain extent in a production system. Commercialization of subsistence agriculture can take many different forms. Commercialization can occur *on the output side* of production with increased marketed surplus, but it can also occur *on the input side* with increased use of purchased inputs. Commercialization is not restricted to just cash crops: The so-called traditional food crops are frequently marketed to a considerable extent, and the so-called cash crops are retained, to a substantial extent, on the farm for home consumption, as, for instance, groundnuts in West Africa. Also, increased commercialization is not necessarily identical with expansion of the cash economy when there exist considerable in-kind transactions and payments with food commodities for land use or laborers. Finally, commercialization of agriculture is not identical with commercialization of the rural economy. The deviation between these two processes becomes all the more obvious when off-farm nonagricultural employment already exists to a large extent in a certain setting.

At the household level we may thus specify forms of commercialization and integration into the cash economy from at least three different angles and measure the extent of their prevalence at the household level with the following ratios:

(1a) Commercialization of agriculture (output side) $= \dfrac{\text{Value of agricultural sales in markets}}{\text{Agricultural production value}}$

(1b) Commercialization of agriculture (input side) = $\dfrac{\text{Value of inputs acquired from market}}{\text{Agricultural production value}}$

(2) Commercialization of rural economy = $\dfrac{\text{Value of goods and services acquired through market transactions}}{\text{Total income}}$

(3) Degree of integration into the cash economy = $\dfrac{\text{Value of goods and services acquired by cash transactions}}{\text{Total income}}$

Specific characteristics of cash crops of a certain land may, under defined circumstances, imply certain household food security and nutritional effects (Longhurst 1988, 34).

Basic Relationships[1]

The effects of commercialization on income, consumption, and nutrition are mediated through complex relationships at household and intrahousehold levels. Generally speaking, the improvement of the status of a food-deficient and malnourished person has to come about by an improvement in the ability to acquire more food or better quality food, or both, hence, through the growth of income. An expected increase of production capacity and income motivates a household or individual household members to enter the exchange economy and become more commercialized. Thus, insofar as increased sale of produce, purchase of inputs, and off-farm employment occur on a voluntary basis, and insofar as the responsibilities and preferences within a household ensure sharing of gains, it can be expected that commercialization contributes to a household's food security. In other words, food consumption benefits are assured for all when markets do exist and intrahousehold conflicts do not. The relationship is more complex when it comes to the real world of rural households and thin and volatile rural markets, often characterized by structural imbalances and institutional constraints.

In spite of dynamic interdependencies of causes and effects, it may facilitate the analysis if exogenous factors that determine commercialization are separated from endogenous factors that tend to affect the influence of commercialization on income and nutrition. Figure 2.1 describes major relationships between both groups of factors.

As far as the exogenous determinants of commercialization (left-hand side of figure 2.1) are concerned, among the most important

1. The sections on basic relations and theoretical foundations draw on von Braun, de Haen, and Blanken 1991.

FIGURE 2.1 Commercialization at the household level: determinants and consequences for income, consumption, and nutrition

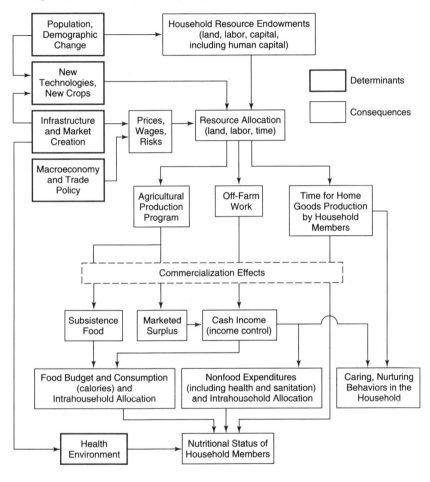

Source: Derived from von Braun, de Haen, and Blanken (1991).

driving forces are population change, availability of new technologies, infrastructure and market creation, and macroeconomic and trade policy. Some of these factors, which are briefly discussed below, may have more immediate effects on farmers' decisions to become more integrated in the market, whereas others may only have long-term effects. Figure 2.1 cannot capture the respective time subscripts.

Demographic change is certainly a key long-term determinant of commercialization. It may facilitate or impede commercialization, depending on the availability of resources. If an expansion of the cultivated area is still possible, and if the marginal labor productivity exceeds the marginal subsistence requirements, population growth may in fact enable an increase of the marketable surplus. However, this situation has certainly become rare. With no concurrent change in the preferences for a high degree of self-sufficiency in staple food (due to perceived food security risks) on the one hand, population growth might lead to a reduced volume of marketed surplus in relative or even absolute terms in regions with deficient market connections. On the other hand, an increased person-land ratio might lead to an increased demand for off-farm employment in order to generate cash income, of which a high proportion will be spent on food.

The availability of new technologies, such as improved seeds and agronomic practices, and investment in infrastructure and policies for market creation are key factors that facilitate the commercialization process. Increased commercialization can occur without technological change in agriculture, but technological change without increased commercialization seems unlikely because the increased use of purchased inputs and specialization are inherent elements of most technological innovations in agricultural production. Policies for the promotion of commercialization and technological change may focus on either one or—in a more complex, dynamic fashion—on both. Technological change implies increased total factor productivity. Policies that generate technological change focus on human capital improvement, research and extension, and related institution building. In order to have a sustainable effect on the food security of the poor, the income streams resulting from technological change must reach them, directly or indirectly, through employment expansion, returns to their resources, or favorable food-price effects. Commercialization implies increased market transactions for capturing the gains from specialization. Policies that foster commercialization focus on facilitating an open international and domestic trade environment, improving hard and soft infrastructure for opening up new market opportunities, and ensuring legal security.

Ideally, policies to speed up commercialization and technological change move jointly in a reinforcing way. Policies for increased commercialization facilitate the generation and diffusion of new production technology. The latter reinforce the gains from specialization. However, there exists much concern with the potential risks of commercialization for the food security of the poor. These potential risks are derived from a host of potential market failure and policy failure problems[2] relating not

---

2. Comprehensive reviews of the arguments are provided in von Braun and Kennedy (1986) and *IDS Bulletin* (1988).

only to food and cash crop markets, but to deficiencies in land and financial markets, and to absent insurance markets as well. Addressing these deficiencies promises high payoffs for policy-oriented research.

Commercialization can also be enforced by direct government action, namely, by various forms of compulsion related to the establishment of plantations, execution of certain management practices and input use, or forced procurement of produce. Examples of such aberrations are presented in Part V.

It is now well understood that nutritional improvement is often constrained by the health and sanitation environment, which is partly a function of related hard infrastructure (for example, water) and soft infrastructure (for example, health centers). These factors are endogenous to government policy and development but are treated as exogenous to our discussions of the household decision-making process.

The endogenous consequences of commercialization for consumption and nutrition are also indicated in figure 2.1. They relate to three different but linked types of decision making within the households. One affects the allocation of income for food and nonfood expenditures and how household members spend their time. It may be hypothesized that a reduced share or a reduced absolute volume of subsistence production will motivate a rise in the volume of purchased food and vice versa. The second decision level relates to how the available food budget is actually spent, that is, which types and which quantities of food are purchased and how these purchases are distributed intertemporally. The third decision level concerns how the available food and other consumption items are distributed among household members.

To understand how these decisions may be affected by the commercialization process, the other indirect consequences of commercialization, such as changes in the time allocation of men and women and in the control over household resources and cash income, need to be fully considered. For instance, when men's involvement in market production or off-farm work increases, women may have to spend more time in own-farm production and so may have less time for child care and home-based work and less control over their household resources. Since men and women and younger and older people have different preferences in the allocation of household income (for example, for health care and nutrition), commercialization may differently affect the welfare of various family members, depending on how work responsibilities and control over income within a household change.

Finally, caring or nurturing activities within the household may be critically important in influencing child health. This encompasses activities such as breast-feeding and weaning practices, child care, and other nurturing activities, all of which may be affected if the commercialization of agriculture increases or reduces demands on the time of household members, particularly that of women. Time allocation studies indicate that, on average, women in developing countries put in more hours per day in nonleisure activities than do men. Not only are women actively engaged in own-farm production and wage-earning activities, but a substantial amount of their day is devoted to home production activities such as food preparation, child care, cleaning, and water and fuelwood collection. These home production activities may be a key factor influencing child health. Few studies have looked at the link between the modernization of agriculture and caring. A number of the chapters in this book will fill some of these gaps.

Theoretical Foundations

The complexity of the relationships just described suggests that a comprehensive model of the rural household would be essential in deriving hypotheses about the process of transition from subsistence to more market integration.

Household and Intrahousehold Decisions

Since Tschajanow (1923) first developed a theory of subjective household equilibrium, many researchers have refined the model of the peasant household. According to Tschajanow, a peasant family does not try to maximize a monetary profit, but a subjective utility. Maximum utility is reached when the marginal drudgery of family labor in various activities is equated with the marginal goods and services gained from the labor input. Nakajima (1970, 1986), stimulated by Tschajanow, developed a set of much more sophisticated subjective equilibrium models that postulate the same basic behavioral rules, with and without exchange with the external labor market. Not only did Nakajima specify a more formal mathematical structure that made it possible to trace the consequences of external changes, such as variations in wages, prices, and productivities on household labor allocation, he also specified certain properties of a household's indifference curves with a lower limit of income ("minimum subsistence"), below which leisure has zero marginal utility, and an upper bound ("achievement standard of living"), above which income generated from further work has a marginal utility of zero. While Nakajima's models identify the specific factors that determine the decision of household members to be engaged in wage employment or to employ hired labor in the farm household, they unfortunately do not explicitly specify the factors that influence a household's allocation of resources between subsistence and market production. Implicitly, his subjective equilibrium models assume a fully commercialized farm where a price can be imputed to all commodities.

Modeling the commodity side of the transition process requires (1) introducing the distinction of subsistence and market production at the level of resource use (including labor), (2) specifying the underlying causal determinants, such as risk aversion, preferences for tasks, and habits, that may motivate a household to maintain a certain degree of self-sufficiency even at the cost of market income forgone, and (3) assigning a common nonmonetary utility index to nonmarketable household goods and services as well as market goods.

The specification of a household's utility function in nonmonetary terms is one of the strengths of the modern theory of household economics, originating from Becker (1965) and Lancaster (1966). Models based on this theory postulate that a household's utility function is directly specified by a set of goods and services acquired in the market or produced at home. These so-called Z-goods are produced using assets owned by the household, in combination with the time input of household members. Maximization of a household's utility subject to a full-income constraint is then equivalent to minimizing the costs of producing a set of Z-goods, including leisure.

Figure 2.2 portrays the basic structure of Evenson's (1978) model of the peasant household. The composite Z-good is measured along the vertical axis, whereas the horizontal axis measures the working time, with the remainder of the full-time capacity (OH) being leisure. Curve s traces the production function for home goods, and curve m describes the combined production function of the household where agricultural production is added on the home production function. The basic assumption is made that the composite Z-good can be produced at home or purchased in the market.[3] Purchased goods might not be identical but would be close substitutes for home-produced Z-goods. Thus, line d measures the opportunities in terms of Z-goods offered by the labor market. Its slope is defined as the wage rate divided by the goods price, indicating the purchasing power of the off-farm wage in terms of Z-goods (d' is the parallel to d). Finally, curve u shows the indifference curve in terms of Z-goods and leisure.

At equilibrium the household would have LH leisure time and LG Z-goods for consumption. It would spend OF units of time (and correspondingly household resources) for home goods production, FM units of

---

3. It is possible to include the part of the subsistence production from household's resources that cannot be used to produce market goods. This would normally include house and shelter, cooking facilities, and maybe a small home garden.

FIGURE 2.2 Allocation of household time between home goods production, farming for the market, wage earning, and leisure

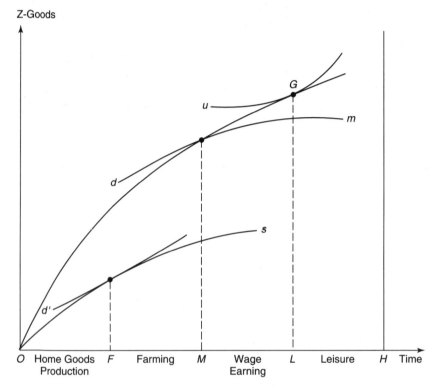

Source: von Braun, de Haen, and Blanken (1991).

time for farm production, and ML units of time for wage earning. Thus the model postulates principally the same equilibrium conditions as the Nakajima model (Nakajima 1970, 1986): the marginal productivities of time in various activities in and outside the household are equated to the off-farm wage rate. But in addition, the ratio of marginal utility of leisure over the marginal utility of Z-goods is set equal to the wage rate over the price of the Z-good. For the one-person household (as reported in figure 2.2; or for a household where everyone is similar in terms of productivities and preferences), the equilibrium condition may be stated as

$$\frac{\text{Wage}}{P_Z} = \frac{MU_{\text{LEISURE}}}{MU_Z} = \frac{MP_{\text{FARM}}}{P_Z} = \frac{MP_{\text{HOME}}}{P_Z}.$$

Under the assumption that leisure is a normal good, if income increases for any reason while the marginal cost of leisure remains constant (that is, a constant wage), then leisure will increase.

The matrix below (table 2.1) gives the patterns of time allocation effects under different exogenous changes, which may result from commercialization promotion.

Illuminating conclusions can be derived from this simple model:

- Increasing the wage rate raises the opportunity costs and hence motivates a reduction of the volume of home as well as farm production. It increases the incentive to seek wage employment and, depending on the position of the indifference curve, may also affect the overall time allocation between work and leisure.
- Increasing the price of the Z-good reduces leisure time (income goes down; leisure is assumed to be normal good), and more time is spent in the work force. The wage and marginal productivity in farm and household work remain equated if time does not change in the farm and household work.
- Increasing the productivity of farm work causes an upward shift of the overall production function. It motivates more on-farm work and reduced off-farm work. Time allocated to home production is not much affected.
- Increasing the productivity of home goods production will have a symmetrical effect, increasing time spent in the household and reducing off-farm work; leisure increases, and farm work remains largely the same.
- Increasing the family size will have complex implications for the household, depending on the effect on the labor force and the Z-goods requirements, respectively. The impact on the demand for Z-goods includes needs for additional food, child care, and other household goods and services. This may increase the family's prefer-

TABLE 2.1 Patterns of time allocation effects under different exogenous changes

Exogenous Change/Policy	Direction of Change in Time Allocation			
	Leisure	Wage Labor	Farm Production	Home Production
A. Increase in wage	?	+	−	−
B. Increase in P_Z	−	+	0	0
C. Increase in farm productivity	+	−	+	0
D. Increase in income	+	−	0	0

ences for Z-goods instead of leisure. Also, if additional employment cannot be found or can only be found at a reduced wage in areas under population pressure, the household would perceive reduced opportunity costs of labor and intensify the time spent in home and farm production.

While a number of conclusions can be drawn from this simple model, some of the aforementioned aspects of commercialization cannot be easily incorporated. Essentially the model assumes a complete separation of a household's resources for home production from those for farm production. Only the household's labor is being allocated between different types of activity. The model does not explain how land and other resources are allocated to market and subsistence production. We therefore expand the discussion in two different directions: first, we elaborate the issue of different players in the household (for example, husband and wife) and, second, we discuss the role of risk for subsistence orientation and commercialization.

A simple example for a two-person household in the appendix demonstrates one possible economically driven explanation for why household members tend to specialize in types of activities when new work opportunities arise. The naive example of a household with its members cooperating for maximum income may only reflect a partial reality. Conflicts may arise over "fair" work burden-sharing by task and—not addressed above—the sharing of benefits. The issue of balance between cooperation (adding to total availabilities) and conflict (dividing total availabilities among members of the household) constitutes bargaining problems. For the case studies, we particularly address the issue of how women and children fare under different socioeconomic and cultural circumstances in these "cooperative conflicts" (Sen 1985).

The Role of Risk

Subsistence food production is essentially an insurance and credit market substitute.[4] It is, thus, much a function of (perceived) *risks* of food markets (prices, availability), of factor markets, and of related income streams. Static concepts cannot properly capture those relationships. In order to derive preliminary hypotheses as to what motivates a household to adjust the share of resources determined for market production, the risks involved would at least have to be accounted for. Figure 2.3 portrays a simple model addressing this issue.

The household's production possibilities are indicated in section I of the figure. Curve a shows the transformation between Z-goods, namely

---

4. It also avoids the margin between farmgate and retail prices, which may be substantial.

FIGURE 2.3 Resource allocation to market versus subsistence production under risk

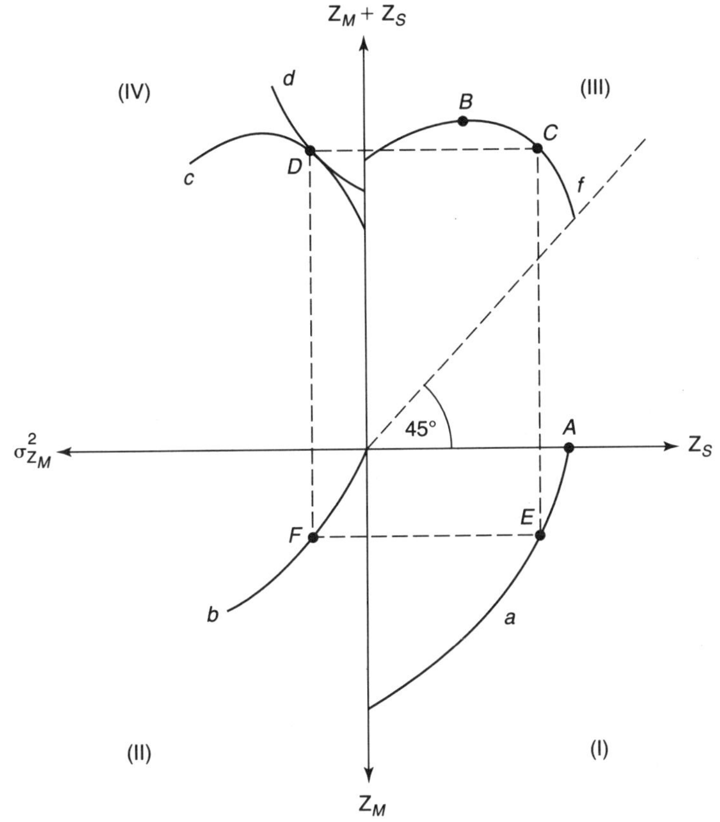

Source: von Braun, de Haen, and Blanken (1991).

food, that the household can directly produce for subsistence (Z_S), and Z-goods that the household can obtain by producing cash products and exchanging them for goods from the market (Z_M). Hence the curve's slope is not only determined by the physical production functions for subsistence and cash products, but also by the net price (gross margin) of the marketed crop divided by the net price of the Z-good.

Curve b (section II) indicates the market risks associated with an increased intensity of market exchange. Note that only those risks related to fluctuations of prices or availability of quantities are indicated here. Production risks are assumed to be identical in market and subsistence production. They do not enter the theoretical analysis here but are considered in the empirical analyses in this volume.

Curve f (section III) represents the aggregate availability of Z-goods from subsistence and market exchange, as a function of the volume of subsistence production. The aggregate quantity is equivalent to the household's income. As the transformation curve is nonlinear, the aggregate has a maximum (B) somewhere between 0 and A—that is, at a degree of subsistence (subsistence divided by the total aggregate income) between 0 and 1. Obviously, any extreme specialization on either subsistence or market production would be suboptimal due to declining marginal productivities (curve a).

Finally, section IV portrays how the household might arrive at a decision concerning resource allocation for subsistence and market production by equating the marginal utility per unit of incremental aggregate production (curve c) with the marginal disutility due to the additional risks involved (curve d). A series of additional hypotheses can thus be derived about a household's decision making once market risk is introduced:

- Risk-averse families may tend to keep subsistence production beyond the maximum income point (B)—say, at C or E—in order to keep the risk of market integration low (at F).
- A reduction of marketing risks—say, by improved infrastructure—(downward shift of curve b) and an increase in the profitability of marketing (downward turn of curve a through A) would both reduce the preference for a high degree of subsistence.
- Increasing a household's total resources (right shift of curve a) would most likely motivate a decline in the degree of subsistence, probably going along with an increased absolute volume of subsistence production.

The hypotheses advanced above have been studied in the literature.[5] Finkelshtain and Chalfant (1991) and Fafchamps (1992) use the agricultural household model framework (Singh, Squire, and Strauss 1986) to show that risk-averse semisubsistence households may produce more of the risky subsistent good (as compared to production under the no-risk situation) under certain conditions. Both studies show that this situation holds when the consumption effect exceeds the income effect. The consumption effect is defined as the amount by which expenditures would have to change after a shock to keep marginal utility constant. It is shown to depend on the share of risky crop out of total consumption, income elasticity of demand of risky crop, risk preferences, and covariance between the risky crop's consumption price and its revenue. The

---

5. Kene Ezemenari contributed substantively to this section on risk and subsistence orientation.

income effect is defined as the change in income due to a shock. It is shown to depend on the proportion of income of risky good out of total income and degree to risk (covariance between prices and revenue for the risky crop) (Fafchamps 1992; Finkelshtain and Chalfant 1991). When the consumption effect and the income effect are equal, then the household is made risk neutral with respect to the shock. Therefore, households with enough resources at their disposal are able to diversify risk and maintain a risk-neutral attitude toward shocks. If the consumption effect is greater (less) than the income effect, the household produces more (less) of the risky staple (Finkelshtain and Chalfant 1991).

Fafchamps (1992, 93) explains the intuition behind this result. The result depends on the share of the risky crop in consumption and on the covariance between price and revenue from that crop: the larger the share and the larger the covariance, the more likely a more risk-averse producer is to shift production toward the risky crop. This is because when consumption prices and crop output are correlated positively, growing a particular crop serves as insurance against consumption price. A household's decision to commercialize depends, therefore, on the sum of the consumption effect and income effect.

It should be noted, however, that other factors, such as food habits, and agronomic conditions may also be reasons for farmers to retain some subsistence production.

Wiebe (1992) examined the impact of change in wealth (given risk, risk preferences) within a lexicographic framework and found that changing attitude toward risk without improving the resource endowment of semisubsistence households is unlikely to have a favorable effect on commercialization.

Given the above discussion, a simple model of the decision to commercialize can be estimated. Following Chavas and Holt (1990), the proportion of land allocated to various crops can be expressed as a function of the key variables identified above, which may influence commercialization decisions:

$$l_{it} = \alpha_0 + \alpha_{it} \cdot W + \Sigma_j K^j_{it} \cdot P^j_{it} + \Sigma_k \beta^k_{it} \cdot R_{it} + \gamma_{it} \cdot S_{it}. \qquad (2.1)$$

The subscript t may represent either a time series of aggregate variables for a given country or region or a cross section of households at a point in time, where

l_{it} is the proportion of crop i allocated to land;
W is wealth;
P^j is the expected return per unit of cropped land for the jth crop ($j = 1 \ldots i \ldots N$);
R_{it} is the interaction terms between income elasticity of demand for crop i, subsistence orientation in consumption for

tth household, proportion of crop i income out of total household income, and the covariance between price and total revenue, and coefficient of risk aversion;

S_{it} is socioeconomic factors such as proportion of dependents in household, age, education, and years of farming of head of household, and so forth.

The above may be estimated as a system of equations for the $i = 1 \ldots M$ crops. Identification of the system depends on the restriction that the proportion of land allocated to each crop sum to 1.

While commercialization is largely being dealt with as exogenous in chapter 3, this is not the case in several of the case studies (Part V), where approaches of the above design are employed to explain commercialization processes. Some of the hypotheses derived from the preceding analysis will be subject to empirical tests in the case study chapters in Part V. While time use and risk are two of the most widely researched topics in the general area of household resource allocation, other more complex conceptualizations of the household decision-making process have been formulated. Examples include incorporation of individual health production functions as constraints (such that better nutrition becomes an input into increasing household productivity, not merely a desirable outcome), intrahousehold bargaining (who within the household earns income affects how resources are allocated), and allocation of household resources across extended time periods (the manner in which inheritance customs will result in a transfer of resources to young children several years hence may affect present period resource allocation). Each of these improvements, while resulting in models that are seemingly more realistic, increases the complexity of interactions and, so, the amount of information and effort required to understand the nature of the allocation process (Singh, Squire, and Strauss 1986).

In all of these models, allocation decisions/outcomes (such as what to produce, what to consume, intrahousehold distribution of food, and individual health outcomes) are treated as simultaneously determined. Econometrically, a common practice is to estimate a set of reduced-form equations with an extended list of exogenous explanatory variables that affect any of the structural relations. This approach is not followed in this book, in part because of data limitations. More important, because it is not possible to identify structural coefficients from these reduced-form estimates, it is not possible to draw any firm conclusions about the specific impact of crucial variables in the system at each particular link. Thus, it is difficult to gain an understanding of the process through which nutrition is affected by changes in the production system, or to identify the key factors that drive that process, which is the objective of this book. As outlined below, the approach, rather, is to focus on selected structural relationships (which combine exogenous and endogenous variables as explanatory variables) that will illuminate these key factors and processes.

Approach and Issues

The actual analysis at the household level is carried out as sketched in figure 2.1 in an attempt to trace the relevant exogenous forces of commercialization to their effects on resource allocation, patterns of commercialization, consumption, and nutrition.

Agricultural Production/National-Level Food Availability

Many developing countries are pursuing a policy of stressing the increased production of export crops as well as food crops for domestic consumption. Indeed, an emphasis on export crops has typically been one component of macroeconomic policy reforms. However, critics of accelerated export crop production argue that national-level food security will deteriorate because of falling domestic food supplies.

The extent to which national food availability will be affected depends on the competition for scarce resources—land, labor, water, and capital—between food crops consumed domestically and crops for export. If land that has been devoted to basic staples is replaced by nonedible cash crops, food availability may drop. However, even under this scenario, national-level food supplies need not be affected; if the foreign exchange generated from the sale of export crops is used to increase food imports, national-level food availability may remain unchanged or even increase. This topic is taken up again in chapter 6, where analysis of data from 90 countries shows that in the majority of countries studied, an increase in nonfood production was accompanied by an increase in food supply.

However, national-level food availability is a poor predictor of food security at the community or household level. The Malawi case study (chapter 20) indicates that, while the country has historically had an enviable record of achieving national-level food self-sufficiency, this aggregate picture masks widespread food insecurity at the household level. This phenomenon can be generalized to a large number of countries. Countries that have achieved food self-sufficiency often have a significant proportion of their populations with inadequate food intakes because these households do not have access to the available food (von Braun, Bouis, Kumar, and Pandya-Lorch 1992).

The analyses in this book focus primarily on the effects of cash crop production on producers and nonproducers in the areas undergoing commercialization. However, households outside of those areas are also likely to be affected by the commercialization process because of linkages between the various sectors of the economy. For example, foreign exchange earnings from agricultural exports make possible investments in entirely different regions and sectors of the economy, which can stimulate their growth and employment. Any complete evaluation of the income, employment, health, and nutrition effects of the commercialization process would require construction of economy-wide models disaggregated by employment and income groups. This type of analysis is beyond the scope of this book.

Community-Level Food Availability

The availability of food at the national level is a necessary but not a sufficient condition for food security at the community and household levels. Some of the most vocal criticisms of an export-oriented food policy have been related to its perceived negative effect on local food supplies.

Here, again, there is no inherent reason why cash crop production should have a negative effect on local food supplies. Expanded cash crop production is likely to influence local food supplies in one of two ways. First, to the extent that land is shifted out of basic staples and into crops for sale outside the community, the volume of marketed food supplies could decrease, which could exert an upward pressure on food prices. These higher local food prices could be offset by movement of food supplies from other regions of the country or from food imports. However, many countries restrict the movement of food supplies from one area to another and, thus, a sufficient inflow of food into the affected region may not occur to offset the rising food prices.

Second, if incomes of agricultural laborers increase as a result of more commercialized production, the demand for food will increase in the local area. This increased demand for food may occur simultaneously with declining marketed food supplies. Infrastructure's key role comes into play in this context. Higher food prices induced by increased nonfood cash cropping may also stem from deficient rural infrastructure, including poorly developed transportation systems (chapter 8).

Employment and Income Effects

The discussion thus far has assumed that most of the effects of cash cropping will be on the households participating directly in commercialization. However, there may be major positive or negative effects of commercial agriculture on employment opportunities within the communities. The effects will depend in large part on whether the new crop is more or less labor intensive than the crop it replaces. The commercialization of agriculture may have a substantial effect on the demand for labor in a given area. If cash crop production increases, the need for hired labor and, hence, the incomes of landless laborers may increase. Increased production of labor-intensive crops is an attractive way of reaching the landless poor who are often not reached by other development projects. The net employment effects of a range of cropping strategies are summarized in chapter 3.

In the conceptual framework, the primary link between agricultural production patterns or income-generating strategies, on the one hand, and household food intake on the other hand, is through household income. Proponents of cash crop production assume that household income will increase as a result of the transition to a more commercialized agriculture. Each of the case studies in Part V evaluates the income effects of cash crop production, and chapter 3 provides a synthesis of this information.

Household Food Availability and Consumption Effects

A primary concern of the research described in this book was to document the links between the transition to commercial agriculture and household-level food security. The decisions about how to allocate household resources—land, labor, time, and capital—have implications for the ultimate impact of commercialization on household food supplies.

The potential negative impacts of commercialization on household food security can be short term or long term. In the short- to medium-term, the decision to allocate land to a cash crop—particularly a nonfood cash crop with a long growing cycle—can decrease the food supplies available to a household. If the household has other sources of off-farm income available, this money could be used to supplement food purchases. The worst-case scenario is one where a household allocates a disproportionate share of available farmland to a nonedible cash crop with a long gestation period and is trapped when other income sources become less available and the terms of trade for the cash crop develop unfavorably. This volume contains examples of cash crops that would fall into the category of long gestation periods—sugarcane in Kenya (chapter 16) and the Philippines (chapter 13) and tree crops in Sierra Leone (chapter 21).

If expected real income gains materialize, the ability of the farmer to acquire food should be higher under cash crop production. However, the income–food consumption relationship is influenced by more than just total household income. The form of income (lump-sum versus periodic, cash versus kind) and the control of income within the household may be as important in understanding the ultimate effects of income on household food consumption. A number of the case studies in Part V show the influence of these income-related variables on household consumption, and chapter 4 provides a synthesis of these consumption-related effects.

Semisubsistence agriculture frequently produces a rather constant flow of income in the form of food and some cash, whereas income from cash crops, such as sugarcane, often comes in one lump-sum payment. In the absence of well-integrated rural financial markets, income in the form of lump-sum payments may be used in a different manner than a smaller, more continuous flow of income. Lump-sum payments, typically, are associated with the purchase of consumer durables, whereas continual forms of income are more likely to be spent on food (von Braun and Kennedy 1986). In addition, who controls the income within the household may partly explain why lump-sum payments may be used differently than some periodic forms of income.

Income is one of the key determinants of household food consumption. Therefore, one would expect that increases in household income associated with cash crop production would result in increases in household food consumption. However, caloric intake at the household level may increase more slowly than income for two reasons. First, there is a tendency as incomes increase for households to move toward the purchase of more expensive calories, although this trend is less pronounced in the African case studies. Second, as incomes increase, there is also a movement toward the purchase of more nonfood items. How total household income is spent may be influenced by who within the household earns the income. Intrahousehold earning patterns may change with commercialization. Similarly, how consumption items are distributed within the household may again vary with who controls income.

Much of the research analyzing the effects of agricultural policies and programs on food consumption tended to limit analyses to macronutrients—mainly calories and, to a lesser extent, protein. The assumption was that these were the nutrients most limiting in the diet and that an improvement in energy consumption would lead to a concurrent increase in intake of other nutrients. However, this may not always be true or, at least, not for all nutrients. For example, in Kenya, some recent evidence suggests that as preschooler energy consumption increased, there was a decrease in vitamin A intake, due to shifting dietary patterns (Kennedy and Oniang'o 1993). Beta carotene–rich foods were being replaced by higher status foods.

The case studies presented in this volume tend to concentrate on calories. However, it is clear that other measures might also be appropriate to capture the full range of dietary changes associated with the commercialization of agriculture.

Health and Nutritional Status of Women and Children

As evident from the richness and complexity of the household model, caloric intake is but one link through which the commercialization of agriculture can potentially influence an individual's nutritional status. For preschooler nutritional status, other factors such as morbidity patterns—and how these are influenced by changes in the health and sanitation environment, breast-feeding, and weaning practices—as well as allocation of time and other resources to the child can be as important as the diet in affecting the overall welfare of the child.

Gender issues have rarely been included in the analysis of commercial agriculture schemes. However, even where they have been included, much of the analysis has been limited to the impact of cash cropping on women's income or on women as the primary caretakers of children. Virtually no information exists on the direct effects of agricultural commercialization on women's health and nutritional status. Until recently, much of the evidence related to the commercialization of agriculture and its effects on women has been anecdotal. In contrast, some of the case studies on commercialization of agriculture reported in this volume allow a gender-disaggregated assessment of the effects of commercialization on women's nutrition.

There are complex linkages between households and women. Figure 2.4 assumes that the entry of a household into commercial agriculture can affect women—either positively or negatively—through its impact on their income, time, or energy expenditures. These factors can ultimately influence women's health and nutritional status, which, in turn, can also influence children's health.

A number of the case studies (Part V) provide detailed information on the microenvironment affecting preschoolers. Many constraints to the production of satisfactory children's health and nutrition are context specific. However, it is clear from almost every case study presented that the causes of malnutrition are multifaceted; relieving one factor associated with malnutrition does little unless other causal factors are also simultaneously addressed. The net effects of commercialization on energy consumption, morbidity, and other factors affecting women's and

FIGURE 2.4 Commercialization of agriculture: potential pathways through which women's and preschoolers' health and nutritional status may be influenced

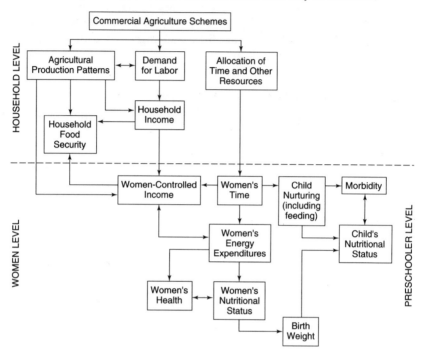

children's health and nutritional status are summarized for a number of the case studies in chapter 5.

APPENDIX

An example for the two-person household demonstrates one possible (economically driven) explanation for why household members (husband and wife) tend to specialize in certain broad types of activities when a new work opportunity arises, be it a cash crop available for men because of gender-biased extension or market institutions or male off-farm work opportunities at higher wage rates for men than for women.[6] We observe frequently that men work on cash crops or in the off-farm labor force, while women perform domestic chores. Certainly there are other reasons why this occurs, but the point is to show, among other

---

6. The monetary values shown in this appendix are developed solely for the purpose of illustrating the examples. They do not refer to any case study in this volume.

Conceptual Framework 31

TABLE 2.2 Assumed production function per plot

Labor Inputs Per Day	Value of Net Output (cents)	Implied Marginal Value Product Per Additional Hour (cents)
2.00 hours	30	
2.75 hours	47	
3.00 hours	50	20
3.75 hours	58	11
4.00 hours	60	10

NOTE: Values for labor inputs and net output are expressed on an average daily basis over the production cycle.

things, one set of economic forces that is operating. Several assumptions are made:

- Husband's wage in the labor force is 20 cents per hour; wife's wage in the labor force is 10 cents per hour.
- Husband's and wife's labor in agricultural production and home production are perfectly substitutable.
- Household has two plots of farmland that are identical, with the production function for each plot shown in table 2.2.
- Household has two sets of domestic chores that have identical production functions (table 2.3).
- Objective of the household (husband and wife) is to maximize total income.
- Each agrees to work eleven hours a day.

Stage 1

No specialization: husband and wife agree to split the types of work. Each gets one plot of farmland and one set of household chores to do. In essence, these are two one-person households. Husband and wife individually set their marginal products in all three types of activities equal to one another. Equilibrium is reached as shown in table 2.4.

TABLE 2.3 Assumed production function of domestic chores

Labor Inputs Per Day	Value of Net Output (cents)	Implied Marginal Product Per Additional Hour (cents)
0.00 hours	0	
0.75 hours	17	
1.00 hours	20	20
1.75 hours	28	11
2.00 hours	30	10

TABLE 2.4 Farming and household chore equilibrium (Stage 1)

	Time					
	Labor Force	Plot 1	Plot 2	Chores 1	Chores 2	Total
Husband	7 hours (20)	3 hours (20)		1 hour (20)		11 hours
Wife	5 hours (10)		4 hours (10)		2 hours (10)	11 hours

NOTES: Marginal product is in parentheses (in cents).
Husband's income (in cents per day) = (7 hours × 20) + 50 + 0 + 20 + 0 = 210
Wife's income (in cents per day)   = (5 hours × 10) + 0 + 60 + 0 + 30 = 140
Total income (in cents per day)   =   350

Stage 2

The wife is working for 10 cents per hour; yet she could be more productive by switching some of her time to plot 1 and chores 1. The husband tells his wife: "You put in two hours on plot 1 and two hours in chores 1 and reduce your time in the labor force by these four hours. I will reduce my time in plot 1 and chores 1 by one hour and put in two extra hours in the labor force. We both still work eleven hours a day, but we will be better off." The result is shown in table 2.5.

Stage 3

The wife says to her husband: "You're beginning to catch on, but you haven't quite got it right. Why don't you put in all your time in the labor force, and I will take care of all the farm work and household

TABLE 2.5 Allocation of time for improved profitability (Stage 2)

	Time					
	Labor Force	Plot 1	Plot 2	Chores 1	Chores 2	Total
Husband	9 hours (20)	2 hours (10)				11 hours
Wife	1 hour (10)	2 hours (10)	4 hours (10)	2 hours (10)	2 hours (10)	11 hours

NOTES: There is a total profit of 40 cents from plot 1, which both husband and wife have earned together. Marginal product is in parentheses (in cents).
Husband's income (in cents per day) = (9 hours × 20) + 30 + 0 + 0 + 0   = 210
Wife's income (in cents per day)   = (1 hour × 10) + 30 + 60 + 30 + 30 = 160
Total income (in cents per day)   =   370

TABLE 2.6 Labor specialization and income maximization (Stage 3)

	Time					
	Labor Force	Plot 1	Plot 2	Chores 1	Chores 2	Total
Husband	11 hours (20)					11 hours
Wife		3.75 hrs (11)	3.75 hrs (11)	1.75 hrs (11)	1.75 hrs (11)	11 hours

NOTE: Marginal product is in parentheses (in cents).
Husband's income (in cents per day) = (11 hours × 20) + 0 + 0 + 0 + 0 = 220
Wife's income (in cents per day) = (0 hours × 10) + 58 + 58 + 28 + 28 = 172
Total income (in cents per day) = 392

chores. As things stand now (Stage 2), you are not using those two hours on plot 1 as productively as you might." The final result is shown in table 2.6.

As a result, by agreeing to cooperate, the husband and wife have increased total household income, which is maximized by specialization. The real world is, of course, much more complex, as is our conceptual framework of study of commercialization effects for women (see below). The above analysis assumes that both husband and wife will continue to work eleven hours a day even though income is higher. This is not a realistic assumption, as is later shown by the empirical studies as well. Again, abstracting from considerations of leisure, note that it is the wife who will reallocate her time when there is a change in farm productivity or household productivity (see Low 1986 on this issue). If some utility is derived from work in the labor force (by the wife), or from work on one's own fields (by the husband), the Stage 3 solution will not be reached.

PART III

Household-Level Effects: A Synthesis

3 Production, Employment, and Income Effects of Commercialization of Agriculture

JOACHIM VON BRAUN

The Policy Issues

The three chapters in Part III report synthesis findings from the microlevel IFPRI research in The Gambia, Guatemala, Kenya, the Philippines, and Rwanda, as well as from the other case studies presented in Part V. Any attempt to synthesize and generalize on the basis of the detailed case studies runs the risk of excessively extrapolating from special circumstances and of losing insights gained from these case studies, whose strengths are the detailed assessments of the commercialization-production-income-consumption-nutrition chain and the important feedbacks from these elements. This chapter, on the first elements of the commercialization chain, is therefore to be seen in the context of the following two chapters, and all the three synthesis chapters together are to be seen in the context of the rich insights from the individual studies discussed later. Furthermore, the microlevel experiences are also to be seen in an economy-wide context that is addressed in Part IV.

Many of the theoretically possible problems of commercialization for household-level food security and nutrition derive from "if statements," such as: if food crops are replaced by nonfood cash crops in the production program, and, if markets are not well integrated, and, if landless farm laborers are replaced by less labor-intensive production, then the resulting employment and price effects may have adverse impacts on the food security of this population group. Or, if more gross sales of food crops are induced by improved market access, and, if the resulting cash returns are controlled by male heads of households, and, if their propensities to consume are inclined toward nonfoods, and, if women lack income-generating alternatives, and, if these households were close to food insecurity to begin with, then the changes induced by the increased commercialization may have negative impacts on household food security (or on selected subgroups in the households). Empirical analysis is required to address at least the more relevant *if*s.

Following the subsets of the conceptual framework laid out in chapter 2, this chapter addresses the production, income, and employment effects of commercialization, and their food security implications. It specifically focuses on the following policy questions relating to increased commercialization:

1. To what extent are resources for market production drawn from subsistence food production under different real-world circumstances?
2. Is there a major reduction in subsistence food availability at the household level?
3. Is there a reduction in the real incomes of the poor (or of selected subgroups of the poor) temporarily or over the long term?
4. Is there an increase in seasonal or irregular fluctuations in food availability, prices, and income?
5. Is there a reduction in labor demand for landless laborers?
6. Is there a change, in disfavor of the poor, in access to land?

Obviously, the answers to these questions as well as the resulting policy implications are quite determined by the nature of the commercialization process and by the economic, sociocultural, and structural characteristics of the affected area. Moreover, commercialization effects at the household level cannot be comprehensively assessed in a vacuum of time and space: the historical context matters.

Generalizations from this comparative look at specific cases must, therefore, be fairly broad and need to be reassessed in the context of the specific comprehensive case studies. This chapter and the following two comparative chapters capture, only to a small extent, the rich policy-relevant findings from the case studies that prove that commercialization-production-income-consumption-nutrition linkages are often quite location and program–design specific. Table 3.1 gives an overview of the case study settings and synthesizes, in broad terms, findings regarding production, employment, and income effects of commercialization.

Research Design and Survey Protocol

The survey protocols used for the various studies, although similar, were not identical. The survey design and data collection protocols were guided by the conceptualization outlined in figure 2.1. Since a synthesis of results from the various IFPRI studies was envisioned from their inception, the studies were designed so that certain key pieces of information would be available from the individual surveys. This information included household income by source, household expenditure patterns, household energy consumption and/or preschooler caloric consumption,

TABLE 3.1 Overview for case study settings of commercialization features and production, employment, and income effects

Country	Commercialization Scheme	Main Subsistence Crops/Commercial Crops	Did Subsistence Food Production Decline?	Did Employment Increase?	Did Income Increase?
Guatemala	Export vegetable-producing cooperative of Cuatro Pinos	Maize, beans/snow peas, cauliflower	No, despite decline in area cultivated, higher yields meant greater per capita maize production.	Yes, overall increase of 21 percent in agricultural employment. Labor input up by 45 percent on commercial farms.	Yes, substantially, especially among recent participants and smallest farmers. Farm incomes up more than enough to offset reduced off-farm incomes.
Philippines	Bukidnon Sugar Company	Maize/sugarcane	Yes, substantially, by about 50 percent compared to nonparticipants.	Not much, about 10 percent. Labor input on corn and sugar farms similar; but use of hired labor on sugar farms doubled and, of family labor, women's employment fell sharply.	Yes, substantially, except for laborers.
Papua New Guinea	Karimui Spice Company cardamom plantation	Root crops, sago, bananas/cardamom	Perhaps, wage households produced considerably less food than nonwage households.	Yes, when KSC was set up, but as it declined, most laborers were laid off.	No; in fact, wage rates of male laborers were reduced. Total income similar in wage and nonwage households.

continued

TABLE 3.1 Overview for case study settings of commercialization features and production, employment, and income effects (Continued)

Country	Commercialization Scheme	Main Subsistence Crops/Commercial Crops	Did Subsistence Food Production Decline?	Did Employment Increase?	Did Income Increase?
India	Karnataka Dairy Development Project	Milk	Information not available.	Information not available.	Yes, to some extent. Average expenditures up by 8 percent in villages with dairy co-ops.
Kenya	South Nyanza Sugar Company	Maize/sugarcane	No. Participants devoted as much land to food crops as sugar farmers. New entrants even increased subsistence income.	Information not available.	Yes, significantly, for both new entrants and sugar farmers.
Kenya	Two rice schemes: Ahero Irrigation Scheme and West Kano Irrigation Scheme	Maize, sorghum/rice, sugarcane	Yes, for those tenants who reside at the schemes, but not for nonresident tenants who have access to much more land.	Information not available.	Yes, for individual rice growers and nonresident tenants, but not for nonrice growers and residents. Controlling for household size, there were barely any differences among the four groups.
Rwanda	Potato production in the Gishwati forest area	Peas, beans, sweet potatoes, maize, sorghum/potatoes, tea	No, no change in crop composition as modern potato production exclusively	Yes, significantly. Sixty-four man-days of wage labor hired per hectare per season	Yes, to some extent. But income control within the household affected as men's greater

Country	Project	Crops			
Zambia	Technological change in maize	Maize, sorghum, finger millet/hybrid maize	on the Gishwati land. The increased potato production raised food availability.	to help in potato production.	participation in potato production leads to their greater control over that output. Yes, significantly. Per capita income about 25 percent higher in adopting households.
			Information not available. Note, though, that only those farmers who could expand cultivated area beyond subsistence food requirements adopted hybrid maize.	Yes, household labor input up by 46 percent in high-adopting areas, for both males and females.	Yes, but with wide variations. Income of tobacco specialists much higher but that of small tobacco households slightly less than of nontobacco households.
Malawi	Commercialization of maize and tobacco	Maize, legumes/maize, tobacco	No, strong desire to produce as much maize as possible for food-security reasons. On average, tobacco did not displace maize.	Information not available.	No, in fact, it was about 12 percent lower among new adopters, and about the same for established plantation farmers and subsistence farmers.
Sierra Leone	Bo-Pujehun Rural Development Project-Tree Crop Promotion	Upland rice, vegetables, roots/coffee, cocoa, oil palm	Not clear, less food crop area is cultivated but land productivity is much higher, which may offset the area substitution effects.	Information not available.	
The Gambia	Jahally-Pacharr Smallholder Rice Project	Millet, sorghum, rice/rice under modern irrigation, groundnuts	No, new rice technology increased rice yields substantially.	Yes, a 56 percent increase.	Yes, for households as a whole, but women generally benefited less than men.

NOTE: For details, see chapters in Part V.

weaning and child-feeding patterns, and preschooler morbidity and nutritional status. The other case studies contained in this volume were also guided, in large part, by the conceptualization outlined in figure 2.1.

A variety of techniques were used to collect data from the community, household, and individual levels, with the approaches tailored to the context. In each study site attention was given to choosing a representative sample of participant and nonparticipant households and applying the same survey methods to the two groups. Thus, appropriate comparisons could be made between participant and nonparticipant households, using the same survey techniques. The question of determinants of participation is addressed at the level of the case studies (Part V), as local circumstances play an important role.

Characteristics of the Study Areas

The nature and source of agricultural commercialization differ among the study settings. For example, in Guatemala, the introduction of export vegetable crops spurred commercial production. In The Gambia, fully water-controlled double-crop rice production provided a new income-earning opportunity. In Rwanda, potato and, to a lesser extent, tea production led to increased commercialization. In both Kenya and the Philippines, sugarcane production was the source of agricultural commercialization; both regions are similar in that they were primarily engaged in maize production before the construction of sugar mills permitted extensive sugarcane production.

In most of the study settings, commercialization of agriculture occurs jointly on the output and input sides of agricultural production (see definitions in chapter 2). An aggregate measure of "commercialization of the farm household" would be the ratio of net-marketed surplus over income. The emphases, however, vary in the different study settings. While commercialization occurs more in terms of sales over total production value, that is, on the output side, in the Philippine and Kenya sugarcane cases, commercialization on the input side is particularly pronounced in the Gambian case of double-cropped rice. On the other hand, the overall degree of commercialization of the rural economy is particularly far advanced in the Guatemalan setting because of much off-farm employment. That is also the case in Rwanda, but here in-kind transactions are still considerable, and thus the degree of integration into the cash economy is lower, although the overall degree of commercialization of the rural economy is substantial due to off-farm employment.

This chapter carries the comparative analysis along the lines of the conceptual framework, tracing the commercialization effects up to income and employment, from where the effects on consumption, health, and nutrition are picked up by chapters 4 and 5. The relationships between income and nutritional improvement are certainly complex. Although we later actually disentangle these complexities, there may be a risk of losing the broad perspective. Therefore, before studying the processes and interactions outlined in the conceptual framework, an aggregate analysis of commercialization-nutrition links is presented first.

The Aggregate Commercialization-Nutrition Relationship

Increased household income theoretically permits households to respond in a number of ways that may favor nutritional improvement: more food may be acquired; workloads may be reduced and, thereby, child care improved; household sanitation and housing environments may be enhanced and, thereby, exposure to infectious diseases reduced; water availability, both quantity and quality, may be improved; and effective demand for health care, both preventive and curative, may be strengthened. Furthermore, when household resources are less constrained, the ability of households to respond to existing or new knowledge for nutritional improvement may be increased.

For an income-based aggregate model to explain nutritional status, all other potentially income-related determinants of nutritional status, such as food consumption, diet composition, health and sanitation, and so forth, are excluded, since we hypothesize that these determinants may also be driven, at least partially, by household income. The model, therefore, includes only income, income composition-related variables, and child demographic variables (age, sex, birth order, and, where applicable, duration of breast-feeding of the child). The reduced form model is thus:

Nutritional status of child = f(Per Capita Income, Per Capita Income squared, Income Share from Cash Crops, Child Demographics).

The results of this analysis for six of the study environments, taking weight-for-age Z-score values for children as the dependent variable, are presented in table 3.2. In all study areas, except Kenya, a positive and significant effect of increased income on nutritional improvement is identified. The income effect on nutritional improvement is decreasing at the margin, which is captured by the negative parameter for the income-squared variable in the model, and is significant in all cases except Kenya (chapter 16) and Papua New Guinea (chapter 14).

In none of the study areas is there a significant negative effect of an increased share of cash crop income on children's nutrition. A separate

TABLE 3.2 Regression analysis of effects of income on nutrition of children for six study areas undergoing commercialization

Explanatory Variable	Guatemala		The Gambia		The Philippines	
	Parameter	t-Value	Parameter	t-Value	Parameter	t-Value
Income[a]	8.231E-04	3.20	0.0749	2.91	5.853E-03	5.76
Income squared[a]	-3.930E-07	-2.74	-1.487E-05	-2.00	-1.391E-05	-3.87
Income share from cash crops	0.1569	1.88	0.267	1.62	0.0214	0.20
Male off-farm income share	0.1613	2.00	-0.4649	-2.59	—	—
Female income share	0.4953	2.20	-0.5762	-0.99	—	—
Age (months)	-3.489E-03	-0.85	2.3634	6.16	8.1101E-03	1.33
Age squared	3.497E-05	1.14	-0.01151	-3.57	-3.355E-05	-0.40
Sex (1 = male, 2 = female)[b]	7.083E-03	0.12	8.8169	1.57	-0.3938[b]	-11.55
Birth order	1.706E-04	0.01	—	—	3.241E-03	0.40
Breast-feeding (months)	9.663E-05	2.03	—	—	5.421E-03	2.45
Constant	-2.1165		-282.409		-1.8614	
R^2	0.032		0.111		0.086	
F-value	3.59		20.23		25.26	
Degrees of freedom	785		1,227		2,065	

Explanatory Variable	Kenya Parameter	Kenya t-Value	Malawi Parameter	Malawi t-Value	Rwanda Parameter	Rwanda t-Value
Income[a]	8.224E-05	0.67	9.03-03	3.85	4.9602E-05	2.13
Income squared[a]	-1.357E-08	-0.75	-1.63-05	-2.78	-1.539E-09	-2.16
Income share from cash crops	9.741E-03	2.63	-0.211	-0.58	0.4573	1.42
Male off-farm income share	—	—	-0.115	-0.59	—	—
Female income share	—	—	0.283	1.39	-0.3439	-1.67
Age (months)	0.01185	1.56	8.27E-04	0.09	-0.0116	-1.20
Age squared	-1.0184E-04	-1.08	4.41E-05	0.43	1.0994E-04	1.18
Sex (1 = male, 2 = female)[b]	0.08643	1.22	2.98	2.92	0.2464	3.43
Birth order	7.6203E-03	0.53	—	—	-0.0452	-2.16
Breast-feeding (months)	—	—	—	—	0.0165	3.68
Constant	-1.6025		-2.27		-1.1940	
$\bar{R}^2$	0.009		0.089		0.054	
F-value	2.15		4.47		5.249	
Degrees of freedom	941		374		662	

NOTE: Dependent variable is weight-for-age Z-score values of children. The age range of child population covered differs in the study settings: Guatemala, 6–120 months; The Gambia, 6–120 months; Philippines, 6–60 months; Kenya, 7–80 months; Rwanda, 6–72 months. Dependent variable in The Gambia is the Z-score value multiplied by 100.

[a] Annual total expenditure per capita (in national currencies) is used as an income proxy.
[b] In the Philippine and Kenya surveys, male = 1, female = 0.

exercise on Sierra Leone (chapter 21), however, found a significant negative effect on children's nutrition of a higher income share from cash crops, controlling for income level. This would perhaps be explained by the source of the cash crop income, which is tree crops that for the first five to seven years do not yield a cash return and hence cause a net income loss. In three of the study areas—The Gambia, the Philippines, and Rwanda—the effect of cash crop share is positive but statistically not significant. In Guatemala, and even more so in Kenya, a positive significant effect results from an increased share in total income from cash crops on children's weight-for-age. In the Guatemala study setting, this cash crop income share variable may be capturing some of the social programs associated with commercialization in the export vegetable cooperative described in chapter 12. The positive effect in Kenya, in the context of commercialization, remains somewhat puzzling. It may be that households with more entrepreneurial attitudes who may use household resources more efficiently, including for child welfare, joined the sugarcane outgrowers' scheme. In another commercialization project in West Kenya, this time, rice (chapter 17), the composition of income appeared to be much more important than the level of income for determining household nutrition—nutritional indicators generally improved when incomes were more diversified. An important point of this analysis is that there is generally no evidence for an adverse effect on child nutrition from increased commercialization, even when income is held constant.

In table 3.3, parameter estimates are used to evaluate the effect of a

TABLE 3.3 Effect on children's nutritional status (weight-for-age) of a 10 percent increase in income of the poor (at US$100 per capita)

Country	Effect of a 10-Percent Increase in Income	
	Level of Z-Score	Percentage Change of Z-Score
Guatemala	+0.019	+1.06
Philippines	+0.017	+1.13
Kenya	n.s.	n.s.
Rwanda	+0.015	+2.46
Malawi	+0.068	+4.90
The Gambia	+0.023	+1.92

SOURCE: Derived from table 3.2.
NOTE: n.s. = not statistically significant.
ᵃHolding income constant.

10 percent increase in income on children's nutrition, measured in terms of weight-for-age, at a uniform income of US$100 per capita (thus moving households from $100 to $110) in some study settings. The elasticity of nutritional improvement with respect to income ranges between 0.1 and 0.49, which translates to between 1.0 and 4.9 percent for a 10.0 percent increase in income. It is highest in Malawi and Rwanda, followed by The Gambia, and is substantially lower in Guatemala and the Philippines. The level of these elasticities suggests that major increases are required in the income levels of rural poor households to actually have a major nutritional improvement effect. Also, the long-term effect of increased income for nutrition is likely to be higher.

Commercialization of Agriculture and Staple Food Production

The nature and extent of competition versus complementarity between substance crops for own consumption and crops for sale in the market are central elements of the forces that drive the food security effects of commercialization. The discussion that follows centers on the reallocation of land and labor resources that occurs with commercialization, as well as the profitability and productivity of the commercialized crops as compared with the subsistence food crops.

Commercialization of agriculture, as noted earlier, takes a variety of forms. Therefore, any classification of households and farms into commercialized or noncommercialized has its shortcomings. Households are therefore separated in each study setting (within farm-size classes) into two groups, according to their participation or nonparticipation (or very little participation) in the commercialization scheme. A broad grouping of this nature does not, of course, satisfactorily capture indirect participation in commercialization schemes—for instance, through labor market effects. We therefore augment the presentation of results in various parts of the analysis beyond the participant/nonparticipant categories. The detailed case studies in Part V look at production relationships in a more disaggregated way.

Commercialization and Changes in Production Patterns

PARTICIPATION OF THE SMALLEST FARMS. A pertinent question is, "Did the small farmers effectively get equal access to commercialization schemes?"[1] Sampling was basically done randomly among participant and nonparticipant households in and around the respective schemes,

---

1. It should be noted, however, that the overlap of "small farms" with "poor farm families" is rather bad (Lipton 1989), and therefore the above question addresses only a segment of the poverty-commercialization relationship.

which permitted an assessment of the relationship between scheme participation and farm size. While the smallest farms in each study setting do participate in the commercialization schemes, it is only in The Gambia that their representation is more than proportional (table 3.4). In the other four settings, particularly the Philippines, the smallest farms participate disproportionately less (proportional participation would mean a 33.3 percent figure in column 3, table 3.4).

While there is an element of choice in adopting new crops or technologies in the scheme, it is noteworthy that scheme participation in all study settings is not just a matter of choice by farm households. There are cases in which farmers, one way or another, were provided rationed access for cash crops: a major attempt at providing women and poor people access to land in the rice scheme was made in the Gambian case, and access to the new potato-growing area in Rwanda was a matter of bureaucratic procedure apparently not resulting in equal access.

STAPLE FOOD PRODUCTION BY COMMERCIALIZED FARMS. With the exception of Guatemala, the average-size farms in the respective survey areas (using location-specific mean values) are self-sufficient in staple food supplies if a rule-of-thumb figure of 170 kilograms of cereal equivalents per capita per annum is applied. In the case of Rwanda, this is, however, only barely met, and only for the average, leaving many below this level.

Participants in commercialization schemes maintained a considerable part of their area allotment for staple food production (that is, of subsistence crops) (table 3.5). Comparisons are made for the middle tercile of farm size in each of these study areas in an effort to control for differences in the respective distributions. Scheme participants in the Philippines and Guatemala allocated the highest proportion of their crop area to new cash crops or crops using new technology, 52 percent and 44 percent, respectively, whereas participants in The Gambia allocated the lowest production to these crops.

TABLE 3.4 Average farm size and food production in study areas

Country	Average Farm Size (hectares)	Staple Food Production (kilogram/capita cereal equivalents)	Share of Participants in Commercialization Schemes Among Bottom Farm-Size Tercile[a] (percent)
Guatemala	0.8	110	19.4
Philippines	4.3	324	15.7
Kenya	4.5	233	22.3
Rwanda	0.7	172	29.0
The Gambia	3.9	344	41.3

[a] 33.3 percent would be equal participation.

TABLE 3.5 Change in cropping pattern with adoption of new cash crops and crops under new crop technology (averages for the middle tercile farm-size groups)

Country	Degree of Participation in Schemes	Area with New Cash Crops or Crops Under New Technology in Schemes	Area with Staple Food Crops (Subsistence Crops) (percent of) total farm area)	Area with Other Crops (Including Fallow, Traditional Cash Crops)
Guatemala	Participating	43.9	48.2	7.9
	Not participating	0.0	88.4	11.6
Philippines	Participating	52.0	32.4	15.6
	Not participating	0.0	66.1	33.9
Kenya	Participating	37.6	49.7	12.7
	Not participating	2.6	45.3	52.1
Rwanda	Participating	17.0	75.9	7.1
	Not participating	1.9	85.4	12.7
The Gambia	Much participation	14.3	55.0	30.7
	Little participation	2.5	58.2	39.3

NOTE: The middle farm-size tercile of each sample was chosen to exclude farm-size effects.

The difference between participants and nonparticipants in crop area allocated to subsistence crops is large in the Philippines and Guatemala; nonparticipants in these study settings did not allocate any land to cash crops. A large difference is also observed in Sierra Leone, where farmers who have adopted tree crops have an area under food crops that is just half the area cultivated by subsistence farmers; this is surprising, considering the long gestation period for tree crops before they yield cash returns. In Kenya, however, participants in the sugar scheme use an even higher share of their land to grow subsistence crops (mainly maize) than do nonparticipants: sugarcane cultivation has cut mainly into fallow land held by participants (table 3.5, col. 3). The long-term implications of this practice on soil fertility, unless mitigated by fertilizer use and erosion control, may be of concern.

While it was observed earlier that the smaller farms were less represented among the scheme participants (table 3.4), it is also true that the smallest farmers who actually did join the schemes allocated a larger share of their land resources to scheme participation than did the larger farmers in the participant group. An interesting adoption pattern is observed: the smallest farm-size terciles in each of the schemes' participant groups, except Guatemala, adopted the new crop the most, that is, they converted the largest proportion of their land to the new crops (table 3.6). In Guatemala, a rather equal distribution across farm-size terciles is found.

A key question is how implementation of the commercialization schemes and participation in these schemes affected staple food production. Despite the reallocation of land to the new cash crops, which was substantial in some cases, staple food production per capita is maintained or even increased. The exception again is in Sierra Leone, where tree crop farmers reduced the area under food crops. This high level of staple food production is not surprising in the Gambian or Rwandan cases, since the commercialization projects focus on rice and potatoes, respectively. The Guatemalan case is, however, surprising: the difference in staple food production between the participant and nonparticipant groups is small, despite the large reallocation of land to the new cash crops by participants (table 3.7). The drop in staple food production by participants in the Philippine study case is substantial, about 50 percent compared with the nonparticipant group. Note that the gross marketed surplus of staples by the Philippine farmers is largest in this comparison across study sites, showing them to be the most market-integrated farmers in staple foods. We thus have here a case of shifting from production of maize for the market and home consumption to production or sugarcane for the market, and not simply a shift from "subsistence" to "cash cropping."

The expansion of food crop area (where fallow land is available) or increases in yields of staple foods (where technology is available) permits either a generally high level of staple food production to be maintained or a less than expected decline in that production. Yield increases were particularly important in Guatemala, where participants' yields increased by 34 percent, and in the Philippines, where there was a 28 percent increase in yields (table 3.8). The lack of such a tendency in

TABLE 3.6 Degree of participation of scheme participants in commercialization schemes by farm-size terciles

	Percent of Land Use for Selected Commercialization Crops		
Country	Bottom Tercile	Middle Tercile	Top Tercile
Guatemala	38.0	43.9	38.5
Philippines	68.1	52.0	53.8
Kenya	45.3	37.6	18.8
Rwanda	30.8	17.0	22.5
The Gambia	31.8	14.3	9.7

TABLE 3.7 Staple food production per capita with new cash crops and crops under new crop technology (averages for the middle tercile farm-size groups)

Country	Staple Food Production	
	Participants	Nonparticipants
	(kilogram/capita)	
Guatemala	87	108
Philippines	193	306
Kenya	238	225
Rwanda	153	132
The Gambia	469	179

NOTE: The middle farm-size tercile of each sample was chosen to exclude farm-size effects.

Rwanda is disturbing and may be attributed to the absence of a yield-increasing technology in cereals, which stimulates farmers who are under land pressure to seek increased calorie output per unit of land by shifting toward roots and tubers, for example, sweet potatoes. The situation appears different in Kenya, where excess land permitted staple food production through area expansion. Yield increases are critical in the land-scarce study settings, and their existence points at opportunities for joint growth in cash crops and staple food production for home consumption.

TABLE 3.8 Yields of major cereal crops of farms participating and not participating in commercialization schemes

	Participation in Commercialization Schemes	
Country/Crop	Much	Little or None
	(tons/hectare)	
Guatemala		
Maize	2.19	1.63
Philippines		
Maize	0.97	0.76
Kenya		
Local maize	1.33	1.31
Hybrid maize	1.33	1.42
Rwanda		
Average maize[a]	1.03	1.06
The Gambia		
Fully water-controlled rice	5.33	1.21
Swamp rice	1.36	1.21
Millet, sorghum	0.38	0.73

[a]In maize (grain) equivalent basis in maize monoculture and maize mixed cropping.

Generally, but not always (Sierra Leone being an exception), farm households, where possible, expanded staple food production *with* cash crop production, at least on a per-unit-of-land basis. This microlevel finding coincides with observations from 78 developing countries (von Braun and Kennedy 1986).

Why do farm households respond the way they do with their food production in the context of commercialization? Where there are no technical or contractual constraints on expanding the production of cash crops and where these crops are more profitable, such a food production response by farmers relates to market risks and production risks, as discussed in chapter 2.

The potential gains from specialization are certainly not fully exploited by the small farmers. Actually, farmers are willing to pay a price to maintain household food security based on own production of food crops. This insurance approach, for instance, cost small farmers in Guatemala 6 cents per kilogram of corn produced on the farm, because of deviation from profit-maximizing resource allocation. Yet this deviation from full specialization makes sense from a social security perspective, when insurance markets are largely absent. Such substitutes for insurance markets can be effectively supported as a second-best policy option by rapid technological change in staple food production: yield-increasing technology, which reduces cost of production per unit of output, brings down the "insurance cost" paid by small farmers for their own food security, permits more rapid adoption of crops with higher payoffs that would be economically desirable and thereby permits enhancement of household food security because of resulting increased income.

Profitability and Productivity

Small farmers tend to allocate their resources efficiently and to respond to incentives, but they are conscious about taking production and price risks into account. This general pattern, however, does not preclude small farmers from making management mistakes, especially when they are in the early stages of adopting new production technologies or new crops, of whose market and price risks they are not yet fully aware. For households operating close to the borderline of food insecurity and hunger, management mistakes can be disastrous. Again, most small farmers seem conscious of this and therefore adopt new crops and technologies only if the margin of increased profitability against that from the old systems and subsistence crops is large and there is insurance against additional risks.

The returns to land and labor are, in general, substantially higher for

TABLE 3.9 Net returns to land and family labor (gross margins) of new cash crops or crops under new technology and of staple foods (subsistence crops), in 1984–85 U.S. dollars

Country	Cash Crops or Crops Under New Technology/ Subsistence Crops	Returns to Land		Returns to Family Labor	
		Cash Crops and Crops Under New Technology	Subsistence Crops[a]	Cash Crops and Crops Under New Technology	Subsistence Crops
		(US$/hectare/year)		(US$/labor-day)	
Guatemala	Snow peas/maize	736	52	2.19	1.04
Philippines	Sugarcane/maize	246	124	(3.45)[b]	2.08
Kenya	Sugarcane/maize	181	190	3.53	1.05
Rwanda	Potatoes/maize	226	115	0.36	0.46
The Gambia	Fully water-controlled rice/ swamp rice	593	207	2.46	1.12

[a]The respective subsistence crop returns are adjusted to the multiyear land use situation of sugarcane in the cases of the Philippines and Kenya. (Two maize harvests per year compete with 12 months of sugarcane on the field.)
[b]Returns to total labor are US$3.45. Only 36 percent of labor input in sugarcane is from family labor in the Philippine case, the rest being from hired labor.

the new crops or the crops grown under new technology.[2] The returns to land at least doubled in all cases, except Kenya, and, in some cases, increased severalfold (table 3.9). The most dramatic case was that of export vegetables in Guatemala.

Furthermore, labor productivity in the new crops is substantially higher than in the subsistence crops, in general, except in Rwanda, which should not come as a surprise. In Rwanda's labor-surplus environment, returns per day of work are roughly equal in both the new crops and the traditional subsistence crops.

Government market interventions have a major impact on farm-level profitability of crops in some study settings. For instance, rice producers in The Gambia and sugarcane producers in the Philippines and Kenya are protected by government price and trade policies and benefit from substantial subsidies. Under an agricultural price policy that was oriented more towards long-term international price ratios and their changes, the competitiveness of sugarcane production in the Philippines and in Kenya would be less significant and returns would be more unstable. Similarly, labor productivity in fully water-controlled rice pro-

2. The profitability of crops is expressed here in terms of private returns at farm level.

duction in The Gambia would come close to labor productivity in upland cereal production. Farmers producing commercial crops do not always benefit from government policy actions; for instance, export vegetable production in the Guatemala case was taxed due to exchange rate regulations.

Employment Effects and Women's Work

Commercialization and diversification of agriculture can affect the structure and level of employment. Changes may take place in the use of hired labor versus family labor, the distribution of family labor by gender, and the level of labor input for field operations and for processing.

Increased field-labor demand stimulated employment creation on a particularly large scale in the Guatemalan export vegetable case; employment expanded by 45 percent on participants' farms. Employment expansion was also large in the Gambian case of technological change in rice—56 percent. However, in the case of sugarcane production in the Philippines, employment expanded by only 10 percent. Similarly, the increased processing of crops in rural areas substantially increased employment in the case of export vegetables in Guatemala, but not in the case of sugarcane, whose processing is more capital intensive.

An assessment of the income and employment benefits of commercialization of agriculture is not complete if only the farm household situation is evaluated. The increased income stream for hired labor is an indirect benefit that goes beyond the effects on directly participating farms. In almost all case studies, there was a large expansion in the use of hired labor, which is indicative of a form of commercialization of the rural economy, that of the labor market. In The Gambia, this increase in hired labor use is particularly significant, but it started from a very low base. In the Philippines and Guatemala, the share of hired labor in total labor was already quite high, and there was a large increase from 36 percent to 63 percent in the Philippines, and from 21 percent to 26 percent in Guatemala. Increases in the use of hired labor in these settings create employment for the rural poor.

Not only can labor input patterns in terms of family versus hired labor change in the context of agricultural commercialization, but so can labor input by gender within households (table 3.10). There is great heterogeneity relating to gender-specific crops, work tasks, and seasonal work distribution, not only between study areas but also within them.[3] Changes in cropping patterns and crop technology may affect any of

---

3. Details on these issues are described in the specific case study reports in Part V.

TABLE 3.10 Change in women's labor use when agriculture is more commercialized

	Women's Family Labor in Percent of Total Labor[a]			
Country	Cash Crops/ Crops Under New Technology		Staple Food (Subsistence Crops)	
Guatemala	Export vegetables	21.5	Maize	6.1
	Traditional vegetables	19.5	Beans	18.0
Philippines	Sugarcane	2.5	Maize	9.1
Kenya	Sugarcane	1.2	Maize	50.5
Rwanda	Potatoes	29.5	Maize intercrop	69.9
			Sorghum intercrop	56.5
			Beans	63.6
The Gambia	Groundnuts	21.9	Millet, sorghum	2.1
	Fully water-controlled rice	31.2	Swamp rice	64.5

[a]The labor share of women's family labor reported refers to the observed mean values in the scheme participant groups in the first column and in the nonparticipant groups in the second column, respectively.

these elements. It was found that, in general, women work less on the more commercialized crops than do men or hired laborers, who are also mostly men. Women generally work much more on subsistence crops than they do on commercialized crops, with the possible exception of women in Guatemala. Thus, at least in terms of direct labor input, the cash crops and cash-intensive new technologies have largely become "men's crops."

It is expected that favorable wage rate effects from increased employment would spread the benefits of the increased labor demand in agriculture across a broad spectrum of the rural economy. While these general equilibrium effects may be substantial, they are not traced here.

Marketed Surplus and Price Risks

It could be hypothesized that while the switch to cash crops leads to an overall increase in commercialization of the farm in terms of aggregate output sold, less of the remaining food crops is sold, both in relative and absolute terms. Contrary to this hypothesis, we find that the proportion of total staple food production that is sold (that is, the gross marketed surplus) tends to be higher among scheme participants than among nonparticipants (table 3.11). Even in farms where per capita production of food declined, the marketed share did not decline. The farms that have joined the commercialization schemes are apparently more integrated into the exchange economy, in general, as sellers and buyers of

TABLE 3.11 Marketed surplus (gross) of staple foods when agriculture becomes more commercialized

Country	Crops	Participants	Nonparticipants
		(percent sales out of production)	
Guatemala	All staple foods	6.1	3.9
Philippines	All staple foods	68.0	70.0
Kenya	Maize	15.0	12.6
Rwanda	All staple foods	14.7	3.7
The Gambia	All cereals	32.1	25.0

NOTE: The middle tercile farm-size groups of each sample are presented here to exclude farm-size effects.

commodities, or they have developed greater market participation in staple foods in the context of scheme participation.

Note that marketed surplus in the two study areas with the smallest farms and highest population densities, Guatemala and Rwanda, is very low—only 4 percent of staple food output was sold by nonparticipants (table 3.11). In these locations, it could be hypothesized that even small changes in marketable surplus affect local food prices a lot. Yet, two very different situations apply here and determine potential price effects of changes in local marketable surplus. In Guatemala, the cereal trade functions rather freely interregionally and, despite various government interferences, the local cereal market is reasonably tied to the international one. In Rwanda, on the other hand, the situation is quite different due to the country's landlocked position and deficiencies in local infrastructure and transportation. Here, price fluctuations, due to local supply variations, can be very large—in 1985, for instance, staple food prices were up to three times higher than in 1986 in the study area. The low and cautious adoption of nonfood cash crops such as tea by smallholders in the study area can partly be explained by this situation.

The technical characteristics of crops impinge on the ability of farm households to respond to changing price ratios in the short run. The characteristics of sugarcane, for instance, give it a much less short-term ability to respond to price changes—even if contractual arrangements would allow it—than do, for instance, potatoes, export vegetables, or rice in the other study environments. Sugarcane, with its potentials for harvesting ratoon crops, represents a semifixed factor situation to a farmer who has switched to it. If, after the switch and investments are made, the terms of trade between sugarcane and a competing crop (say, maize, in the Philippines and Kenya) shifts in favor of the competing crop, then moving out of sugarcane is constrained in the short run. Production will continue as long as variable costs are covered. It is therefore of interest to look into the evolution of the terms of trade to assess if farmers who opted for sugarcane production got trapped in a disadvantaged situation because of adverse terms of trade developments. Figure 3.1 suggests that this was not the case, in general, in the Kenyan and Philippine situations. In the Philippine case, however, terms of trade were far from stable. Quota regulations and protection-cum-stabilization of the domestic sugar price played an important role in stabilizing favorable returns from sugar for farmers for some time. The sustainability of such a policy in the long run is, however, questionable.

Price risk is a key consideration for farmers who adopt more commercial crop mixes. However, while a careful consideration of price risk is called for in the commercialization schemes, the existence of alternative income risks to small farmers, if they do not choose to reallocate labor and land resources into more commercialized agriculture, should also be taken into account. Frequently it is not a steady in-kind income

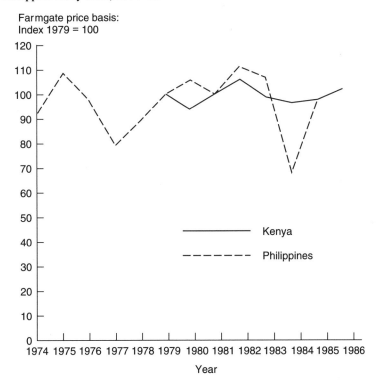

FIGURE 3.1 Local terms of trade of sugarcane/maize in the Kenyan and Philippine study areas, 1974–86

stream from subsistence food production, but rather an uncertain and risky income stream from off-farm work, that is the appropriate point of reference for comparing with the risks of cash cropping. In Guatemala, for instance, the main income alternative to export vegetable production is more off-farm work in urban services and the large-scale plantation sector (seasonal work). The income flow from off-farm employment may hardly fluctuate less and may even be riskier than that from export vegetable production.

Households that adopt more commercial agricultural production are pressed to constantly solve new intertemporal cash management problems. Thus, extension services and savings facilities are important in the commercialization programs to facilitate adjustment to new situations, especially in the short run. Moreover, information acquisition, among other dimensions, may be important to farmers' perceptions of risks and their attitudes toward it (Nerlove 1988). Market information for the small farmer may be of critical importance, especially when interregional trade and specialized crops become more of an option for production.

Income Effects and Implications

With few exceptions, commercialization of agriculture in the study settings has directly generated employment or increased agricultural labor productivity or both. The direct beneficiaries are farm households participating in the schemes and, to some extent, hired laborers. The direct income effects, expected to be generally positive, are further complemented by indirect income effects through forward and backward linkages that are generated by the increased demand for goods and services by the direct income beneficiaries as well as by increased demand for inputs for commercialized agriculture.

The wage rate and employment effects from commercialization are not restricted to the schemes; benefits can be spread across regions and far away from the schemes when family labor from participating households is withdrawn from the off-farm labor market or when hired labor migrates into the scheme areas. The more mobile the labor force, the less a wage rate effect is to be expected locally and the more it is spread across the economy. For instance, both effects appear important in the Guatemalan case, where much of the additional hired labor came from other communities outside the export vegetable cooperative and family labor reduced seasonal outmigration to the large-scale farm sector. The second effect, the spread of the wage rate effect across the economy, played a role, for instance, in The Gambia, especially during the drought year of

1984, when the scheme attracted labor even from neighboring countries such as Senegal and Guinea.

Aggregate Income Effects and Multiple Income Sources

Total per capita income in all study cases—holding farm size constant—is higher among participants than nonparticipants, except in Papua New Guinea and Sierra Leone (table 3.1). However, the relative differences are much less than might have been expected, given the large increases in land and agricultural labor productivity noted before. The incrementally employed labor and land had, of course, opportunity costs. The cash crop share in total income among participants ranges between 11 and 35 percent. If total expenditure (including value of home-produced food) is taken as a reliable proxy for income, the participants of commercialization schemes in the five IFPRI case studies (Guatemala, The Gambia, Rwanda, the Philippines, and Kenya) emerge with increases in income of 17 to 25 percent,[4] while returns per day to family labor in the new crops about doubled in most cases (Rwanda being the exception).

The aggregate income effect is much lower than the agricultural income effect because of the limited share of income from crops in total income and because of substitution between on-farm and off-farm work. The change in agricultural income due to commercialization is much greater than the change in overall income because in all study settings farm households depend on a wide range of multiple income sources, both farm and nonfarm (see figure 3.2). In relative terms, off-farm income is highest in the most land-scarce settings (Guatemala and Rwanda), as should be expected. Staple food income, even among project participants in all settings except Guatemala, remains higher than cash crop income (figure 3.2).

Specific Gainer-Loser Situations

It would probably be idealistic to assume that an agricultural development and growth process that fosters the transition from semisubsistence to commercial agriculture could be designed without any relative losers in the process. However, program and project design has to take the complex gainer-loser patterns into account and consider appropriate ways of short-term compensation and long-term income generation for the potential losers. At the household level, losers certainly include those farmers who are displaced by the introduction of commercialization

---

4. More refined model analyses than these comparisons of mean values in similar farm-size classes confirm this range of net income gains due to scheme participation.

FIGURE 3.2 Income and income sources of scheme participants and nonparticipants

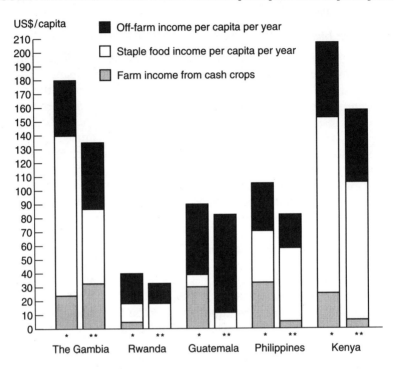

Note: Only the middle tercile farm-size groups are represented in the figure to exclude farm-size differences.

* Participants
** Nonparticipants

schemes and not fully compensated. Particular attention has to be paid to those households that may lose not only relatively but also absolutely, to the extent that commercialization poses a food security problem and nutritional risk. Frequently the losers are only small groups but, nevertheless, they require attention.

The gainer-loser patterns in each study setting are complex, and the following examples may shed light on the diversity of the pattern.

THE GAMBIA. Pastoralists lost grazing grounds in the area that was taken over by the fully water-controlled rice scheme. Individual women farmers lost rice land, most of which became a communally farmed area for the compound as a whole, under the control of male household heads. Women, in general, lost out to men in terms of production and income.

GUATEMALA. The increased returns to land from export vegetable production may have put upward pressure on land rental values and thereby made more costly the provision of food security for those households that obtain most of their cash income from off-farm sources and want to maintain a certain level of "food insurance" on the basis of own production on rented land.

KENYA. Some households were displaced by the sugar factory. The relocated households own substantially less land than before the creation of the scheme, and even less than that owned by the nonsugar producers. The income per capita of the relocated households is slightly lower than the non-sugar-producing comparison group, but more striking is the significantly lower calorie intake per adult equivalent in the relocated households. Also, the incidence of calorie-deficient households is significantly higher in the relocated group. However, there are no differences, in general, in the average nutritional status of children and the total amount of time ill for children from the two groups.

RWANDA. Households that adopted smallholder tea production experienced returns to tea that were not competitive with crops such as potatoes and cereals. Legal regulations did not permit them to completely abandon the tea fields and convert them back to subsistance crops.

Some farm households were displaced by tea factories. Seventy-two percent of the surveyed displaced households reported smaller farm sizes than before. These households also worked much more for income from off-farm sources. On average, however, the relocated households were not found to consume less food (calories) on an adult-equivalent basis than other sample households. Entitlements to food were maintained, despite the reduced farm resource base, via off-farm employment opportunities.

PHILIPPINES. Fifty percent of the households that were primarily engaged in sugar production were landless. A large proportion of these landless households identified themselves as corn tenants before the establishment of the sugar mill. Some were even former corn landowners. Thus, a substantial number of former corn farmers experienced a decline in tenancy status as a result of the introduction of sugar.

Conclusions

Commercialization of agriculture is a reality in many developing countries and an important part of their development strategies. Frequently, however, countries adopt a mix of policies that combines direct and indirect promotion of cash crops with discrimination against commercialization, the net outcome of which may vary a lot. There is considerable evidence from several developing countries that export crop producers are heavily taxed, mainly through marketing boards and overvalued exchange rates (see, for example, Bautista 1987; Tshibaka 1986; Oyejide 1986; Krueger, Schiff, and Valdés 1988).

The more specific policy findings from the comparative analyses of the case studies highlight the program-level challenges of a balanced strategy for promotion of agricultural commercialization. They suggest that considerable potentials exist to address the food security problems of the poor with promotion of commercialization of agriculture. The following conclusions relate to program-level issues, acknowledging that appropriate trade policies are a precondition for tapping the long-term benefits from commercialization.

- Smallholder producers make a conscious effort to maintain subsistence food production alongside the new cash crops. They do this despite higher returns to land and labor from the cash crops in the schemes studied. This reliance on food from own production under household control is a response to market, employment, and production risks and can be viewed as an insurance policy by farm households in a risky income environment. Theoretically, this strategy is a second-best option compared to full market integration, since related benefits of specialization are forgone. However, in risky economic environments, maintenance of own food supplies is certainly a sensible strategy. Agricultural policy can effectively support this strategy by promoting technological change in staple (subsistence) foods. This also provides further room for specialization at farm levels and thereby permits capturing of further gains by the economy from commercialization and market integration of smallholders. Rapid technological change in food production that increases yields per unit of land and output per unit of labor must play a parallel role in agricultural and rural commercialization, especially in African settings.
- The smallest farm households participate less than proportionately in the commercialization schemes, but when they participate, they tend to be the more radical adopters of the new cash crops. Scheme design and management can enhance integration of the smallest farms into the schemes, especially where access is rationed.
- The employment effects for the poor that result from commercialization are generally large but are very crop specific and a function of the local labor market and the new technologies introduced. Choice of crop and technology, therefore, has a major implication for the actual outcome of the employment effects. This applies not only to on-field employment creation, as exemplified by the substantial

employment increases in vegetable production in Guatemala and potato production in Rwanda, but also to the processing and trading employment that results from commercialization. The commercialization of agriculture generally entails a substantial expansion of the demand for hired labor. To the extent that hired-labor households rank among the malnourished poor, this employment effect is expected to be of particular benefit.
- Positive income effects of commercialization programs and projects were generally observed for scheme participants, but not necessarily for all households or for all components of the commercialization process. Although substantial, the net income gains were generally much less than the gross income from the new cash crops because of major substitution effects within agricultural production and between agricultural and off-farm employment. The latter was particularly notable in the Guatemalan case: off-farm income earnings were reduced when the labor-intensive export vegetable production drew family labor to the farm fields. In The Gambia, double-cropped irrigated rice production gained to a large extent at the cost of upland crops such as groundnuts and millet.
- In the short run, some households lost income because of the commercialization schemes. This group of losers is rather small and heterogeneous across the study settings. In the Philippines, the rapid expansion of sugarcane production contributed to the creation of a landless class of households that used to be tenants growing corn on rented land before the introduction of the sugarcane. An important contributing factor to the consolidation of landholdings was a long-run decline in corn productivity, which discouraged smallholders, tenants, and landowners from continuing to produce corn, and which resulted in declining incomes of the poor before the introduction of sugarcane. Careful ex ante assessment of possible creation of absolute losers is required. General employment expansion cannot be relied upon to reach out to these groups in the short run.
- It is common for women's work in cash crops and new crop technologies and for women's direct control over income from the new cash crops to be much less than that of men and frequently even to be disproportional to their labor input into these crops.

In none of the commercialization schemes studied did women play a significant role as decision makers and operators of the more commercialized crop production line, not even where typical "women's crops" were promoted (rice in The Gambia) or where the agricultural production environment was largely female dominated (potato with modern inputs in Rwanda). These findings, however, should not be interpreted as if women did not benefit indirectly from the income and employment gains provided through commercialization and technological change in agriculture. Judging distribution of benefits only from the production and labor side may be misleading. This becomes clear when spending patterns of income in the study settings are reviewed.

In sum, there is not much supportive evidence for a pessimistic position vis-à-vis commercialization of smallholder agriculture, given the production, employment, and income effects found in the large majority of the researched real-world examples. Where adverse effects emerged, these were due largely to policy and program design. The positive aggregate income effects for nutritional improvement were highlighted up front; further insights on process and on cause and effect linkages are addressed in the following two synthesis chapters.

4 Consumption Effects of Commercialization of Agriculture

HOWARTH BOUIS

Introduction

The commercialization of agriculture has, in many diverse circumstances, led both to an increase in household income and to changes in the way household resources are organized to earn that income. Have these changes meant that food intakes are more nutritious and that health and sanitation conditions are improved? This chapter addresses three central questions: (1) to what extent are increments in income spent on nonfood items, in particular, health-related items; (2) to what extent are increments in income spent on food, and (controlling income) does the switch to commercial production alter the marginal propensities to spend on food; and (3) to what extent do increments in food expenditures lead to greater calorie intakes, both at the household level and at the individual preschooler level?

To answer these questions thoroughly requires collecting household survey information on incomes by family member, on expenditures by type of item, and on food intakes by family member. This is a time-consuming and expensive process, with the most effective methods varying by cultural setting.

While it was not feasible to collect all the desired information in all the study areas, in all case studies, expenditure surveys were undertaken that recorded food and nonfood expenditures over varying time periods for a disaggregate list of items.[1] Per capita total expenditures are used as a proxy for income because it is difficult to measure income directly and because expenditure data are thought to be a better reflection of "perma-

---

1. Information was collected on food items consumed out of own production, earned as in-kind wages, or borrowed. The value of this nonmarket food consumption was added to market purchases of food and nonfood items to estimate total expenditures.

nent" income. In most case studies, a direct measure of income was also derived.[2]

Table 4.1 provides a summary of the detailed findings of all eleven case studies with respect to the effect of commercialization on total food expenditures, calorie consumption, and income control at the margin as it affects food-acquisition behavior. These topics are discussed in more detail below.

It was in the collection of calorie-consumption information that the greatest divergence in survey methods occurred. In part, this divergence was dictated by cultural factors related to eating habits. In some case studies, for instance, in Rwanda and The Gambia, a seven-day recall of foods cooked (divided by persons present for these meals) was conducted to give an estimate of household calorie intake.[3] In other case studies, for example, in Guatemala, Kenya, and the Philippines, food quantities from the expenditure information were converted to calories to give household calorie availability.[4] In Kenya, the Philippines, Papua New Guinea, and Sierra Leone, in addition, a 24-hour recall of foods consumed by the household was administered. This not only produced an estimate of household calorie intake as a cross-check of the household calorie availability information, it also gave estimates of calorie intakes of individuals within the household. Furthermore, it provided an estimate of family food consumption that was surveyed independently of total expenditures. This separation of surveyed information means that a potential source of upward bias is avoided when using regression techniques to measure the effect of income on food intakes (Bouis and Haddad 1992).

Among the six study sites reviewed below, the highest per capita total expenditures are found in The Gambia. Average expenditures are somewhat lower, but roughly comparable, in Guatemala, Kenya, and the Philippines, lower still in Rwanda, and lowest by far in Malawi. The largest difference between the more commercialized and less commercialized farmers is found in the Philippines; this large difference is partly

---

2. Farm profits were computed from detailed farm production information and were then added to wages and salaries earned off the farm, profits from off-farm business activities, and income transfers. In some instances, it was possible to compute the share of total household income that was earned by the wife or that was earned from commercial production. These variables are used to test for changes in the marginal propensity to buy food while controlling for changes in total household income.

3. During this seven-day recall, information was also collected on snack foods, food prices, and expenditures in order to compute household food expenditures used in the total household expenditure calculation described above.

4. An estimate of calories from meals taken by nonfamily members was subtracted from this total.

TABLE 4.1 Overview for case study settings of food expenditure and consumption effects

Country	Commercialization Scheme	Was More Spent on Food?	Did Calorie Consumption Improve?	Did the Source of Income (Women's/Cash Crop) Affect Food Expenditure?
Guatemala	Export-producing cooperative of Cuatro Pinos	Yes, about 18 percent more per capita on average. More expensive calories were acquired. The budget share to food was slightly lower.	Yes, significantly with income, but the effect was decreasing at the margin. No significant changes in diet composition.	Scheme participation and women's off-farm income share did not affect food budget share. Increases in male nonagricultural income decreased food budget share. Less of the marginal income was spent on food, perhaps because export crop income is male controlled.
Philippines	Bukidnon Sugar Company	Yes, especially on meat whose consumption increased dramatically as incomes rose. More expensive calories were purchased. Food budget share declined.	Yes, for both households and household members, since, at the margin, calories were distributed fairly equally. greater variety in diets.	Information not available.
Papua New Guinea	Karimui Spice Company cardamom plantation (in adverse times)	No, in fact, slightly less was spent. More expensive calories were bought. Food budget share increased as employment opportunities dwindled.	No, there was slightly less food energy available, especially when employment dried up and access to purchased foods diminished.	The share of wage income had a negative but insignificant effect on energy availability, whereas the share of cash income to females positively and significantly affected food availability.

continued

TABLE 4.1 Overview for case study settings of food expenditure and consumption effects (Continued)

Country	Commercialization Scheme	Was More Spent on Food?	Did Calorie Consumption Improve?	Did the Source of Income (Women's/Cash Crop) Affect Food Expenditure?
India	Karnataka Dairy Development Project	Information not available.	Energy consumption slightly increased, but there was a highly significant decrease in consumption of dairy products.	Information not available.
Kenya	South Nyanza Sugar Company	Yes, more expensive calories were acquired, particularly from meat and fruits.	Yes, not only did new scheme entrants consume significantly more calories, but a smaller proportion of the group consumed less than 80 percent of requirements.	Nonfarm income has a negative effect on household calorie consumption, whereas women's income has a beneficial effect. These effects are related to control of income within the household.
Kenya	Two rice schemes: Ahero Irrigation Scheme and West Kano Irrigation Scheme	Information not available.	For the resident tenants, it did not. For nonresident tenants, to some extent it did, but calorie consumption of individual rice growers was highest. No changes in the staple diet.	Women's income from casual labor and local sales of paddy and bund crops increased per capita food expenditures.

Rwanda	Potato production in the Gishwati forest area	Yes, more expensive calories were acquired, both from meat and more preferred food staples, such as potatoes.	Yes, significantly. On average, potato production in Gishwati provided an additional 40 percent calorie production, making these households surplus-calorie producers.	Among poor households, female-headed households consume more staple foods per capita.
Zambia	Technological change in maize	Information not available.	Yes, per capita calorie consumption is higher for both small and large farms in high-adoption areas.	Information not available.
Malawi	Commercialization of maize and tobacco	In absolute terms, no. But the higher income is spent on more expensive food such as meat and fruits than on grains and vegetables.	Yes, especially for those in the top third of the income distribution.	More of women's expenditures go to food purchases than men's, 25 percent versus 13 percent, respectively for maize, for instance.
Sierra Leone	Bo-Pujehun Rural Development Project-Tree crop promotion	No. In fact, less was spent absolutely and relatively by new adopters who suffered from a cash flow problem.	Yes, slightly, despite temporary income reduction for the new adopters.	No, an increased share of tree crop income in total income has no significant impact on calorie intake.
The Gambia	Jahally-Pacharr Smallholder Rice Project	Yes, more expensive calories were acquired, particularly from meat and fruits.	Mixed. As women lose control over rice production, calorie consumption is reduced. At the same time, an increase in total household income favorably affects calorie consumption.	A reduced share of cereals from women's production significantly reduces calorie consumption.

due to the larger landholdings of the commercialized farmers there, who are more likely to adopt sugar.

Nonfood Expenditures

What nonfood items do households purchase at the margin when incomes increase? Comparable information from six case studies on expenditures for health, housing, clothing, education, and transportation indicate that, in general, expenditure elasticities (as suggested by the ratios of expenditures of the top and bottom terciles) are highest for housing and lowest for education and clothing (table 4.2). Nonfood budget shares tend to be higher in Guatemala and the Philippines for almost all categories, particularly health and transportation. Education expenditures stand out as particularly important for the Philippines, and clothing expenditures as relatively important for the African countries. The relatively low budget shares for health and housing suggest that such expenditures do not yet constitute a major link in any observed improvement of nutritional status as income increases. In fact, such low expenditures may be important constraints to the realization of improved weights and heights of preschoolers at higher incomes. If more and better health facilities were made available in rural areas and if rural populations were provided with health and nutrition education, this could result in significant household budget reallocations.

TABLE 4.2 Expenditure elasticities and budget shares for selected nonfood items

Item	The Gambia	Guatemala	Kenya	Malawi	Philippines	Rwanda
		(ratio between top and bottom tercile)				
Health	3.0	24.9	1.6	2.1	6.3	4.2
Education	1.5	2.3	1.4	3.8	5.0	1.6
Housing	9.3	—	14.1	4.0	22.8	7.7
Clothing	2.3	4.6	4.1	3.6	5.0	2.5
Transportation	3.4	7.9	5.9	10.7	12.5	1.2
Total expenditure budget share (average)			(percent)			
Health	0.9	2.9	0.3	1.4	2.1	1.3
Education	0.3	0.3	0.9	0.3	2.3	0.9
Housing	1.0	2.3	1.0	0.4	2.0	2.8
Clothing	5.8	4.5	2.7	10.1	4.3	7.5
Transportation	2.5	5.3	2.7	0.9	3.5	0.5

SOURCE: Data sets of respective case studies.

Food Expenditures

In many of the case studies, food budget shares are well above 50 percent and are close to 80 percent in Rwanda and Kenya. Food budget shares generally decline with rising incomes, although more money is spent on food. In several instances, more expensive foods such as meat were consumed. This is especially the case in the Philippines. Comparable information from The Gambia, Guatemala, Kenya, Malawi, the Philippines, and Rwanda suggests that, controlling for income, the four African countries tend to have higher food budget shares than Guatemala and the Philippines. For example, food budget shares in Kenya exceed 75 percent, whereas at similar income levels in the Philippines, food budget shares are about 65 percent. Food expenditures did increase rapidly with income in all six countries, but much more so in the African countries. This difference in expenditure behavior may be due to the relatively easy access to consumer goods in Guatemala and the Philippines compared with Africa.

As incomes more than double in Africa, the food budget share declines only slightly, especially in The Gambia and Rwanda. This contrasts with the much more rapid decline in Guatemala and the Philippines. These patterns indicate that households in the four African study countries spend a much higher proportion of their incremental income on food. Direct estimation of food expenditure elasticities with respect to total expenditures gave values of 0.94 and 1.00, respectively, for The Gambia and Rwanda, but a lower value of 0.84 for the Philippines.[5] These elasticities are evaluated at the mean of the data. Estimations for Guatemala and the Philippines (not shown) indicated that these elasticities are not constant across income groups but decline at higher income levels.

Differences in the Marginal Propensity to Spend Cash Crop Income

The discussion thus far has centered on how increases in income, from any source, were spent. But when income is controlled, do marginal propensities to spend for food differ across groups that are more or less commercialized? These marginal propensities could differ for several reasons. For instance, production and consumption decisions may not be separable; a household's decision to produce a subsistence crop may well affect the implicit price it pays for that subsistence crop if marketing margins are substantial. Or certain household members (for example,

---

5. An alternative estimate for the Philippines, obtained with predicted income (using two-stage least squares) in place of total expenditures, gave a lower estimate of 0.65. A value of 0.84 probably overestimates the relationship because of common errors in measuring food expenditures and total expenditures.

women) may participate strongly in subsistence production but not in commercial crop production, which may alter their control over income earned or change household energy requirements by transforming time allocation patterns, or both. In The Gambia, Kenya, and the Philippines, for instance, the share of income from cash crops did not significantly affect the marginal propensity to spend on food.[6] In Guatemala, on the other hand, it was estimated that an increase in the share of cash crop income from 0 to 50 percent led to a 1.2 percent decrease in the share of expenditures on food. Although statistically significant, this effect, in practical terms, is small. These results may be contrasted with those from Rwanda, where a 10 percent increase in the share of cash crop income led to a 4.8 percent decrease in the food budget share, suggesting that cash crop income was treated quite differently from other forms of income in the intrahousehold allocation process.

Food expenditure effects of income earned specifically by women were positive in most of the case studies where such information was available (table 4.1). In The Gambia, a shift away from production of the traditional women's crop led to reduced calorie consumption, holding income constant; however, the increased income due to commercialization more than compensated for this effect. In Rwanda, after controlling for income, female-headed households in the bottom income quartile consumed 17 percent more calories than male-headed households. Likewise, in Kenya (in both case study areas), Malawi, and Papua New Guinea, income going to women increased calorie intake significantly. Only in Guatemala was the effect of female income share on food expenditures not found to be statistically significant. With the possible exception of Rwanda, cash crop income generally had little effect on food expenditures, controlling total household income.

Household Calorie Availability

To what extent do higher food expenditures lead to higher calorie consumption at the household level? It is widely observed that, as incomes increase, more expensive calories are purchased so that a given percentage increase in food expenditures will lead to a lower percentage increase in calorie availability at the household level.

Cost per Calorie

It is evident from available data from The Gambia, Guatemala, the Philippines, and Rwanda that cost per calorie increases between the

---

6. For Kenya, the dependent variable was household calorie intakes instead of food expenditures. The effect of income from sugar was insignificant for the subsample that excluded nonagricultural households.

TABLE 4.3 Calories purchased per U.S. dollar

Country	Average		Ratio Between Top and Bottom Terciles[a]	
	Participants	Nonparticipants	Participants	Nonparticipants
	(thousands)			
The Gambia	9.55	10.37	1.53	1.56
Guatemala	12.16	15.07	1.66	1.66
Philippines	7.76	8.64	1.48	1.41
Rwanda	10.25	11.28	1.17	1.20

SOURCE: Data sets of respective case studies.
[a] Food expenditures roughly double from bottom to top tercile, except in Guatemala and Rwanda, where food expenditures increase by somewhat more, 150 percent.

bottom and top expenditure terciles (table 4.3). The smallest increase is observed for Rwandan sample households, whose incomes are the lowest of the four countries in table 4.3, but the largest increase in cost per calorie from the bottom to the top expenditure tercile is observed in Guatemala, where average household expenditures are lower than in The Gambia but where there is a wider variation in expenditures. After controlling for income, costs per calorie hardly differ between scheme participants and nonparticipants.

Direct estimation of household calorie consumption with respect to total expenditure (which combines the effects of "leakages" due to lower food budget shares and more expensive calorie sources) resulted in elasticity estimates of 0.48 and 0.50 at the high end of the scale for The Gambia and Rwanda, respectively, and 0.31 and 0.34 for Guatemala and the Philippines, respectively, at the low end of the scale. These results suggest that a doubling of income results in a 50 percent increase in household calorie consumption in The Gambia and Rwanda (and that the divergence from a calorie increase of 100 percent is almost entirely due to the purchase of more expensive calorie sources), but only a 30–35 percent increase in Guatemala and the Philippines. Although there is reason to believe that all of these elasticities are upwardly biased, their relative magnitudes indicate significant differences in food consumption behavior, again between the African and non-African study sites.[7]

7. A similar relationship was estimated for Kenya and the Philippines, using household-level calorie intake data derived from a 24-hour recall of foods consumed. These resulted in much lower elasticity estimates, 0.17 and 0.11 for Kenya and the Philippines, respectively. Use of calorie and food expenditure data (to construct total expenditures) from the same survey information can lead to upwardly biased estimates. Also, if there is a strong correlation between total expenditure, meals taken in other people's homes (low-income households), and meals provided to guests and hired workers (high-income households), and care is not taken to record these meals for nonfamily members accurately, there will be an additional upward bias (Bouis and Haddad 1992).

How do diets change as incomes increase? What types of food do higher-income households purchase that are expensive calorie sources? In Guatemala and the Philippines, meat, which is one of the most expensive sources of calories, accounts for at least 20 percent of food expenditures, even for the lowest expenditure tercile. The budget share for meat increases with income especially in the Philippines, with a consequent decline in the budget share of staple foods, which are the least expensive calorie sources. In the African countries, by contrast, the budget share for meat does not reach 20 percent even for the highest total expenditure tercile—although it should be noted that the share of fish in the Malawian food budget is quite high. The extreme case is that of Rwanda, where the budget share of staple foods is 70 percent and of meat only 6 percent, even for the highest total expenditure tercile. However, in Rwanda, as incomes increase, there is a shift away from sweet potatoes, a cheap calorie source, to potatoes and sorghum, which are more expensive calorie sources. The tendency for the budget share of meat to increase and of staples to decrease with increases in income is observed in the four African countries, but it is much less pronounced than in Guatemala and the Philippines.

It is apparent, then, that "leakage" (that is, income increases that are not generating commensurate increases in calorie consumption) is greatest where markets for nonfood items are better developed, and where there are stronger preferences for meat. To the extent that meat and other nonstaple foods improve dietary quality and provide necessary micronutrients, it may be that expenditures on such items are as important in improving nutrition as are further increases in calories.

Other Determinants of Demand for Calories

The calorie income elasticities (evaluated at mean expenditures) are estimated for five countries that have available data (table 4.4). Besides income, the price of the primary food staple, years of formal education of the mother, and household size are three additional variables in the demand for calories. The elasticity with respect to staple price is consistently negative or statistically not significant across the various studies. There is no evidence that formal education of the mother is a significant determinant of demand for calories. The effect of household size is mixed; in some cases, it positively affects calorie consumption and, in others, it has a negative effect.

Market Dependency and Household Food Security

It is often argued that calorie costs may rise with commercialization and that household food security may deteriorate because households may rely less on own-produced food and more on food purchases in the

TABLE 4.4 Selected determinants of household demand for calories

Country	Income Elasticity	Staple Price	Mother's Education	Household Size
The Gambia	0.37–0.48	−	NI	−
Guatemala	0.31	−	NI	+
Kenya	0.17	NI	0	+
Philippines	0.11	0	0	−
Rwanda	0.47	−	NI	−

NOTES: A minus (−) indicates that variable was negative and statistically significant; a plus (+) indicates that variable was positive and statistically significant; a zero indicates that the variable was not statistically significant; and NI indicates that variable was not included in regression estimations. Calorie intake is the dependent variable for Kenya and the Philippines; calorie availability is the dependent variable for The Gambia, Guatemala, and Rwanda. Income elasticities are evaluated at the mean of the data.

retail market. For example, Philippine farmers who grew and consumed their own corn saved a premium of 25 percent on the retail price by avoiding marketing costs. Or, food prices may rise locally if a region shifts from being a net exporter to a net importer (the rise may be even higher if government policy constrained interregional trade of food), which would offset the income gains from commercialization.

Did participation in commercialization schemes significantly affect dependency on market purchases of food? Data from six case studies suggest that, in general, there was not a significant change in sources of food acquisition (table 4.5). At one end of the spectrum, Guatemalan farmers, whether engaged in commercial agriculture or not, were almost totally dependent on market purchases of food. At the other end, the

TABLE 4.5 Sources of food acquisition (in value terms)

Country	Own-Farm Production		Purchases	
	Participants	Nonparticipants	Participants	Nonparticipants
		(percent)		
The Gambia	51	55	49	48
Guatemala	17	14	83	86
Kenya	58	60	42	40
Malawi	55	50	45	50
Philippines	35	48	65	52
Rwanda	62	61	38	39

SOURCE: Data sets of respective case studies.

more commercialized farmers in the four African countries continued to produce more than half of their food (in value terms) on their own farms. In Malawi, tobacco farmers actually depended less on market purchases of food than did farmers who did not grow tobacco; tobacco farmers had larger farms and, thus, were not as dependent on off-farm employment. Hybrid-maize growers in Zambia relied much less on purchased maize meal than households that did not grow hybrid maize (table 4.1).

Households/Individuals Below Calorie Requirements

In general, the calorie consumption of households in the lowest expenditure tercile is deficient, while that of households in the highest tercile is either well above recommended allowances or approaching recommended levels (table 4.6). Where food expenditure survey information was used to compute household calories, increases in income appear to have a strong influence on nutrient consumption. The 24-hour recall data indicate that the influence of income is still positive, but the magnitude of the effect appears to be much smaller. This dichotomy in patterns is reflected in the substantial differences in the calorie-income elasticities cited earlier, which were generated using the various methods for collecting calorie consumption information and total expenditures.

Average adequacy ratios give only a partial indication of the variance in calorie consumption within the survey populations. It is also important to know what proportion of households or individuals fall

TABLE 4.6 Household-level calorie adequacy ratios, by total expenditure tercile

| | Calorie Adequacy Ratio | | | |
| | Total Expenditure Tercile | | | |
Country	Bottom	Middle	Top	Average
The Gambia[a,b]	0.88	1.07	1.25	1.08
Guatemala[a,c]	0.85	1.08	1.22	1.05
Kenya[d,e]	0.80	0.95	1.04	0.92
Preschoolers	0.58	0.56	0.62	0.58
Philippines[e,f]	0.83	0.92	0.97	0.91
Preschoolers	0.71	0.74	0.81	0.75
Rwanda[a,f]	0.72	0.91	1.16	0.93

SOURCE: Data sets of respective case studies.
[a]Recommended intake = 2,800 calories.
[b]Based on food consumption surveys (7-day recalls).
[c]Based on food acquisition survey.
[d]Recommended intake = 2,850 calories.
[e]Based on calorie intake surveys (24-hour recalls).
[f]Recommended intake = 2,580 calories.

below a threshold of calorie requirements, say, 80 percent. A group average that is above the recommended daily allowance (RDA) may mask a substantial number of households that are consuming below the RDA, while, conversely, some low-income households associated with a low group average may be consuming well above the average. In addition, there may be substantial variance across seasons and across individuals within households. For example, in The Gambia, there is considerable seasonal variation in the percentage of households that fall below 80 percent or 60 percent of their caloric requirements. Nevertheless, the decline in the percentage of households that are consuming less than 80 percent of caloric requirements across expenditure groups is much more precipitous than the increase in average calorie-adequacy ratios.

Even if calorie-adequacy levels are above the RDA at the household level, this does not necessarily mean that all household members are consuming their recommended intakes, because calories may not be equally distributed among family members. Data from Kenya and the Philippines show that even though household calorie-adequacy ratios are close to 1 for the highest income terciles, calorie-adequacy ratios for preschoolers are well below 1. The apparent inequitable distribution of calories among household members, however, requires a closer look at recommended calorie intakes vis-à-vis actual requirements due to energy expenditures (activity patterns) and reduced weights and heights as a result of long-run malnutrition. Regression estimations, with preschooler calorie intake as the dependent variable, indicate, for both Kenya and the Philippines, that marginal increases in household calorie intakes are shared equitably with preschoolers. For the Philippines, calorie intake is negatively associated with morbidity, while for Kenya it is positively associated with the number of meals per day that the preschooler is fed.

Conclusion

Preferences for nonfood items and higher-priced calories at the margin as incomes increase result in calorie consumption at the household level increasing much more slowly than income. As incomes increase, food purchases at the margin are directed toward more expensive sources of calories, which may improve the quality of the diet and increase the intake of micronutrients. Calorie intake data from 24-hour recall surveys for Kenya and the Philippines indicate that household-level calorie adequacy does not translate into adequate calorie intakes for preschoolers, although calorie adequacy figures obviously are sensitive to the RDA used to assess adequacy.

In general, commercialization has increased incomes, which, in turn, has led to higher calorie intakes. There is little evidence that commercialization per se has altered behavior patterns in a manner that is detrimental to nutrition. Any negative tendencies to spend less for food, because of loss of income control by women or because of increased involvement in market (cash) transactions, are generally small and are more than compensated for by increased incomes due to commercialization. Greater dependence on the market for food purchases has not led to noticeably higher food costs per calorie consumed, nor has this market dependency resulted in a decline in household food security. However, there appears to be much scope for improvement in terms of redirecting expenditure behavior and intrahousehold allocational decisions toward placing more emphasis on improved nutritional outcomes.

5 Health and Nutrition Effects of Commercialization of Agriculture

EILEEN KENNEDY

Introduction

One of the most contentious issues in the cash crop/food crop debate revolves around the impact of commercialization of agriculture on the health and nutritional status of women and children. This chapter examines the effects of commercialization of agriculture on preschoolers' health and nutritional status. The chapter also assesses the effects on women along the lines of the conceptual framework described in chapter 2.

It is typically assumed that increases in household income will ultimately result in health and nutritional benefits to individual household members. This income-mediated effect on health and nutrition operates through two main pathways. First, increased incomes can be used to purchase either a different mix of goods and services or more of the current market basket, for example, more access to health care, better housing, and so forth. This new or increased market basket could produce a positive health effect. Second, income–food consumption linkages, by improving an individual household member's energy or other nutrient intake, could improve nutritional status, which in turn could improve health. Each of these pathways will be examined in this chapter. Table 5.1 gives an overview on all case study findings as regards health and nutrition effects.

Morbidity and Nutritional Status of Preschoolers

No significant differences are detectable in the percentage of preschoolers who are ill from households that participate in cash crop production and households that do not participate in four of the case study settings: the Philippines, Kenya, Rwanda, and The Gambia (table 5.2). The high rates of illness among preschoolers in The Gambia and

TABLE 5.1 Overview for case study settings of health and nutrition effects

Country	Commercialization Scheme	Was There a Favorable Effect on Health and Nutrition?	Did Source of Income Affect Health and Nutrition?
Guatemala	Export vegetable—producing cooperative of Cuatro Pinos	Not much. The prevalence of weight deficiency and stunting was slightly less. But when households were longer in the cooperative, positive effect increased.	Yes, increased income decreased stunting and weight deficiency, but the effect was reduced at higher income levels. A higher off-farm nonagricultural income share is associated with reduced weight deficiency, with the effect being greater in the case of women-controlled income. Higher export-crop income share did not appear to have adverse effects on children's nutrition.
Philippines	Bukidnon Sugar Company	No, these children are significantly more stunted and experience diarrhea and fever more frequently.	No, it seems that higher-income households purchase nonfood items and higher-priced calories while their preschoolers have less than recommended intakes.
Papua New Guinea	Karimui Spice Company cardamom plantation (in adverse times)	Yes, generally, but those children whose mothers worked at the KSC did not fare as well as other Karimui children. As economic conditions worsened, the generally favorable trend reversed. Frequency of diarrhea was higher.	Availability of household food energy or the share of purchased food had insignificant effects on children's nutritional status. The prevalence of diarrhea had a very negative effect on children's weight-for-age and height-for-age.
India	Karnataka Dairy Development Project	Information not available.	Information not available.
Kenya	South Nyanza Sugar Company	No, there was no significant difference in total time ill or time ill with diarrhea. Also, no difference in the prevalence of stunting, wasting, or weight-for-age. No significant change in total time ill for women.	Income is not a significant determinant of illness for either preschoolers or women, at least in the short run. The health and sanitation environment had the greatest impact on preschooler growth. Children's height and weight are positively determined by preschooler caloric intake and negatively affected by time ill with diarrhea.

Kenya	Two rice schemes: Ahero Irrigation Scheme and West Kano Irrigation Scheme	General improvement in nutritional indicators as incomes get more diversified, but no difference in health conditions.	Yes, in terms of income diversification. Composition of income appears to be more important than level of income for household nutrition. Those farms mainly dependent on specialized income show symptoms of severe nutritional stress.
Rwanda	Potato production in the Gishwati forest area	Yes, the prevalence of malnutrition was lowered in households, and while the net effect on improving children's nutrition was small, it was present.	Yes, there is a positive relationship between income and nutritional improvement. Household health and sanitation conditions strongly determined children's nutritional status.
Zambia	Technological change in maize	Yes, for the children under 5 years, who do better with respect to the shorter-term weight-related measures. Improvements are not so evident for children aged 5–10 years. The prevalence of diarrheal diseases, malaria, and respiratory and other fevers appeared to be lower.	Information not available, but there does not appear to be a close relationship.
Malawi	Commercialization of maize and tobacco	Not much. There was no significant difference in nutritional status between children of tobacco growers and of nontobacco households.	Yes, income was positively associated with nutritional outcomes, with per capita income and per capita expenditures being the most powerful correlates of child stature. However, morbidity was largely independent of income and wealth. There was no significant effect of cash cropping on height-for-age of children. Levels of stunting were lower in female-headed than male-headed households.
Sierra Leone	Bo-Pujehun Rural Development Project-Tree crop promotion	No, children of commercialized farmers seem worse off than children of subsistence farmers.	Off-farm income had a positive effect on children's weight-for-age, but the share of tree crop income had a negative effect.
The Gambia	Jahally-Pacharr Smallholder Rice Project	Yes, especially for those children who were in a particularly bad nutritional situation, but there were intrahousehold biases. Women's fluctuations in weight are reduced and seasonal imbalances are leveled out.	Yes, a strong effect, via increased levels of and more stability in food consumption.

TABLE 5.2 Percentage of preschoolers in participant and nonparticipant households who were ill

	Percentage of Preschoolers Who Were Ill	
Country	Participants	Nonparticipants
Guatemala[a]	10.6*	16.1
Philippines[b]	31.5	28.0
Kenya[b]	89.6	87.7
Rwanda[c]	33.6	28.3
The Gambia[c]	78.8	80.3

SOURCE: Data sets of respective case studies.
NOTE: Includes all-round average for each child.
[a]Based on 72-hour recall.
[b]Based on two-week recall.
[c]Based on one-month recall.
*Significant at $p < .05$.

Kenya is due largely to malaria. However, the children of cooperative members of the smallholder export vegetable scheme (that is, cash crop–growing households) in Guatemala have a noticeably lower rate of illness than do the children of noncooperative members. This difference is due in part to a package of social and health services provided as part of the scheme. The positive health effect of participation in the agricultural cooperative will be explored later in the multivariate analyses.

Table 5.3 looks at the total time ill (all symptoms) and total time ill with diarrhea alone for preschoolers from participant and nonparticipant households. There is no definite pattern in the length of illness for preschoolers from participant or nonparticipant households.

Also, no definite pattern is observed for preschoolers in either the total time ill or the time ill with diarrhea alone when households are stratified by terciles of income. Preschoolers from the highest income category appear as likely to be sick, on average, as preschoolers from the lowest income category.

The effect of wealth on health patterns is also of research interest. Since all communities studied are overwhelmingly engaged in agriculture, landholdings per capita was used as a proxy for wealth. In this case, the pattern is not as clear-cut as was the income/morbidity stratification. In The Gambia, preschoolers in the top tercile of landholdings per capita are less likely to be sick when compared to preschoolers from the lowest category. However, this pattern does not exist in the other case studies: in Kenya, the Philippines, and Rwanda, increasing landholdings per capita is not associated with decreasing total morbidity.

When the morbidity data for preschoolers are disaggregated into age categories, children aged between 7 and 24 months are, with few exceptions, most likely to be sick. There are several reasons for this. This period corresponds to the period when children are weaned, first partially and then totally. Even if the breast-milk substitute is nutritionally comparable, the additional steps involved in preparing the food often introduce new pathogens into the children's environment. The fact that preschoolers also become much more mobile during this period further exposes them to a wider range of pathogens.

Commercialization can potentially influence preschooler nutritional status through a number of pathways. One of these is through the impact on child-feeding patterns. If increased demands are placed on a mother's time to provide agricultural labor for a specific cash crop, early weaning or the early introduction of solid foods for preschoolers, or both, can occur. However, no significant differences are found in the weaning age between participant and nonparticipant households (table 5.4). In all study countries, breast-feeding occurs for an extended period of time. Similarly, the age at which solid foods are first introduced to infants does

TABLE 5.3 Percentage of time that preschoolers are ill with any illness and with diarrhea in particular, participant and nonparticipant households

	Preschoolers		
Country/Group[a]	Sample Size	Any Illness (percent)	Diarrhea
Philippines[b]			
Participants	543	12.9	0.5
Nonparticipants	1,016	12.5	0.8
Kenya[b]			
Participants	291	28.2	4.8
Nonparticipants	446	28.5	3.8
Rwanda[c]			
Participants	311	12.4	5.6
Nonparticipants	279	12.3	5.7
Malawi[b]			
Participants	148	17.7	1.9
Nonparticipants	294	21.4	2.1
The Gambia[c]			
Participants	209	15.2	3.9
Nonparticipants	174	15.9	6.0

SOURCE: Data sets of respective case studies.
[a]The Guatemala case study had data only on the incidence of morbidity, not the total time ill.
[b]Based on two-week recall.
[c]Based on one-month recall.

TABLE 5.4 Age at which infants are totally weaned, participant and nonparticipant households

Country	Participants (months)	Nonparticipants
Guatemala	16.2	16.2
Philippines	14.2	13.7
Kenya	19.1	20.2
Rwanda	23.5	21.2

SOURCE: Data sets of respective case studies.
NOTE: Age in months weaned refers to the age when mother stopped breast-feeding completely. If child never breast-fed, age in months weaned equals 0. Data not available for The Gambia or Malawi.

not differ between the two groups. It is normally recommended that four to six months after birth, breast milk complements be added to an infant's diet. With the exception of Rwanda, where children receive solids later than in other countries, children in the study countries receive solids within this period. Thus, as table 5.4 suggests, entry of households into cash cropping schemes has not had detrimental effects on child-feeding patterns.

In general, the nutritional status indicators signal an improvement in the nutritional status—expressed in Z-scores[1]—when income increases. However, differences between participants and nonparticipants were not statistically significant in any case study setting. Neither were there any significant differences identified in weight-for-age or weight-for-height Z-scores between the two groups, stratified by income terciles.

Most of the case studies in this volume present anthropometric data for preschoolers of all ages combined into one group. It may be that the most pronounced effects of the commercialization process are on children born after the households have benefited from the increased income and/or on older children whose families have been involved in cash crop production for some time. The Kenya case study examined some of these issues by stratifying children into various age groups, and the results are similar to what is reported here (Kennedy 1993). More disaggregation of nutritional status data by age group would be useful in elucidating the process by which income-generating schemes do or do not have effects on health and nutritional status.

1. $\text{Z-score} = \dfrac{\text{(Actual measurement} - 50 \text{ percentile standard)}}{\text{Standard deviation of the standard}}$

Based on National Center for Health Statistics Growth Standards.

Prevalence rates for stunting (less than 90 percent of height-for-age), wasting (less than 90 percent of weight-for-height), and malnutrition (less than 80 percent of weight-for-age) are shown in table 5.5. With the exception of The Gambia, there are no significant differences between participant and nonparticipant households in any of these indicators. In The Gambia, preschoolers from participant households are less stunted. Moreover, there is more wasting than stunting in The Gambia, unlike other countries. This is true for preschoolers from participant as well as nonparticipant households.

Z-scores for each nutritional-status indicator were also stratified by landholdings per capita for participant and nonparticipant households. In The Gambia, Guatemala, Kenya, and Rwanda study settings, there is no consistent pattern. In the Philippines and Malawi cases, however, children from the highest landholding tercile—whether cash cropping or not—have better Z-scores on all three nutritional status indicators when compared to children from the lowest landholding tercile.

Determinants of Morbidity and Nutritional Status of Preschoolers and Links to Commercialization

So far, the results on morbidity and nutritional status have been purely descriptive. However, the multivariate analysis conducted on the case studies reinforces the descriptive results. Table 5.6 presents the

TABLE 5.5 Prevalence of stunting, wasting, and malnutrition among preschoolers, participant and nonparticipant households

Country	Stunting[a]		Low Weight-for-Age[b]		Wasting[c]	
	Partici-pants	Nonpartici-pants	Partici-pants	Nonpartici-pants	Partici-pants	Nonpartici-pants
			(percent of preschoolers)			
Guatemala	66.7	72.8	47.2	49.8	6.3	6.9
Philippines	32.2	36.3	45.7	51.8	22.4	27.1
Kenya	24.3	25.3	20.1	23.3	12.3	16.3
Rwanda	18.6	24.0	11.3	11.5	4.5	3.9
Malawi	55.2	52.7	52.4	46.6	16.6	19.7
The Gambia	9.4	17.4	27.5	27.0	29.0	28.7

SOURCE: Data sets of respective case studies.
[a] Less than 90 percent below height-for-age standard.
[b] Less than 80 percent below weight-for-age standard.
[c] Less than 80 percent below weight-for-height standard.

TABLE 5.6 Selected coefficients for relationship between program participation, socioeconomic variables, and preschoolers' morbidity (all preschoolers)

	β Coefficient on Preschoolers' Total Time Ill					
Independent Variable	The Gambia	Guatemala	Kenya	Malawi	Philippines	Rwanda
Household income[a]	6.06-05	−0.00025	−1.60−05	6.89−03	7.10−04	1.18−03
	(−0.899)	(−0.87)	(−0.20)	(0.50)	(1.67)	(1.60)
Mother's schooling	n.a.	n.a.	−0.25	−8.84−04	−2.28−03	−0.59
	n.a.	n.a.	(−1.18)	(−0.04)	(−0.99)	(−1.27)
Age of child (in months)	−2.54−03	−0.00051	−0.16	−0.20	−2.43	−0.21
	(−6.42)	(−0.073)	(−4.52)	(−5.41)	(−7.21)	(−4.00)
Household size	n.a.	n.a.	−0.36	5.88-0.5	−2.05	n.a.
	n.a.	n.a.	(2.35)	(0.0)	(−0.98)	n.a.
Participation dummy (1 = participants)	5.31−03	−0.432	0.29	1.80	3.22	4.1
	(0.33)	(−2.86)	(0.20)	(1.33)	(0.29)	(1.73)
Sample size	561	477	994	425	n.a.	585

NOTES: *t*-values in parentheses. The variable for morbidity equals all-round average for total time ill, except for Guatemala, where analysis is based on a probit analysis of incidence of illness.
[a]Total household expenditures used as a proxy for income for The Gambia, Guatemala, the Philippines, and Rwanda.
n.a. = Not available.

morbidity model for preschoolers in the six study settings of The Gambia, Guatemala, Kenya, Malawi, the Philippines, and Rwanda. Current income is not significantly associated with total illness in any of the six study countries. The same finding is obtained for morbidity for children less than 36 months of age. These findings are consistent with the absence of an income-morbidity relationship shown in the descriptive statistics above.

Age, however, is a significant determinant of illness in all study settings except Guatemala (table 5.6). As children get older, they are likely to be sick for a shorter period of time. This finding, again, is consistent with the descriptive analyses presented earlier. Note, however, that many of the very sick children have died, and, in a sense, an aging sample of preschoolers is biased toward "healthy" children.

Mother's schooling is not a significant predictor of preschooler morbidity in any of the case studies where data are available. However, the average level of education for women in all study areas was low (three to six years of schooling), with many women having no formal education. More variation in schooling may be needed in order to see a beneficial effect on children's health.

A separate dummy representing participation in commercialization schemes was included in the morbidity model. There are no negative effects of participation in commercialization schemes on children's health. In fact, in Guatemala, membership in the export crop–producing cooperative has a beneficial effect on children's health. This finding reinforces the data on incidence of illness presented in table 5.2. Planners of the cooperative in Guatemala attempted to maximize the welfare effects of commercialization through implementing a package of health and social services, partly funded out of cooperative profits. Table 5.6 suggests that the health of children of cooperative members has benefited partially as a result of this package.

The implications of the morbidity patterns of preschoolers for their nutritional status are explored in child-growth models. Tables 5.7 through 5.9 present results of the growth models for the various case studies. Given that household incomes had increased in all of the case studies partly as a result of the specific commercial agriculture schemes, one research interest was to trace through some of the effects of this increased household income on child nutrition.

In three of the five case studies for which caloric intake data are available, household and/or child energy consumption is a major determinant of linear growth. Similarly, household energy consumption is a significant determinant of gains in weight-for-age in Rwanda and The Gambia. In Kenya, Malawi, and the Philippines, child calorie consumption is a significant determinant of weight-for-age. In general, household

TABLE 5.7 Selected coefficients for preschooler growth model for height-for-age

	β Coefficient on Z-Scores for Height-for-Age					
Independent Variable	The Gambia	Guatemala	Kenya	Malawi	Rwanda	Philippines
Sex (1 = boy)	−0.29	−0.056	0.090	0.072	0.326	−0.03
	(−2.661)	(−0.52)	(1.09)	(0.95)	(3.30)	(−0.34)
Age (in months)	0.009	0.025	0.025	2.08−03	1.62−03	n.a.
	(3.26)	(4.67)	(8.10)	(0.88)	(0.589)	—
Calories[a]	0.069−03[b]	n.a.	2.42−04	1.30−04	1.35−04	2.1−04
	(1.02)		(1.96)	(1.32)	(2.79)	(3.35)
Mother's height	0.019	n.a.	0.026	0.022	0.02	0.038
	(2.41)		(3.69)	(2.97)	(2.39)	(5.53)
Total time ill (percent)	−0.796	n.a.	−0.02	9.28−03	−1.26−04	0.32
	(−2.63)		(−3.36)	(−3.06)	(−0.059)	(1.24)
Prior Z-score of preschooler	n.a.	0.58	0.605	0.862	n.a.	n.a.
		(11.48)	(20.67)	(25.72)	—	—

[a] Household calories used for The Gambia and Rwanda; child calories used for Kenya, the Philippines, and Malawi.
[b] A strong positive calorie-nutritional status relationship was identified for preschoolers weight-for-age and in analysis for populations, including children up to age ten also for height.
n.a. = Not available.

TABLE 5.8 Selected coefficients for preschooler growth model for weight-for-age

Independent Variable	β Coefficient on Z-Scores for Weight-for-Age					
	The Gambia[a]	Guatemala	Kenya	Malawi	Rwanda	Philippines
Sex (1 = boy)	6.019	0.048	0.03	0.146	0.269	−0.40
	(0.53)	(0.41)	(0.40)	(1.87)	(3.56)	(−7.07)
Age (in months)	1.799	0.29	0.01	−9.62−04	5.31−03	n.a.
	(0.78)	(5.71)	(6.21)	(−0.39)	(2.53)	—
Calories[b]	0.0161	−9.0−07	2.02−04	3.16−04	9.49−05	2.27−04
	(2.22)	(−0.23)	(2.13)	(3.08)	(2.57)	(5.93)
Mother's height	n.a.	n.a.	7.80−03	0.018	0.013	0.022
			(1.46)	(2.46)	(2.04)	(5.93)
Total time ill[c]	−3.16	n.a.	−0.014	−9.98−03	−5.90−03	−0.425
	(−1.79)		(−3.03)	(−3.29)	(−3.59)	(−2.66)
Prior Z-score	0.172	0.388	0.55	0.813	n.a.	n.a.
of preschooler	(4.19)	(8.05)	(20.09)	(19.21)	—	—

[a] In this model, the Z-score values are multiplied by 100; thus, the scale of parameters other than the prior Z-score is to be adjusted to compare with the other parameters in the table.
[b] Household calories used for The Gambia, Guatemala, and Rwanda; child calories used for Kenya, the Philippines, and Malawi.
[c] Time ill reflects diarrhea only in The Gambia model.
n.a. = Not available.

TABLE 5.9 Selected coefficients for preschooler growth model for weight-for-height

	β Coefficient on Z-Scores for Weight-for-Height					
Independent Variable	The Gambia	Guatemala	Kenya	Malawi	Rwanda	Philippines
Sex	0.058	0.178	−0.03	0.087	0.086	−0.067
	(0.75)	(1.80)	(0.47)	(0.98)	(1.34)	(−1.22)
Age (in months)	0.004	0.021	1.27−03	5.73−04	6.54−03	n.a.
	(2.15)	(4.46)	(0.51)	(0.20)	(3.68)	
Calories[a]	−0.011−03	−4.0−05	1.10−04	3.37−04	4.46−05	1.52−04
	(−0.23)	(0.99)	(1.07)	(2.91)	(1.43)	(4.03)
Mother's height	0.0072	n.a.	6.3−03	4.67−03	7.69−03	9.44−03
	(1.29)		(1.09)	(0.58)	(1.41)	(2.28)
Total time ill	−0.32	n.a.	−7.09−03	−7.56−03	−6.69−03	−0.74
	(−1.48)		(−1.46)	(−2.12)	(−4.82)	(−4.68)
Prior Z-score of preschooler	n.a.	0.169	0.39	0.717	n.a.	n.a.
	—	(3.48)	(10.64)	(13.43)	—	—

[a]Household calories used for The Gambia, Guatemala, and Rwanda; child calories used for Kenya, Malawi, and the Philippines.
n.a. = Not available.

calorie consumption is less significant for improving short-term growth, that is, weight-for-height. However, in Malawi and the Philippines, child calorie consumption is a significant determinant of weight-for-height.

Prior height and weight information on preschoolers was available for four case studies.[2] For each growth model, it is clear that prior nutritional status of preschoolers is a major predictor of their growth in the later period. Preschoolers who were not doing well nutritionally in the earlier period continued to exhibit patterns of inadequate growth. The data from these longitudinal studies give credence to those who advocate growth faltering as one criterion for identifying at-risk children.

Increments in household income result in improvements in preschooler nutritional status (as measured by growth) via the income–household-calorie–child-calorie route. However, even where these linkages are significant, the magnitude of the effect is often small.

In the Philippines and Kenya case studies, information on preschoolers' caloric consumption was available, which was used to examine the household-income–child-calorie–child-growth linkages. In the Philippines, a doubling of household income would increase a preschooler's caloric intake by 9 percent and, in turn, would improve the average weight-for-height Z-score by 3.6 percent, from -0.63 to -0.61, on average. Even with a very substantial increase in income in the short run, in this case a doubling of income, the ultimate physiological effect on growth via the income–household-calorie–child-calorie route alone is weak.

Kenyan data suggest similar, weak income–child-calorie–growth linkages. A doubling of household income in this setting results in an approximate 4 percent increase in preschooler caloric intake, which would result in an approximate 7 percent improvement in the weight-for-height Z-score. Thus, as in the Philippines, these household-income–household-calorie–child-calorie–child-growth linkages, although statistically significant at each point, result in very small changes in growth.

These weak linkages are of concern for those preschoolers who are moderately and severely malnourished. A doubling of household income would not be sufficient to move these children from severe to moderate malnutrition or even from moderate to mild malnutrition.

As the regression results in tables 5.7 through 5.9 show, a major determinant of nutritional status in each case study is the morbidity patterns of the children. The lack of a significant income-morbidity relationship has already been discussed. Therefore, these data suggest that in order to enhance the income–child-growth effect, at least in the

---

2. The studies in Guatemala, Kenya, and Malawi were longitudinal studies; in The Gambia, such information was available from retrospective data.

short to medium term, the health infrastructure and sanitary environment must be addressed along with agricultural policies and programs.

Health and Nutritional Status of Women

One of the most volatile topics in the commercialization debate is the impact of cash cropping on women. A flavor of this controversy is revealed from a review by Irene Tinker:

> a recurring theme in all these studies [in her review] of new technology for cash crops is that, while cash incomes may have increased, nutritional levels tend to fall. The primary reason for this seemingly contradictory phenomenon is the fact that this income belongs to men (Tinker 1979).

The general theme of much of the literature is that the effect of commercialization of agriculture on women is negative.[3]

There is no significant difference in the mean body mass index (BMI) or the total time ill between women of different income categories, whether their households are participating or not in cash cropping (table 5.10). This finding may not be surprising since, as chapter 3 indicated, in each case study the specific cash crop was seen as a men's crop and the proceeds from that crop were viewed as men's income. Multivariate analysis for women's illness and women's BMI in Kenya, which found that household income was not a significant determinant of either morbidity or nutritional status (Kennedy 1989), reinforced this finding.

A more surprising finding from the Kenya case study is that, as women's own income increased, there was a significant decrease in their BMI (Kennedy 1989). This appears counterintuitive until the linkages between household income, women's income, and women's nutritional status are examined. For many Kenyan women in agricultural households, energy expenditures increase with an increase in their income. For many women, this increase in the energy intensity of activities is greater than the concurrent increase in their caloric intake.

Results from a number of the cash cropping case studies (chapters 16, 18, and 22) indicate that both the total amount and the control of income are important influences on household caloric intake. In Kenya, female-controlled income has a significant positive effect on household caloric intake; this effect is above and beyond the effect of household income on household food consumption. A number of other case studies reported in this volume also show an effect of female income on household caloric consumption that is over and above the pure household income effect. For example, in Rwanda, when income is held constant,

---

3. See von Braun and Kennedy (1986) for some of the pros and cons of this literature.

TABLE 5.10 Women's body mass index (BMI) and percent time ill, by income per capita tercile, participants and nonparticipants

Country[a]	Income Tercile	BMI		Total Time Ill (percent)
		Participants	Nonparticipants	
Philippines	1	21.0	21.0	3.1
	2	21.0	21.0	7.5
	3	22.0	21.0	7.9
Kenya	1	22.4	22.1	19.9
	2	21.5	22.1	23.5
	3	22.3	22.0	23.3
Rwanda	1	23.4	23.1	23.8
	2	23.4	22.8	20.4
	3	23.1	22.4	29.2
Malawi	1	20.2	20.6	n.a.
	2	21.0	21.4	n.a.
	3	21.8	21.0	n.a.
The Gambia	1	21.5	20.8	11.6
	2	21.1	20.3	17.0
	3	20.4	20.6	15.5

SOURCE: Data sets of respective case studies.
NOTE: Body mass index (BMI) = weight (in kilograms)/height (in meters) squared.
[a] Data not available for Guatemala.
n.a. = Not available.

female-headed households consume 377 calories per adult-equivalent per day more than male-headed households. This effect is most pronounced in the lowest income groups. Similarly, in The Gambia, the share of cereal production under the control of women adds 322 calories per adult-equivalent per day to household energy consumption.

The influence of female-controlled income on household caloric intake seems to be most consistent and dramatic in the African case studies.[4] This may be because it is within the African context that income streams and expenditure responsibilities are still most highly differentiated by gender. In Malawi, a detailed comparison of the expenditures of household heads showed that most females allocated a higher share of their budget to food and a lower share to both drinks (between one-quarter and one-half of the share allocated by male heads) and agricultural inputs, mainly fertilizer. It appears that when total

---

4. The percentage of female-controlled income was not significant in the household consumption function specified for Malawi.

income is low, female-headed households tend to allocate a higher share of their expenditures to food than do male-headed households.

Control of income seems to be related more to the gender of the household head than to cash cropping per se. For example, in Kenya, the percentage of income that is female controlled is similar in male-headed sugarcane-producing households (43 percent) and in male-headed non-sugarcane-producing households (42 percent). Female-headed households, on the other hand, whether sugarcane- or nonsugarcane-producing, control approximately 63 percent of total household income.

What are the implications of gender of head of household for child-level outcomes? Earlier, we saw that there were minimal differences in the health and nutritional status of preschoolers from cash-cropping and non-cash-cropping households. What if households are disaggregated by the gender of the household head? This exercise can be performed for the Kenya and Malawi case studies.

Table 5.11 looks at child-level outcomes for households disaggregated by their participation in cash-cropping schemes and by the gender of the household head. In Kenya, the female-headed household category has been subdivided into de jure (the legal head of household is a woman) and de facto (in practice the head of household is a woman; the male head of household is absent). On many measures of child health and nutrition, children from male-headed households, whether or not these households are participating in cash-cropping schemes, have similar outcomes. However, there are differences in child health and nutrition outcomes in female-headed households when these households are disaggregated by their cash-cropping status. If it is assumed that many of the positive effects of development on child welfare are income mediated, particularly if the income is controlled by women, it would be predicted that higher-income households would do better in producing child health or child nutrition. In these cases, the assumption would be that female-headed cash-cropping households should have a preferential preschooler health and nutrition status. This assumption only holds true in certain cases.

Preschoolers in Kenya from sugar-producing, de facto female-headed households have a lower prevalence of total illness and diarrhea than do children from non-sugar-producing, de facto female-headed households. Some of the lower prevalence of illness of preschoolers from de facto female-headed sugar producers is associated with their higher household incomes. It is curious that this effect is not observed in the de jure female-headed sugar-producing households. One reason is that the female heads of de jure households are typically not the mothers of the preschoolers. In Kenya, and to a lesser extent to Malawi, many of these de jure female household heads are the grandmothers. It is therefore not

TABLE 5.11 Time ill, caloric intake, and nutritional status of preschoolers by gender of head of household and cash crop production, Kenya and Malawi

	Kenya									Malawi			
	Male		All Female		De jure Female		De facto Female			Male		Female	
Item	Sugar Farmers	Non-sugar Farmers	Sugar Farmers	Non-sugar Farmers	Sugar Farmers	Non-sugar Farmers	Sugar Farmers	Non-sugar Farmers		Tobacco Farmers	Non-tobacco Farmers	Tobacco Farmers	Non-tobacco Farmers
Total time ill (percent)	28.66	28.77	29.01	27.32	30.83	26.64	26.96	28.66		18.8	23.8	26.1	22.3
Time ill with diarrhea (percent)	4.67	3.57	3.66	4.83	5.26	3.93	1.85	6.58		1.8	1.9	1.8	1.9
Child caloric adequacy (percent)	59.92	57.34	52.78	59.47	51.21	57.16	54.51	63.71		67.9	72.2	66.6	70.4
Percent weight-for-age Z-score less than −2	19.6	19.0	15.7	15.9	20.4	14.1	10.4	19.5		27.5	29.0	27.3	23.1
Percent height-for-age Z-score less than −2	41.0	41.0	36.0	31.9	37.0	33.3	34.8	28.9		63.7	61.3	45.5	57.1
Percent weight-for-height Z-score less than −2	2.3	1.9	4.0	5.0	5.6	3.7	2.2	7.9		2.2	0.0	0.0	0.0
Household expenditures per capita[a]	2,872.74	2,517.02	2,729.57	2,343.96	3,011.24	2,482.12	2,331.92	2,059.00		80.42	81.51	61.55	72.74

SOURCE: Data sets of respective case studies.

[a] In Kenyan shillings and Malawian kwacha, respectively.

necessarily the case that the increased income received by the household and/or the female head will benefit the individual preschooler. In fact, in Kenya, preschoolers in de jure female-headed sugar-producing households are ill for longer periods of time and have longer bouts of diarrhea. The increased income associated with sugarcane production has not benefited the children. This finding reinforces the earlier discussion that income-morbidity links tend to be weak.

In Kenya, there are no significant differences in the prevalence of stunting, wasting, or malnutrition between cash-cropping and non-cash-cropping male-headed households. This finding is consistent with the analyses presented earlier. Comparing children from all female-headed households in Kenya to children from male-headed households, the preschoolers in the female-headed household group do better on the longer-term measures of nutritional status. A different pattern emerges, however, when we examine different types of female-headed households in Kenya. The de jure female-headed households, that is, those households in which women are considered the legal head, show differences in anthropometric indicators between sugar and nonsugar producers. Preschoolers from de jure female-headed non-sugar-producing households have lower rates of stunting (low height-for-age) and malnutrition (weight-for-age less than -2 Z-scores) than children from sugar-producing de jure female-headed households. In contrast, preschoolers from de facto female-headed sugar-producing households have lower rates of malnutrition based on weight-for-age than preschoolers from non sugarcane-producing de facto female-headed households. Further analysis for Kenya showed that children in de facto female-headed households capture a larger share of household calories than children from male-headed or de jure female-headed households.

In Malawi, because of small sample sizes, it was only possible to compare male-headed households to all female-headed households combined; preschoolers from female-headed households—whether producing tobacco or not—have lower rates of stunting than preschoolers from male-headed households.

The results of table 5.11 suggest that it is not female headedness per se that imparts a nutritional benefit to preschoolers but, rather, it appears to be a complex interaction of female headship, expenditure patterns, and time use that leads to different health and nutrition outcomes in preschoolers (Kennedy and Peters 1992).

Conclusions

None of the case studies reported here shows a clear negative effect of the commercial agriculture schemes on children's health and nutritional status. Not only were negative effects not observed in the rather food-oriented Gambian and Rwandan commercialization schemes, but they were also not observed in the very nonfood, cash crop-oriented Kenyan, Malawian, Philippine, and Guatemalan cases. This point is emphasized, since there exists a belief that cash cropping tends to have adverse effects on children.

However, the increases in income that are associated with participation in the various commercial agriculture schemes are not found to decrease child morbidity, at least in the short and medium run captured by the studies. It may seem counterintuitive that very significant increases in income do not translate into a decrease in the total length of illness. These data need to be interpreted cautiously. As already discussed (see chapter 4), there is a tendency in some case studies for more commercialized households to spend, at the same income levels, a slightly higher proportion of their income on nonfood expenditures. The impact of these new expenditure patterns will depend on the items purchased. There are two broad categories of nonfood expenditures: those with no expected health or nutrition benefits, such as certain types of consumer goods (jewelry, radios, alcohol), and those with potential health impacts. The second category can be further subdivided into those expenditures with short-term effects and those with long-term effects. Obvious expenditures, such as for deworming or other preventive health items, could be expected to have an immediate impact on health, observable even in the very short term. However, increased income in participant households is often spent on items such as improved housing and education. While these expenditures may produce health benefits in the long term, in the short term, they are not associated with changes in morbidity patterns. The present case studies may not have fully captured longer-term benefits. A scenario can be envisioned in which increased expenditures on education—particularly of girls—in the longer term would be likely to result in changes in fertility patterns, which in turn would influence neonatal outcomes and ultimately result in improved infant health. Because of the relatively short time frame of the case studies reported in this volume, these sorts of linkages could not be identified.

In each of the study sites, the health and sanitation environment is poor, and infant mortality and malnutrition rates are high. While increases in income would be expected to bring about improvements in overall health and welfare in the longer term, in the short term it appears that increases in income must be combined with public action to improve the health environment in order to have a significant effect on preschooler morbidity. This is not to argue that income is not important, but rather that when agricultural policies and programs are being planned, attention should be given simultaneously to health and sanitation conditions in rural areas. The complementaries between increased income and an improved health and sanitation environment should be stressed so that the potential effects of commercial agriculture schemes on overall welfare can be enhanced.

Even in households where food availability is greater than energy requirements, preschoolers often consume well below their apparent energy requirements. The reasons for inadequate preschooler energy consumption in situations where income and household food supplies seem not to be constrained are not addressed directly by the case studies. However, two explanations are plausible. First, the communities in which the commercial agriculture schemes have been implemented are ones in which malnutrition is endemic. There may not be an awareness on the part of households that malnutrition is in fact a problem, since their children look like most other children in the community. If primary caretakers do not perceive a nutritional need, then there would be no reason for them to assume that their children need more food.

Second, most of the case studies concentrated on evaluating the nutritional status of children aged up to six years. However, two of the case studies, Guatemala and The Gambia, also included a cohort of children up to ten years of age. Growth models indicated that in Guatemala there was no significant association between household energy intake and child growth, but when growth data for children between the ages of six and ten years are analyzed, the opposite finding emerges. In this group of older children, increases in household caloric intake result in positive impacts on child growth (von Braun, Hotchkiss, and Immink 1989). Why, then, do we not see the same household-calorie–child-growth association in younger children? It is very likely that chronically sick preschoolers, because of a generalized anorexia, feel satiated before their "true" caloric needs have been met. It is also not surprising that parents would not assume that children need more food if the children have indicated that they have had enough food.

The schemes, as implemented, are not associated with increased preschool malnutrition. Therefore, accusations that the introduction of cash-cropping schemes is usually associated with a deterioration in nutritional status are not borne out by data from the case studies included in this chapter. However, a recently completed study in Guatemala, following the same households and the same children surveyed in 1984–85, indicated that children from households participating in an export vegetable cooperative in the Western Highlands did significantly better on some measures of nutritional status than children from the noncooperative households (Immink et al. 1992). These data indicate that some gains in nutritional status associated with participation in the cooperative were maintained in the longer term.

Household income is a key determinant of household caloric consumption. However, beyond total household income, female-controlled income has a positive and significant effect on household food intake in many cases. Female-controlled income is more likely than male-controlled income to be spent on food. Therefore, household food security, in a number of case studies in this volume, appears to be influenced by both the total amount of household income and the proportion of income controlled by women.

Female headedness, per se, does not impart a nutritional benefit, since not all types of female-headed households show lower rates of preschooler malnutrition when compared to male-headed households. The better nutritional status of children in de facto female-headed households, particularly the non-cash-cropping group, appears to be due to a preferential treatment in the allocation of incremental calories to children. The increased allocation of household calories to preschoolers is due to something other than increasing household income. In fact, in an analysis for the Kenyan case, the household income variable had a marginally significant but negative effect on the proportion of calories going to children. This would reinforce a statement made earlier that, as household income increases, preschooler food intake does not increase as rapidly as household food intake.

Increased household income in rural areas, based on agricultural growth (which also includes the smallest farmers and the landless), can make a major contribution to solving the hunger problem. However, it cannot in itself provide a complete solution in the problem of preschooler malnutrition.

Hunger and malnutrition are not synonymous. While insufficient food intake at the household level and, in particular, the child level, is one precipitating cause of malnutrition, it is not the only cause. The results from the six case studies presented in this chapter indicate that morbidity patterns are a key cause of malnutrition in preschoolers. In order for income-generating policies such as commercialization of agriculture to have dramatic effects on improving health and reducing malnutrition in the short to medium term, attention must be given to health, sanitation, and environmental issues as complementary components of agriculture policies and programs.

PART IV
The Driving Forces of Commercialization

6 Commercialization of Agriculture and Food Security: Development Strategy and Trade Policy Issues

NURUL ISLAM

Introduction

The choice between subsistence food crops, on the one hand, and cash crops, especially nonfood cash crops predominantly meant for exports, on the other hand, is a subject of considerable debate among policymakers as well as development specialists. The debate raises issues not only at the level of farming households but also at the level of national and international policies, including macroeconomic policies such as trade and exchange rate policies. This chapter reviews and focuses on those aspects of principal relevance in the context of an overall agricultural development strategy and food security.

The controversy regarding cash or export crops versus food crops is part of a bigger debate relating to the commercialization of agriculture in developing countries. Commercialization can broadly be defined as a rise in the share of marketed output or of purchased inputs per unit of output. A shift from basic food crops, which are produced and predominantly consumed on the farm, to cash crops, which are produced mainly for sale in the market, is therefore viewed as part of the commercialization of agriculture process. It is associated with an extension of the market or of the exchange economy, leading to increased specialization and division of labor. Cash crops can be both food and nonfood crops. Even though nonfood crops have been the principal component of cash crops, their relative importance has been on the decline recently.

Choice Between Food Crops and Nonfood Cash Crops

To put the debate at the national policy-making level in its proper perspective, it is important to stress a few salient factors influencing the choice between food crops and nonfood cash crops. First, there are constraints on allocation of land and other resources between various crops; not all lands are agro-ecologically suitable for all crops. Second, different crops are frequently grown in different seasons and are, therefore, grown in combination over the year. Mixed farming, hence, is often common. Third, mixed farming, including a combination of different crops, is often preferred because monoculture leads to soil degradation or, at least, does not contribute to renewal or maintenance of soil fertility without a supply of additional nutrients. Therefore, cash or export crops and food crops can both, in certain agro-ecological environments, be necessary components of a crop rotation.

Fourth, food crops, including basic staples (cereals, roots, tubers, and pulses) and cash food crops (sugar, fruits, vegetables, oils, and oilseeds), constitute the largest share of the value of total crop output in developing countries. In 93 developing countries, excluding China, food crops of both types constituted 89 percent of the total crop output in 1983–85, whereas the share of nonfood cash crops was 11 percent (table 6.1). Basic food crops constituted 50 percent of the total crop output in the same period; Asia and Africa had the highest share, 59 and 57 percent, respectively. Africa and Latin America had the highest share of

TABLE 6.1 Distribution of total crop output between basic food crops and cash crops, developing country regions, 1961–63 and 1983–85

Region/Year	Basic Food[a]	Cash Crops	
		Food[b] (percent)	Nonfood[c]
Sub-Saharan Africa			
1961–63	54	30	16
1983–85	57	27	15
Near East and North Africa			
1961–63	42	49	9
1983–85	38	54	8
Latin America			
1961–63	36	40	24
1983–85	34	51	15
Asia (excluding China)			
1961–63	59	31	9
1983–85	59	33	8
Developing countries (excluding China)			
1961–63	51	35	14
1983–85	50	39	11

SOURCE: FAO production data tapes.
[a]Cereals, roots and tubers, and pulses.
[b]Sugar, vegetables, bananas, fruits, vegetable oils, and oilseeds.
[c]Tea, tobacco, coffee, cocoa, jute, cotton, fiber, rubber, and so forth.

nonfood cash crops, but this was no more than 15 percent. Over the last 25 years, the proportion of basic food crops in the total crop output has hardly changed. However, within the group of cash crops, the share of food cash crops has expanded from 35 to 39 percent, whereas that of nonfood cash crops has declined from 14 to 11 percent.

Fifth, over the last 25 years—that is, during the 1961–85 period—basic food crop output increased at a rate of 2.7 percent per year, a much higher rate than the 1.4 percent rate of growth of output of nonfood cash crops (FAO production data tapes). However, food cash crops grew at 3.4 percent per year during this period, and the annual rate of growth of all cash crops was about 2.9 percent. During the 1960s, the output of basic food crops increased at a faster rate than cash crops (both food and nonfood), but this situation was reversed during 1976–85, mainly because of an acceleration in the growth rate of food cash crops from 3.0 percent to 3.8 percent per year.

Sixth, is there an association between growth in food production and growth in cash crop production? Do countries that do well in cash crop production also do well in food production? A cross-section study of 78 countries over the 1968–82 period found that growth in acreage under cash crops was positively associated with growth in staple food production, and that, moreover, growth in the proportion of total crop area devoted to cash crops was associated with growth in per capita staple food production (von Braun and Kennedy 1986).[1] Many countries that achieved positive growth rates in basic food production also had positive growth rates in nonfood cash crop production and vice versa. Fifty-seven of the 60 countries that achieved positive growth rates in total nonfood output between 1961–85 also attained positive rates of growth in basic food production. Similarly, of 82 countries that recorded positive growth rates in food production over the same period, 57 countries (70 percent) also attained positive rates of growth in nonfood production. Therefore, on the whole even though high growth rates in basic food production were not necessarily associated with high growth rates in nonfood production, the direction of change was correlated.

Seventh, what is the relationship between growth in production of nonfood crops and growth in overall food supply? Is an increase in nonfood crop production associated with a decrease in total or per capita food supply? In 89 of 90 developing countries for which data are available (FAO data tapes), aggregate food supply (domestic food production

---

1. It should also be noted that in 1982, cash crops—both food and nonfood—occupied more than 3 percent of the cropland in 28 out of 78 developing countries, whereas nonfood cash crops occupied more than 30 percent of cropland in only 3 countries.

plus food imports) increased during the 1961–85 period regardless of change in nonfood production. However, the situation was different as far as per capita food supply was concerned. In the majority of 60 countries that recorded an increase in nonfood production, per capita basic food supply increased; only in 23 countries did per capita food supply decrease.

The foregoing analysis indicates that food crops, including both basic and cash crops, dominate crop production. In the last several years, food cash crops have increased in importance, but total cash crop production has remained unchanged in relative importance due to a decline in the importance of nonfood cash crops. The decline in the relative importance of nonfood cash crops was most significant in Latin America, where the share declined from 24 to 15 percent of total crop output between 1961 and 1985, although it was observed in almost all regions of the developing world. The analysis also indicates that the relative rates of growth in food and nonfood production were not the dominant factors determining overall food supply. Food supply in countries with inadequate domestic food production was a function of access to food imports. Access to food imports was, in turn, a function of food aid, foreign exchange availability, and food import policies of various countries.

At the national development strategy level, a number of arguments have been advanced to explain why specialization in export crops may be undesirable. First, export crops face a long-run adverse trend in the terms of trade. Second, export prices of cash crops are highly unstable and, therefore, have an adverse effect on development, as well as on the stability and availability of food supplies to be obtained in exchange of export crops. Third, long-run dynamic comparative advantage does not seem to favor export crops. Fourth, export crop production habitually tends to be large scale and capital intensive and, therefore, does not promote employment or equitable distribution of income. Finally, an emphasis on export crops as opposed to food crops may detract from household food security in terms of food consumption and nutritional status of poor families. The following sections investigate the validity of these arguments.

Long-Run Trends in Relative Prices in World Markets

It is argued by "food first" advocates that production of basic food crops deserves to be accorded higher priority even if nonfood export or cash crops enjoy a comparative advantage and yield higher returns. The reason usually advanced for such a proposition is that export crops, especially agricultural raw materials and tropical beverages (the principal exports of developing countries), tend to suffer from a long-term decline in the terms of trade and do not have bright market prospects in the future. Although much has been written on long-run movements or trends in the international terms of trade of agricultural exports in general, the evidence is not conclusive about whether a decline, if it occurs, is due to adverse demand conditions or reductions in cost.

A recent comprehensive study (Grilli and Yang 1988) of long-run trends in terms of trade of different groups of agricultural commodities found that from the beginning of this century up to 1986, the real prices or terms of trade of agricultural raw materials, nonbeverage food commodities, and cereals have declined by about -0.82 to -0.77 percent, -0.54 to -0.51 percent, and -0.68 to -0.62 percent, respectively, per year (trend rate of change) (Grilli and Yang 1988). The largest decline was in the real price of agricultural raw materials and, hence, there was a downward trend in the real price of agricultural raw materials vis-à-vis that of food and cereals. However, the annual decline was very small. Furthermore, the real price of beverages rose over the 1900–86 period at about 0.6–0.7 percent per year, with the result that the terms of trade of beverages with respect to agricultural raw materials, nonbeverage food, and cereals moved in its favor over the period. Thus, the long-run movements in the terms of trade of the two groups of export crops vis-à-vis food and cereals crops were very different.

In more recent years, 1962–86, the real prices of nonbeverage foods and cereals have fallen faster than in earlier years (table 6.2). During the 1962–86 period, the rate of decline in the price of cereals was faster than that of agricultural raw materials and other food. Therefore, the terms of trade of agricultural raw materials, tropical beverages and other food all improved vis-à-vis cereals during that period.

TABLE 6.2 Recent trends in terms of trade of different commodity groups, 1948–86 and 1962–86

Commodity Group	Trend Rate of Change Per Year	
	1948–86	1962–86
	(percent)	
Agricultural raw materials	−2.17	−1.83
Tropical beverages	—[a]	—[a]
Other food	−0.64	−1.12
Cereals	−1.85	−2.46

SOURCE: Computed from data provided in World Bank (1988a).
[a]No significant trend.

Neither the long-run decline in the relative price of agricultural raw materials vis-à-vis food and cereals nor the faster decline in cereal price in recent years (1962–86) necessarily provides a signal or guide for long-run investment in any one crop. First, the relative decline in the price of an export crop may have been due to falling costs rather than to declining demand. Second, even if net returns from export crops—for example, agricultural raw materials—declined over time, they may still be higher than those obtained from other crops, such as cereals. A decline in food (cereal) prices in recent years relative to other crops does not suggest reliance upon food imports in exchange for agricultural export crops. The depressed prices of cereals in recent years are mainly due to heavy export subsidies by developed, exporting countries such as the United States and the European Community (EC) that are vying with each other to dispose of the large surpluses generated by domestic price and other support programs. Therefore, the low world food prices of the 1980s, which may not endure with cessation of competitive export subsidies and reduction of surpluses in the developed, exporting countries, do not provide an appropriate signal for long-run allocation of resources between food and export commodities in developing countries.

Also, agricultural exports, which face long-run stagnation or slow growth in world demand, do not provide a viable avenue for specialization on developing countries. Many commodities are faced with the threat of synthetic substitutes; others, especially agriculture raw materials, face a decline in demand due to increased economy in their use per unit of finished output. Trade restrictions in importing countries, mostly developed countries, further constrain the expansion of agricultural exports.

Countries that face inelastic export demand in response to price changes can improve their combined real income by acting in unison to manage or restrict export supply. However, such cooperation among agricultural exporting countries has been rare. In fact, they frequently compete aggressively to secure a larger share of an inelastic and slowly growing export market; the resulting fall in prices tends to reduce aggregate export earnings for all of them. In such situations, the less efficient producers are confronted with the need to diversify their exports to focus on commodities that face an expanding future world demand, while at the same time to strengthen their competitive advantage through cost-reducing innovations and technological progress. While traditional agricultural exports such as sugar and hard fibers do not hold out prospects for rapidly expanding world demand, commodities such as horticultural products, oilseeds, and livestock products show brighter market prospects (Islam 1990).

Relative Prices and Food and Export Crops in Domestic Markets

Even though relative prices of different agricultural commodities in world markets, in relation to their relative domestic costs of production, are expected to provide guidance for selection of an appropriate combination of crops, most developing countries, in fact, intervene through border measures so that domestic relative prices differ, sometimes significantly, from relative prices in world markets. The relative prices of food and export or cash crops confronting producers in their domestic market are of relevance to their decision-making process.

Most developing countries traditionally discriminated against agriculture, including both food and export crops, through a combination of economy-wide and sector-specific policies. Frequently, an overvalued exchange rate, combined with a high level of industrial protection, reduced relative returns from agriculture vis-à-vis nonagriculture, which kept the real prices of both food and export crops low.

The economy-wide policies that tended to discriminate against agriculture were reinforced in the case of export crops by export taxes, the incidence of which often fell on producers, and by export marketing boards, which paid farmers less than export prices or world prices. In several countries, export taxes and export marketing boards were a major source of revenue for the government. At the same time, food price policies pursued through state interventions in food marketing kept prices to farmers low in the interest of urban consumers.

Discrimination against agriculture has been most acute in Africa and, to a lesser extent, in Asia and Latin America. The relative domestic prices of food versus export crops have, however, varied widely among countries, depending on the net result of a multiplicity of policies that impinge on the agricultural sector. In recent years, there has been some decline in the discrimination against agriculture, especially against food crops, in many countries, which can be seen from changes in producer prices of food and export crops and in nominal rates of protection (table 6.3). Discrimination against food crops has declined faster than against export crops. Increases in the nominal rate of protection have been slower for export crops than for food crops (table 6.4). The increase in the nominal rate of protection for cereals has been fastest in Africa, followed by Latin America and the Near East.

A different measure of nominal protection for different crops is expressed in terms of the relative domestic price of each crop vis-à-vis that of nonagricultural commodities, where the world price is converted into domestic price at an equilibrium rate of exchange, rather than at a nominal rate of exchange (Krueger, Schiff, and Valdés 1988). Data from

TABLE 6.3 Changes in farmgate prices in selected developing countries

Crop Group	Index of Farmgate Price (1969–71 = 100)			Percentage Change[a]
	1973–75	1978–80	1981–83	1969–83
Cereals	121	114	120	+18
Export crops[b]	110	117	100	not significant

SOURCE: FAO (1987).
NOTE: Thirty-eight developing countries, including thirteen in Africa, nine in Asia, ten in Latin America, and six in the Near East, are included in this analysis.
[a] The percentage change is calculated on the basis of a trend equation for 1969–83.
[b] Coffee, cocoa, tea, cotton, rubber, jute, soybean, palm oil, and tobacco are included in this group.

16 countries indicate that food crops enjoyed a positive rate of direct nominal protection[2] in 1975–79 and that this continued roughly at the same level, 20 percent, in 1981–83 (Krueger, Schiff, and Valdés 1988). Export crops suffered from a negative rate of direct nominal protection of roughly the same magnitude, − 10 percent, in both periods, 1975–79 and 1981–83. The positive direct nominal protection for food crops was more than offset by the exchange rate overvaluation, resulting in a negative, though small, total protection of about − 5 percent. In the case of export crops, the exchange rate overvaluation aggravated the direct negative protection, so that the total protection was not only negative but also more than five or six times higher than in the case of food crops. Furthermore, the total protection against export crops increased over the years, from − 36 percent to − 40 percent.

Why did export crops suffer greater discrimination in terms of direct as well as total negative protection than staple foods? There were three main reasons. First, as mentioned earlier, developing countries faced with limited sources of revenue found taxes on export crops to be an attractive source of revenue and one that could be relatively easily collected, either directly through export taxes or indirectly through

---

2. The *total nominal rate of protection* is defined as the percentage difference between the relative domestic price (ratio of prices of agricultural commodities to those of nonagricultural commodities) and the relative border price at the equilibrium nominal rate of exchange. The *indirect nominal rate of protection* is the percentage difference between the relative border price at the official nominal rate of exchange and the relative border price at the equilibrium nominal rate of exchange. The *direct nominal rate of protection* is the difference between the total and the indirect nominal rate of protection.

TABLE 6.4 Indices of nominal rates of protection, 1969–71 = 100

Region	1973–75 Cereals	1973–75 Export Crops	1978–80 Cereals	1978–80 Export Crops	1981–83 Cereals	1981–83 Export Crops
Africa[a]	91	89	161	91	170	109
Africa[b]	80	95	134	94	143	101
Asia	69	90	92	100	117	103
Latin America	70	100	117	97	137	105
Near East	74	88	116	91	128	102
Total[c]	74	94	116	96	133	103

SOURCE: FAO (1987), 193–195.

NOTE: Nominal rate of protection is defined as the ratio of farmgate price to border price. Border price, in the case of an export competing crop, is f.o.b. price minus transport, handling, and marketing costs between the border and the farmgate. In the case of an import competing crop, the border price is c.i.f. price plus transport, handling, and marketing costs from the border to the farmgate.
[a]Includes countries that favored higher food prices.
[b]Includes countries that relatively favored higher export crop prices.
[c]Thirty-eight developing countries.

profits of the export marketing boards. In several instances, surpluses siphoned off by parastatals or marketing boards ended up largely paying for the inefficient management and high overhead costs of these bodies. Second, in several countries it was presumed that export demand for such crops was price inelastic, and, hence, taxation of exports, by limiting supply, would contribute towards increased export earnings. For individual countries, it was seldom that price elasticity of export demand was low. Third, in recent years there has been an increasing emphasis on food self-sufficiency and the need to increase domestic food production in the context of rising food imports, as a consequence of increasing population and rising per capita income, especially in middle-income countries. This has led many countries to raise domestic prices either through higher procurement prices of the public marketing agencies or through taxes on imports of food or cereals.

Price and Income Instability: Food Versus Export Crops

The instability of prices of both export and food crops in the world market has been advanced as an additional argument against dependence on the world market for a basic need such as food. Prices of, and earnings from, export crops, which determine the capacity to import food, have been highly unstable. This has been aggravated by fluctuations in world prices (table 6.5) and supplies of food. These two tendencies do not necessarily offset or compensate each other.

Differences in price instability are observed among the commodity groups: agricultural raw materials, in general, seem to have a lower degree of price instability than do cereals and tropical beverages. However, there are variations within each commodity group. For example, price variability around the trend for maize is considerably lower than that for rice. Price variability for sugar is as high as 74 percent around the trend, whereas for soybeans it is 21 percent and for bananas it is 16 percent. Among tropical beverages, the index of instability of cocoa prices is as high as 33 percent. Among raw materials, the index of instability of prices for rubber is as high as 25 percent and for sisal it is 37 percent (UNCTAD 1987).

There is no prima facie reason why the choice between alternative crops at the margin should depend on the relative degree of price instability, even though returns from less stable products are often higher than from more stable products. Risk aversion may lead farmers to discount the higher returns from more unstable crops by a margin determined by their subjective evaluation of risk. Evaluating the risk of instability depends upon the sources of instability. Price instabilities originating from supply fluctuation offset each other and, as supply and income fluctuations are inversely related, stabilize income. This is usually the case with agricultural commodities. Uncertainty or instability of relative prices would make farmers adopt a certain degree of diversification of farm products but would not result in farmers' specializing to the extent warranted by average comparative cost considerations.

However, more pertinent questions may be raised at the national policy level. Should world prices, in view of their volatile and fluctuating

TABLE 6.5 Instability of world market prices, 1962–87

Commodity Group	Percentage Variation from Trend
Food	29
Rice	31
Wheat	23
Maize	20
Tropical beverages	25
Agricultural raw materials	18

SOURCE: UNCTAD (1987).

nature, be used as a guide to long-term allocation of resources among different crops? How should the long-term trend in prices be determined? The pragmatic answer is to rely on three- or five-year moving averages of prices. There is scope for developing-country governments to intervene to stabilize prices around such moving averages to provide appropriate price signals to farmers. Intervention can be in the form of import taxes, import subsidies, or stocks, the latter being more costly. Most countries have taken domestic measures to achieve a degree of stability of domestic prices that is greater than that of corresponding world prices. Such a policy of "exporting" domestic instability contributes to world market destabilization. The policy measures adopted for achieving such domestic price stabilization vary widely in terms of resource costs. The greater the desired degree of price stability, the higher the cost, for example, of the stocks that have to be held, or the higher the loss in efficiency in diverting from relative prices as a guide to resource allocation (Pinckney 1989; Ahmed and Bernard 1989).

Subject to the provision of using long-term or medium-term prices, either comparative costs or domestic resource costs of earning or saving foreign exchange should continue to be used as a guide for resource allocation. Long-term projections of relative prices are needed for making long-term investment decisions such as investment in irrigation or in mechanization projects.

To the extent that price instability leads to instability in export earnings and, hence, in the ability to procure food supplies from abroad, it creates instability in the access to food supplies. Stable food supplies are essential for food security. Food insecurity may result from a rise in world prices of imported food, thus reducing food imports and, hence, food supplies, given unchanged foreign exchange earnings to import food and given typically low short-term supply response.

Even if prices of nonfood exports and food imports move in opposite directions, they do not necessarily compensate each other. Countries confronted with the need to import food or cereals from a highly volatile world market, on the basis of fluctuating agricultural export earnings, tend to emphasize the production of food beyond the point that is warranted by comparative cost considerations. However, reliance on domestic production does not eliminate or necessarily reduce fluctuations in food supply that originate from weather-induced or policy-related fluctuations (Sahn and von Braun 1989). Countries, however, feel more secure dealing with variability in domestic food production than with uncertain supplies and prices in the world market.

The risk of reliance on the world market arising from volatility of export prices or earnings or from food import prices can be reduced or mitigated by an assurance of access to foreign exchange resources to meet rises in food import prices or shortfalls in domestic food production. The Cereal Financing Facility of the International Monetary Fund (IMF) was originally conceived to meet such a need. However, it has been greatly reduced in scope in recent years, partly by its integration with the IMF's Compensatory Financing Facilities for Export Shortfalls and partly by the introduction of "conditionalities" for economic policy reforms. Alternatively, international commodity agreements may be used to stabilize prices in world markets, either by internationally coordinated, nationally held stocks or by supply management by exporting countries through allocation of export quotas among themselves with a view to adjusting supply to expected demand, along the long-run price trend.

The risks of dependence on the world market arise not only from variability of prices but also from disruption of supplies due to war or civil unrest, breakdown of international shipping and transportation arrangements, or export embargoes initiated to overcome domestic shortages in exporting countries or to achieve political and strategic objectives.

The need for assured and uninterrupted access to food supplies has been a powerful motivation behind the search for greater food self-sufficiency and the build-up of large domestic food stocks in both developed and developing countries. In order to encourage developing countries to depend on world trade to an optimal extent, it is necessary to ensure an open and liberal world trading regime so that import restrictions and export subsidies or embargoes do not reduce the level and stability of agricultural export earnings or of world agricultural prices.

Long-Run Dynamic Comparative Costs

The reliance on comparative costs needs to be viewed dynamically by individual countries: relative costs change across countries and commodities in response to technological change fueled by research and development efforts, both national and international. These dynamics raise a difficult question about investment strategy in commodities that are currently in excess supply in the world market—and expected to be in the future—resulting in a long-run downward trend in prices. Countries whose domestic costs of production exceed long-run export or world prices should gradually move out of production of these commodities unless they expect their costs to go down in the future due to cost-reducing innovations and increased productivity. There would be short-term costs of unemployment and income loss on the part of those employed in the unprofitable sector. There is, thus, scope for structural adjustment assistance for the sector in transition.

New producers with comparative advantages may emerge on the world market, either because of their ability to exploit hitherto untapped agro-ecological advantages or because of their success through research and development efforts in achieving cost-reducing innovations. Increasingly, comparative advantage is less a matter of a given endowment of agro-ecological resources and more a matter of technological innovations that improve or modify cost advantages derived from resource endowments. Also, new crops may be introduced in countries that did not grow them earlier due to lack of domestic demand or unfamiliarity with the world market. The rising comparative advantage of East Africa in tea production to the disadvantage of the traditional tea producers in South Asia, the emergence of Malaysia as a low-cost producer of cocoa, and the shift of palm oil production from Africa to Malaysia and then to Indonesia, are all illustrations of shifts in comparative advantage of different commodities among countries. To freeze the pattern of global production of a particular commodity among its traditional producers would be tantamount to sacrificing the advantages of lower costs and cheaper supplies of agricultural commodities in the future.

In recent years, the focus of international and national agricultural research efforts has tended to shift away from export or cash crops toward food crops in response to pressing food needs in many developing countries. This shift in focus is more true of international than of national research expenditures. During colonial times, research on export crops received high priority in most developing countries. Consequently, while a reallocation or shift of emphasis in research efforts was needed in the postindependence period, given the past neglect of research on food crops, the balance may have shifted too far, thus sacrificing opportunities for efficient export crop production. Research on export crops needs to be strengthened to meet competition from synthetic substitutes, expand new uses or markets, and maintain the market share of developing countries in world trade. This is especially important for those export crops that face low or declining demand or price.

The relative importance of research efforts on various crop groups can be observed from the distribution of crop scientists engaged on different crops in different regions (table 6.6). The dominance of food scientists is obvious in all regions during the 1980s. Even among scientists working on cash crops, those working on food cash crops dominate, except in Asia. A smaller share of scientists work on nonfood cash crops in Latin America and the Near East, whereas, in Africa, the share of scientists working on food and nonfood cash crops does not differ significantly.

TABLE 6.6 Distribution of crop scientists working on different crops in different regions, and importance of respective crop groups, 1980–85

	Crop		
Region	Basic Food	Food Cash	Nonfood Cash
		(percent)	
Asia (excluding China)	54	18	28
	(59)	(33)	(8)
Sub-Saharan Africa	50	27	23
	(57)	(27)	(15)
Near East and North Africa	35	47	18
	(38)	(54)	(8)
Latin America	52	37	11
	(34)	(51)	(15)

SOURCE: Oram (1988).

NOTE: The figures in parentheses represent the percentages of total crop output among different groups of crops.

The distribution of scientists does not suggest a relative diversion of scientists from basic food crops to cash crops. Given the inadequate scientific research efforts in agriculture in many countries, what is needed is an increase in total efforts and, at the same time, an increase in emphasis on cash crops in cases of significant past neglect.

If the relative share of different crops in total output is any guide for the allocation of research efforts as indicated by the distribution of crop scientists, there is no serious incongruence, except in the case of Latin America. But, then, it is highly questionable how far the distribution of scientists is a rational criterion; research expenditures, rather than proportion of scientists, weighted by the efficiency or quality of research efforts, would be a better indicator of relative research efforts on different crops. The relative rates of growth in productivity—that is, yield per hectare—could be a measure of the impact of research and development efforts, on the one hand, and of investment, on the other, on different crops. During the 1960s, the rate of growth in productivity for developing countries as a whole was higher for cash crops than for basic food crops (table 6.7). However, in the 1970–84 period, the rate of growth in productivity increased considerably for basic food crops in all regions and was much higher than that for cash crops. In fact, growth in productivity of cash crops declined for the developing world as a whole during 1970–84, compared with 1962–70. Among the regions, it increased

TABLE 6.7 Percentage change in yield value per hectare of different crop groups in developing-country regions, 1962–70 and 1970–84

		Cash Crops		
Region	Basic Food	Food	Nonfood	Total
Sub-Saharan Africa				
1962–70	5.4	0.7	15.9	6.1
1970–84	26.2	−35.7	0.2	−24.8
Near East and North Africa				
1962–70	12.2	−13.1	41.1	0.1
1970–84	27.2	11.6	35.5	20.1
Latin America				
1962–70	−2.5	58.9	−4.9	31.2
1970–84	18.8	−10.4	29.1	7.3
Asia (excluding China)				
1962–70	15.3	11.8	11.9	12.4
1970–84	38.3	21.9	18.7	22.2
Developing countries (excluding China)				
1962–70	10.7	17.5	7.2	14.8
1970–84	32.3	2.2	17.6	9.6

SOURCE: FAO data tapes.

NOTE: Values in dollars per hectare in constant prices of 1979–81 are used for the computations.

only in the Near East and Asia; in Africa, there was a decline, and, in Latin America, growth was less than one-third of what it was in the 1960s. Moreover, there was a considerable divergence in performance between food cash crops and nonfood cash crops in different regions.

Implications of Choice Between Food Crops and Export Crops for Employment, Poverty, and Nutrition

Questions have been raised about whether efficiency considerations or comparative costs are the only criteria for determining choice between food crops and export crops. Are the effects of different choices on income distribution, employment, poverty, and nutrition not relevant constraints, and should they not influence the choice? It is generally agreed that multiple objectives, such as growth, equity, alleviation of poverty, and undernutrition, cannot all be achieved by one single policy instrument such as choice of cropping pattern, and a variety of policy instruments would be necessary.

Whether cash or export crops produce more employment than food crops depends on the choice of technique. An export crop may be more labor intensive than a particular food crop—for example, jute versus rice in Bangladesh. Furthermore, it is often argued that food production is predominantly undertaken by small farmers, whereas export crops are frequently produced on large-scale, plantation-type farms. This is not necessarily so; both crops can be produced on large or small farms, depending on the institutional framework of agricultural production in a country. In many countries, export crops are produced by small farmers. Hence, the choice of crops between food and export crops does not necessarily imply a choice of the scale of production. Rarely are economies of scale specific to a particular crop; often they relate to the precise operations, irrespective of whether the crop is an export crop or a food crop. Similarly, intersectoral linkages—that is, the "spread" effects on the overall economy of the production of a particular crop—depend on the choice of technology that determines the nature and amount of purchased inputs, and that in turn determines the backward production linkages through input-output relationships. Consumption linkages, on the other hand, depend on the extent of increase in income and on who receives the increase in income. If the increase in income accrues to the very poor, it would stimulate demand for food, since the poor spend a higher percentage of their income on food. If the increase in income accrues predominantly to the medium-scale farmers, it is more likely to stimulate demand for nonfood items, such as manufactured goods, as well as for hired labor, leading to a higher level of employment for the landless poor.

The overall impact of commercialization, including the expansion of cash or export crops, on food security depends basically on the nature of technology, institutions, infrastructure, and policies, including macroeconomic policies. Commercialization of agriculture based on large-scale, capital-intensive, plantation-type export crop production, as was the case in the past when it was centered on an urban export enclave without intersectoral linkages to the rest of the economy, does not usually enhance food security. This is especially so if enhanced export production is neither associated with a rise in food imports nor with an increased productivity in food production. Commercialization of agriculture that is based on smallholders and on technological progress in food and export crop production leads to an expansion of income and employment directly as well as indirectly in the rest of the economy through intersectoral linkages and can strengthen food security.

7 China's Experience with Market Reform for Commercialization of Agriculture in Poor Areas

TONG ZHONG, SCOTT ROZELLE, BRUCE STONE, JIANG DEHUA, CHEN JIYUAN, AND XU ZHIKANG

Introduction

This chapter is about the biggest agricultural commercialization "project" of the 1980s: China's economic reforms. It highlights the power of domestic market reform for poverty-alleviating growth, with government remaining an active force in interregional staple food markets, as well as the scope for public policy for poverty alleviation parallel with market reform. As will be shown, China's problem of absolute poverty—a cause for undernutrition in its poor areas—was significantly reduced as a result of these policies. In fact, it may be speculated that China's expansion of its poverty alleviation policy in the 1980s came partly as a result of the increased resource availability that resulted from successful commercialization and partly as a consequence of public demand for regional balance in the growth path.

In the late 1970s and early 1980s, the government of China implemented a series of economic reforms designed to stimulate the rural economy and to form the basis of its drive to create a modern, integrated economic system. Rural reforms of this period may be grouped under four headings: decollectivization, market and price reforms, increased availability of inputs, and agricultural and rural economic diversification. Although the relative contribution of each element is hard to disentangle (Johnson 1989), the 1978–84 period was one of unprecedented high growth rates in nearly every sector of the rural economy, creating a sharp rise in rural household income and a burst of commercial activity. There is little empirical work that addresses the degree of participation in this development boom by impoverished regions where undernutrition was concentrated at the outset of the boom. After 1984, nominal income growth among China's rural residents slowed down and real incomes appear to have declined, yet the broad process of commercialization in rural China has continued.

After briefly summarizing the rural commercialization and income growth process for the country as a whole, this chapter addresses four questions regarding China's poor areas:

1. What are the major arguments or concerns regarding the participation of poor areas in this process?
2. How have they managed during the 1978–84 period of rapid rural development and the subsequent period of high inflation and decline in real incomes?
3. To what extent have they participated in the broad processes of rural commercialization so evident elsewhere in China, and how is their participation or lack thereof related to their income performance?
4. Can their performance be traced to the success or failure of particular public assistance initiatives, economic reforms, or obstacles to their rural development?

Rural Economic Reform, Commercialization, and Development in China

The various elements of China's rural commercialization process since 1978 and how they relate to economic reforms and policy changes since then have been traced in detail elsewhere (An 1989; Perkins 1988; Travers 1984; World Bank 1990). In summary, China's success with commercialization involved several elements:

- rapid growth in foodgrain production based on long-term investments in technical transformation of agriculture and a rapid growth rate of input supplies that were broadly distributed;
- rural reorganization linking family income more closely to productive work and farm-level decisions;
- higher public procurement prices and tax cuts, leaving more cash in farmers' hands and providing incentives for increased farm production;
- liberalization of previously restricted rural income-earning activities and product markets;
- supplementation of autonomous specialization of farm activities with public regionalization plans for agriculture, backed by complementary public market initiatives and increased public investment in transport infrastructure;
- policies supportive of labor-intensive rural industrial development and rural town growth.

During 1978–84 these elements led to rapid growth in farm and rural industrial production; rapid growth of markets, marketing activities and rural income; and some degree of agricultural diversification and specialization. Rural per capita income rose 197 percent in nominal terms between 1978 and 1987, which was partially accounted for, and offset by, the increase in prices.[1] All of the growth in real per capita income seems to have occurred between 1978 and 1984, with average real incomes in rural areas stagnating or even declining between 1985 and 1987. Important causes of this latter stagnation and decline include

- poor integration and bottlenecks in factor market development, especially the lack of a rural financial system and macroeconomic regulatory apparatus appropriate for rapid growth of the rural economy;
- insufficient storage, processing, transport, and communications infrastructure for sustaining rapid development and integration of markets;
- ambivalent public policy toward market integration and privatization;
- a reversal in the favorable terms of trade for major farm commodities procured by public institutions, and excessive taxation by local authorities through remonopolization of input supplies and other fees, taxes, and user charges.

However, the process of rural commercialization, despite retrenching in some regions and sectors after 1984, resumed within a year or two. In particular, commercial sales of foodgrains doubled between 1978 and 1983 and reached even higher levels in 1984 and 1987–88.[2] Rural-to-rural transfers of grain grew from 6 percent of production during 1975–79 to a peak of 15 percent in 1985, subsiding to 7.6 percent by 1987. Rural-to-rural transfers still occurred primarily through public channels that gave priority to urban procurement needs.

Among other farm commodities, the increase in the marketed share of staple foods and edible oils was most significant (table 7.1). Marketing rates for cotton, reflecting state purchasing and industrial-processing monopolies, were already high before the reforms and remained high. Marketing shares for swine and piglets rose and fell in a complex relationship with relative prices and rates of growth and levels of staple food crop production and farmer incomes. The marketed volume of aquatic

---

1. Public purchase prices for farm goods rose 99 percent over the entire period. Prices for consumer goods sold in state stores rose 134 percent between 1978 and 1987. Market prices for the same basket of goods rose 63 percent, but were still some 17 percent above state store prices in 1987 (SSB *Statistical Yearbook of China* 1989). The retail price of grain in rural areas doubled during 1985–89.

2. Foodgrain sales by farmers rose from 51 million metric tons (mmt) to 117 and 121 mmt in 1983–1984 and 1987–88, respectively, in terms of *maoyi liangshi* (trade grain), or partially processed weight (SSB *Statistical Abstract of China's Rural Economy by County* 1990).

products rose, but the marketing rate fell, as infrastructural development was unable to keep pace with rapidly rising production levels.

In part, these trends reflect changes in cropping patterns and production structures during the first decade of the reforms. Throughout China, the proportion of all sown area devoted to grain and green manure crops declined, and the area planted with cash crops increased.[3] With brisk growth in farm input supplies; greater freedom for farmers in utilizing labor, material inputs, and, to some extent, land; rapidly climbing rural incomes and higher public prices; and freer private markets for farm goods, production of many long-suppressed consumer food items grew at phenomenal rates.[4]

Although rural agricultural labor markets and, especially, transactions in land markets are very constrained (Taylor 1988; Rozelle 1991), there was still considerable factor market activity following the reforms. Despite labor market restrictions, the proportion of rural-based labor engaged in off-farm work grew from 6 percent in 1978 to 9 percent by 1983 and to 22 percent by 1988 (SSB *Statistical Yearbook of China* 1984 and 1989). While rural credit via approved channels is still inadequate, agricultural loans quadrupled between 1981 and 1988, supplemented by rapid expansion of credit from unregulated sources. Since the 1950s, public banking and credit institutions have been organized to control flows of real and financial resources between the farm (especially the grain) economy and the government and cities (Byrd 1983; Ishikawa 1986; Stone 1988). Restrictions to achieve this control left credit institutions, on the eve of the reforms, inadequately prepared to meet the needs of prolonged rapid rural development, not only for industrialization and other nonagricultural diversification activities, but even for continuing the rapid expansion in crop production and marketing for which these public institutions were designed. Credit supplementation from unregulated sources shares the blame for accelerating inflation, but it helped to delay and ameliorate the braking effect of the inadequate public credit

---

3. Area planted with cash crops increased from 10 percent of total sown area in 1978 to a peak of 16 percent in 1985 and remained between 14 percent and 15 percent during 1986–89 (SSB *Statistical Abstract of China's Rural Economy by County* 1990).

4. Production of fruit, tobacco, tea, and bast fibers grew by around 50 percent between 1978 and 1984 while production of sugar beets and oil crops doubled and that of cotton tripled. Despite the shift of farmland and labor out of grain cultivation, foodgrain production increased at 5 percent per year (SSB *Statistical Yearbook of China* 1990). The livestock sector's share of gross value of agricultural output (GVAO) increased from 15 percent in 1978 to 22–23 percent during 1985–87. Aquatic production grew from 2 percent of GVAO in 1978 to 6 percent in 1988. Likewise, the share of farm subsidiary output grew from 3 percent of GVAO in 1978 to 7 percent in 1987 (SSB *Statistical Abstract of China's Rural Economy by County* 1990).

facilities and the government's inflation-control efforts (World Bank 1990).

Governmental Concern with Poverty and Links to Rural Reform

How did the commercialization process affect the poorer regions of the country? To understand the concern over the capacity of these regions to participate in this process, a brief digression into the underlying nature of Chinese poverty is useful.

Chinese Poverty Before 1978

The ideological origins of the Chinese Communist Party are grounded in an explicit concern for the poor. Land reform was carried out in areas under party control even before the establishment of the People's Republic and throughout the country within several years thereafter. Land taxes and indirect taxation through the farm goods procurement system were established on a progressive basis. This major redistribution of assets was the basis for important rural welfare improvements and reduction of poverty in the 1950s.

The subsequent formation of agricultural producers' cooperatives and people's communes was undertaken for many purposes, but poverty relief was a major objective. The considerable leveling of wealth that resulted narrowed the dispersion of incomes within particular accounting units. However, the accomplishments in this respect were both limited and temporary (Ishikawa 1968; Roll 1975). Collectivization could not extinguish the remaining sources of poverty, due to geographic differences in endowments among accounting units and disparities in labor endowment among families. The latter were dealt with to a limited extent by maintaining small local and national relief funds to assist the elderly and the dependents of living or dead soldiers (Ahmad and Hussain 1991).

Aside from these rural reorganization efforts, programs and policies specifically aimed at benefiting the rural poor were extremely limited (Lardy 1978; Dixon 1981). The Ministry of Civil Administration provided funds, food, and input supplies to areas suffering from natural disasters, but chronically poor regions were not the principal recipients of emergency resources. To a modest extent, the rural poor were assisted by the already-mentioned small local and national relief funds maintained for the elderly and military dependents. However, the design, targeting, and level of assistance through these programs were not sufficient to improve basic living conditions of the poor.

On the other hand, a number of policies, many of which were aimed at facilitating extraction of agricultural surpluses, tended to impoverish farmers, and often were especially harmful to the poor. Restrictions on labor markets and migration associated with collectivization trapped the poor on unproductive farmlands. Tight controls over cropping decisions, input distribution, and procurement quotas, together with restrictions on income-earning activities in the countryside, kept rural inhabitants excessively focused on grain production, regardless of their comparative advantages. State control of agricultural prices kept farm profitability low (Stone 1988).

These policies were reappraised in the late 1970s, when it was recognized that 30 years of pursuing a Stalinist model of development had left the rural economy undeveloped and characterized by low productivity and low living standards, including undernutrition, with a significant proportion of the rural population unable to maintain a minimum subsistence level.

Rural Policy Reforms and Questions About Poverty

While the policy reform initiatives summarized earlier were successful in raising rural incomes and stimulating rural development in general, official concern has been raised as to whether economic growth and many of the specific reform measures would actually benefit the poorer regions of the country. High surplus quota grain prices and free market sales might not help poor farmers in isolated regions who are unable to fulfill basic quota sales obligations. If higher purchase prices for state farm goods and local public finance initiatives were to substitute for national and provincial revenue sharing and capital construction, where would this leave poor areas often characterized by inadequate local governments and with relatively modest benefits from procurement price increases?

There was concern that agricultural specialization and diversification would be curtailed in poor areas by isolation, poor organization, and chronically low and variable staple food production levels. Could sizable supply increments continue in poor areas where transport costs are high and marketable surpluses low? Were agro-ecological, infrastructural, and credit conditions sufficient to expect productivity increases through agricultural technical transformation in poor counties? There was concern that traditional customs and low education levels would prevent modern economic development in many poor counties, especially those inhabited by minorities. Contrary to these concerns, product market liberalization seems to have been more permissive in poorer counties. Poor areas could benefit most from increased freedom to engage in nonfarm occupations, at least until this policy was extended to encourage collective investment in rural industries (only the wealthier collectives had much capital to invest).

Some of the policies initiated after 1978 were, at least in the beginning, targeted to impoverished areas. The return to family farming under the "production responsibility system" was initiated locally in poor areas where collective farming had clearly not benefited the rural population; it was first officially extended to the poorest one-third of China's counties, and it was not until 1983 that decollectivization was sanctioned as a nationwide policy.

Nevertheless, concern over widening regional income differentials as some areas quickly prospered under reforms, concern that a large portion of China's poorer rural areas was being left behind, and a relaxation of concern over growth in farm production and marketing in the country as a whole led to an increased focus of governmental attention on poor areas during the 1980s.

Special Public Assistance Efforts for Poor Counties

The government's new approach to rural poverty was to provide special assistance to the poor in developing their own capacity to exploit local resources for commodity production and, gradually, to become commercialized. In other words, this new approach replaced the earlier policy of "blood transfusion" with one of "blood creation." Driven by some of the concerns mentioned above, the State Council established in 1985 the Office of the Leading Group for Economic Development in Poor Areas to administer and monitor 4 billion yuan per year in development project expenditures for poor areas. Provincial and regional governments supplement these funds. Poor Area Development Offices (PADO) of provincial and county governments administer funds from both national and provincial sources. This initiative started on a smaller scale in the early 1980s, when the government initiated an effort to address the special problems of economic development in poor areas. An assistance program provided low interest loans to 28 arid, impoverished counties in the Sanxi region of Gansu Province and Ningxia Hui Autonomous Region. This program was used as a model to extend national assistance to 300 counties by 1986, and to an additional 27 pastoral counties by 1988. Some 370 counties were designated for provincial government support in 1986. After adjusting for county consolidation and double counting, some 664 counties received national or provincial assistance in 1986, and 698 counties received public help in 1988.

From 1985 to 1987, poor areas received 2.7 billion yuan in the form of grain, raw cotton, and cloth in exchange for work performed by the poor on infrastructural projects. In 1988, the Office of the Leading Group for Economic Development in Poor Areas initiated the Food and Clothing Supply Project that provided incremental allocations of fertilizer, plastic sheeting, steel, timber, and trucks at 30 percent discount to arid and frigid mountain areas. The state exempted poor counties from agricultural taxes for three to five years beginning in 1985, and from energy and transport construction taxes in 1987, and lowered reserve requirements for their banks in 1987.

While a special economic development program for poor counties marks an important milestone in the evolution of China's public policy, other public finance flows affect poor counties as well. According to the Office of the Leading Group for Economic Development in Poor Areas, the forced sale of low interest bonds (*guokuchuan*) to poor county governments in the late 1980s exceeded PADO expenditures. The resale profit value of fertilizer deliveries at quota prices and heavily subsidized transport to poor counties through the normal contract mechanisms could have totaled several billion RMB by the late 1980s, given the sharp increases in nonquota fertilizer prices.

Reduction in Absolute Poverty, 1978–84

Detailed statistical evidence is lacking for the 1978–82 period, but there is no question that China experienced substantial reduction in poverty between 1978 and 1984: almost one-quarter of the rural population, roughly 180 million people, emerged during this period from income and nutritional levels that were below the absolute poverty line. One exercise, based on estimated food-energy consumption needs, places the number of rural residents living in absolute poverty in 1978 at around one-third of China's total, or 250 million.[5] Using the same poverty standard for 1984, the estimate falls to around 70 million, although, after 1984, rural poverty is estimated to have increased somewhat. Official estimates echo these results, while placing the number of rural poor at somewhat higher levels—102 million in 1985 (State Council 1989). All studies, however, suggest a substantial reduction of poverty during the 1978–84 period.

The remainder of the chapter addresses the economic record of the

---

5. Calculated from survey-based food supply estimates, using 2,150 kilocalories per day as a poverty measure throughout China. While lower kilocalorie estimates are sometimes used in studies for other countries, a high proportion of China's poor are in temperate rather than tropical latitudes and reside in mountainous areas, therefore tending to require more calories for individuals of equivalent age, sex, and stature than poor individuals in tropical climates.

The estimated consumption associated with expenditures on food of about 75 RMB per year, 2,150 kilocalories per day, translates to a poverty line of around 100 RMB in 1978. The 1985 equivalent of the same poverty line would be around 200 RMB. Such calculations inevitably rest on a number of somewhat arbitrary assumptions, and the large decline in the poverty head count is, in part, because of the unusually large concentration of the Chinese population close to the poverty line (Travers and Stone 1991).

poorer part of the Chinese rural population. For conceptual purposes, this population is divided into groups that reside in two areas:

1. "successful poor areas" that operated below the subsistence level in 1978 but were not regarded as poor by national and provincial governments in 1985;
2. "poor counties" designated in 1985 for national or provincial development assistance.

Successful poor areas include the approximately 180 million rural residents of poor villages that were below the poverty line in 1978 but above it in 1985. Large numbers of villages in poor but successful counties are included, as are poor but successful villages located in officially designated poor counties, which included most rural Chinese with average per capita incomes at or near the poverty line in 1985. The poverty line, while based on an estimated relationship between caloric intake and income, was defined in terms of income: 200 RMB per capita.

The 664 poor counties include approximately 300 counties whose annual per capita incomes of rural residents averaged less than 150 RMB (or up to 300 RMB for politically favored counties that are designated for national assistance). The remaining poor counties are designated for assistance by provincial governments; per capita income levels of these counties vary. Some counties with high average income levels still contain many villages and even townships that are quite poor. More than 100 poor counties that are under special military administration in Tibet and elsewhere in western China are excluded from these data.

The reform period may be divided into two subperiods: the rapid growth period of 1978–84 and the period of general rural income stagnation beginning in 1985 and extending into the 1990s. Detailed statistical data on officially designated poor counties span a portion of both periods: 1983–87.

Successful Poor Areas, 1978–84

The poor areas that were able to escape extreme poverty as a result of the initial policy reforms were often those where the geographical or physical endowment was favorable but previous policies had suppressed opportunities for profitable commercialization. These areas included areas that had been forced to devote most of their land to grain, despite having a comparative advantage in cash crops or livestock (Lardy 1983); areas with considerable labor surpluses and easy access to urban centers; and areas where traditional handicrafts or commerce had been well developed but suppressed in recent years. The latter included areas located on historical trade routes of less importance to the socialist economy, border areas with unique forest products (often populated by minority groups), and extremely risk-prone and meager agricultural regions.

Successful poor areas also included grain-producing regions with characteristics not dissimilar from those just on the other side of the poverty line, but which also scored impressive growth in grain yields and production while diversifying the rural economy. Technical change and commercialization of agriculture were key to the success of many successful poor areas. High-yielding variety (HYV) coverage in such areas was significant in 1978 and grew rapidly thereafter. During the 1978–84 period, coverage of hybrid maize rose from 60 percent to 80 percent of China's maize area; coverage of hybrid sorghum rose from 50 percent to almost 90 percent; coverage of short-stature wheat varieties rose to three-quarters of the country's wheat area; and farmland planted with dwarf rice varieties increased from more than 80 percent to 95 percent of China's paddy sown area, inevitably including farmlands occupied by the hundreds of millions of rural residents living in poverty in 1978 (Stone 1990). Irrigation facilities served an important portion of such areas. Coverage of irrigation facilities averaged 54 percent of cultivated area for the 1,300 "nonpoor" counties of China, including hundreds of poor counties that successfully passed over the poverty line during the 1978–84 period (SSB *Statistical Abstract of China's Rural Economy by County* 1980–87).

Some rural areas, relatively impoverished in 1978, made rapid advances in the early 1980s, whereas other poor areas were much less successful. This differential performance was reflected in provincial statistics for growth rates of agricultural production during 1979–82, when, for example, Anhui, Sichuan, Guizhou, and Guangxi (relatively poor provinces) grew rapidly (18–27 percent), but Gansu, Qinghai, Shaanxi, and Ningxia (as poor or poorer provinces) grew very slowly (0–8 percent) (World Bank 1985).

What about the residents of counties still considered poor as late as 1985? Did their incomes remain stagnant or decline during the remarkable period between 1978 and 1984? Were these counties so isolated or so unlike the rest of China? Was China's rural economy so thoroughly dualistic? If not, what was the nature of their progress during the 1978–84 period, and how did they fare during the subsequent period of rural income stagnation across China as a whole? How was any such progress related to the rural economic reforms, public policy, and commercialization process? Empirical work in the remainder of the chapter addresses these questions.

Poor Counties Commercialization and Poverty Alleviation, 1983–87

After the impressive nationwide gains in production and incomes that resulted from the "liberation" of the energies of China's farmers, to what extent have the poor, especially those in impoverished areas, been able to continue to improve their livelihoods during the latter half of the 1980s? Have the poor areas been put on a path to a growing and commercializing rural economy? Has the income gap between the poor and nonpoor been narrowing or increasing over time? Or has the government, in its drive to let "a part of the farmers get rich first" (Deng Xiaoping), totally abandoned its efforts to deal with poverty? Disaggregated data are available for the 1983–87 period for a more detailed examination of the performance of poor counties.[6] This period includes both the end of the rapid growth period (1983–84) and the slower growth period (1985–87) when real incomes in rural China tended to decline.

Location of Poor Counties

Aside from the poor counties under special military administration, the most intractable poverty alleviation problem is centered among the 211 million rural residents of 664 counties, where some 93 million were considered poor in 1986 by the Office of the Leading Group for Poor Area Development (Jiang et al. 1989). The poor counties are distributed across 23 of China's 27 provinces, but 78 percent of them are concentrated to the west of a north-south line that runs through the central mountainous parts of the country from Heilongjiang, Gansu, and Inner Mongolia in the north to Guangxi and Yunnan in the south (figure 7.1). The remaining 146 counties, generally better off among the poor counties, are located in less contiguous islands of poverty in the hills of eastern and southeastern China.

Status in the Early 1980s

Poor counties designated for public assistance are normally characterized as being poorly endowed by geographic location (remote and mountainous) and at a disadvantage in terms of agricultural resources

---

6. The study is based on 664 counties designated as poor counties in 1986, about one-third of all Chinese counties. Thus the successful poor areas discussed in the earlier subsections are generally not included in the sample. Sample data pertain to the years 1983–1984 and 1986–1987. The main data source is unpublished data for agricultural inputs and outputs, village enterprises, grain flows, population, and rural incomes for each of the poor counties. Supplementary data are drawn from national and provincial statistical yearbooks and other secondary sources, especially SSB *Statistical Abstract of China's Rural Economy by County* 1989; *Statistical Yearbook of China* 1990.

FIGURE 7.1 Poor regions of China and their dominant staple food crops

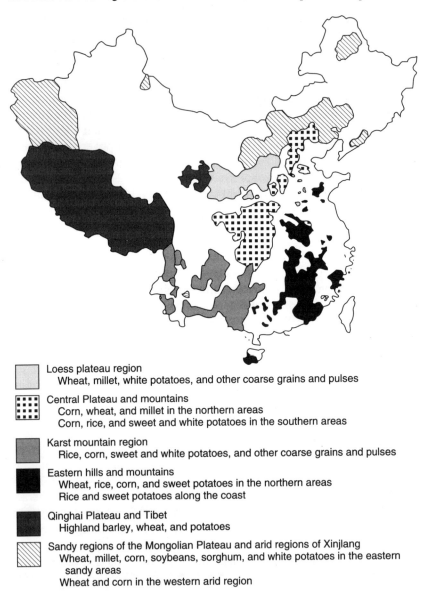

- Loess plateau region
  Wheat, millet, white potatoes, and other coarse grains and pulses
- Central Plateau and mountains
  Corn, wheat, and millet in the northern areas
  Corn, rice, and sweet and white potatoes in the southern areas
- Karst mountain region
  Rice, corn, sweet and white potatoes, and other coarse grains and pulses
- Eastern hills and mountains
  Wheat, rice, corn, and sweet potatoes in the northern areas
  Rice and sweet potatoes along the coast
- Qinghai Plateau and Tibet
  Highland barley, wheat, and potatoes
- Sandy regions of the Mongolian Plateau and arid regions of Xinjlang
  Wheat, millet, corn, soybeans, sorghum, and white potatoes in the eastern sandy areas
  Wheat and corn in the western arid region

Note: Map is based on original map in Dehua (1988, 21).

(such as soil, rainfall, and climate). Many of these areas suffer from severe ecological damage such as deforestation and soil erosion. Counties where staple food crop yields are chronically low or highly variable, or both, as well as counties where yields per hectare are high but there is little arable land and few noncrop activities to support dense populations are included among poor counties as are those counties with poor villages and townships but whose per capita average foodgrain production is surprisingly high (figure 7.2). Infrastructure is generally very poor, particularly in terms of transport links to the outside.[7]

Irrigation in poor counties was available for 31 percent of cultivated area in 1980. Chemical fertilizer sales—an indicator of technology utilization in agriculture—per sown hectare in poor counties in 1980 and 1985 averaged 57–58 percent of sales in nonpoor counties (SSB *Statistical Abstract of China's Rural Economy by County* 1980–87). Large increases in fertilizer supplies were made available to poor counties during the early reform period relatively independently of farm goods sales. Yet farmers in poor counties did not necessarily use all that they received: market resales of fertilizer have been a source of cash income since the 1970s, and especially since the mid-1980s (Stone 1989; Ye 1991).

HYV coverage of Chinese staple food crop area, also in poor counties, increased rapidly during the 1978–84 period (Stone 1990). While irrigation did not appear to increase among poor counties in the early 1980s, fertilizer sales grew by 38 percent between 1980 and 1985, about the same rate (40 percent) as in China as a whole. Consequently, while foodgrain sown area declined between 1980 and 1985, foodgrain production in poor counties increased 22 percent during 1980–84 (17 percent during 1980–85) from 64 million metric tons (mmt) in 1980 to 79 mmt in 1984 (75 mmt in 1985). These figures compare surprisingly well, despite the poorer agro-ecological and water control conditions in these counties, with the 27 percent increase in China as a whole during 1980–84 and 18 percent during 1980–85.

The economic structure of agriculture in poor counties in 1983, though not strikingly different from the national average, reflected the geographic concentration of these counties in mountainous areas: a larger share of output value was derived from forestry and a smaller share from crop production. Despite the pastoral nature of some poor areas, the share of livestock production just equaled the national average (due to greater concentration of pig and poultry production in areas with adequate feed production).

---

7. For example, in Sichuan Province, there are still 792 townships and 50 percent of villages without road access (Chen 1990; State Council 1989).

FIGURE 7.2 Distribution of 664 poor counties by per capita foodgrain production, 1983 and 1987

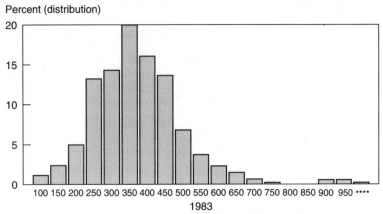

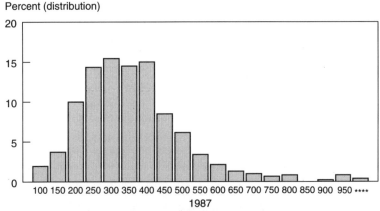

Yet crop production dominated agricultural output value.[8] Foodgrains accounted for 83 percent of sown area during 1980–84. Thus, growth in grain production and any improvement in relative prices are

8. Agricultural output value in 1983 (percent):

	Crops	Forestry	Animals	Fish	Sidelines
Poor counties*	65.6	8.2	17.7	1.1	7.5
National average	70.6	4.6	17.6	2.3	4.9

*The calculations are shares of aggregated poor-county data, more comparable to the national average.

likely to have been important sources of income growth in poor counties during 1978–84.

Poor counties in the mid-1980s were still characterized by semisubsistence economies, with most of the production being for own consumption, although some was sold. About 12 percent of the farming households in these areas earned no income from the sale of farm or sideline produce; 33 percent of the grain producers sold no grain; and 57 percent of the livestock raisers sold no livestock.[9] Thirty-five percent of China's 1984 record grain harvest was marketed, but only 3 percent of the production of poor counties was sold (or 2 percent if government resales are netted out), implying that China's other counties sold 42 percent of their production.

In summary, available information suggests that most poor counties benefited from improved fertilizer and HYV seed availability more or less as the successful areas and wealthier counties did. While it stands to reason that product market, labor, and crop choice decontrol as well as specialization, trade, and crop diversification away from staple food crops (reforms ostensibly available at an earlier date in poor counties) would offer considerable benefits, the share of grain in sown area declined less among poor counties than in nonpoor counties in the 1980–84 period. Economic diversification to noncrop activities would appear to hold greater potential benefits for poor counties where agricultural conditions are, on balance, riskier. While it is clear that noncrop activities were an important and growing source of rural resident income in poor counties by the mid-1980s, data are unavailable for assessing this potential significant contribution to poor-county income growth in the early reform period.

Agricultural Diversification, 1983–87

The principal shift among the five major categories of farm production during 1983–87 in poor counties was an increase in animal husbandry and fishing activities relative to crop farming, forestry, and farm subsidiaries (sideline production).[10] Part of this shift was due to changing

---

9. These figures are based on a sample survey conducted by the Rural Development Institute of the Chinese Academy of Social Sciences of 280 "observation points" distributed in 28 provinces, autonomous regions, and municipalities (Chen 1990). The date of the survey was not specified.

10. Comparison of the structure of agricultural output value in poor counties:

	Crops	Forestry	Animals	Fish	Sidelines
1983 (percent)	58.3	9.6	17.6	0.7	14.0
1987 (percent)	53.8	9.0	26.9	1.1	9.2

relative prices,[11] which would have caused a shift in favor of noncrop production even in the absence of production increases. However, the officially calculated gross value of both livestock and fisheries production increased in real terms by about two-thirds in 1983–87.

While dispersion of average income among counties increased between 1983 and 1987, at the end of the rapid growth period for Chinese agriculture, welfare improvements were not limited to the wealthier portion of the distribution: many of the poorest counties also improved their real income status. However, foodgrain output per capita in poor counties declined in each income stratum between 1983 and 1987.

With production stagnation and rural population growth of 2–3 percent per year in poor areas, production per capita also fell by 5 percent, from an average of 347 kilograms to 330 kilograms.[12] The distribution of per capita foodgrain production among poor counties also changed. Most notable is a large increase in the proportion of counties producing less than 200 kilograms per capita—by government definition, the "poverty line" for per capita food availability—from 8.4 percent to 15.3 percent of counties. Changes in foodgrain sown area in poor counties contributed to this production decline: foodgrain sown area fell by about 2 percent over the period. Average foodgrain yields remained virtually static (at 2.1 tons per hectare), increasing at an annual rate of only 0.4 percent. While foodgrain sown area declined throughout China, yields grew rapidly at an annual rate of 3.8 percent, suggesting that yields in counties not designated as poor grew by around 5 percent per annum (SSB *Statistical Yearbook of China* 1984, 1988).

The decline in foodgrain area, partly due to crop diversification, accounts for falling foodgrain production among the counties with the lowest per capita production. Corresponding increases of 10–20 percent in the area under nongrain crops, mainly oilseeds, account for only a small part of the large increase in nominal income throughout poor counties after 1983. Yet the nutritional significance to the poor of increased oilseed production should not be overlooked—edible oils had

---

11. During 1983–87, the national constant-price deflators for aquatic products and sidelines doubled and that for livestock increased 45 percent, as prices for these products were decontrolled or brought to near-market levels. Crop prices, dominated by grain that was still mainly subject to administered prices, only rose 14 percent by contrast. The constant-price deflators appear to understate price increases, compared with official procurement price indices. The latter also indicate much higher price increases for forestry products, which would typically favor poor areas.

12. According to Chinese accounting convention, foodgrains (*liangshi*) include rice (evaluated at paddy weight), husked corn, wheat, and other cereals (unmilled); soybeans, peas, broad beans, mung beans, and other minor pulses; and potatoes and sweet potatoes (both evaluated at one-fifth fresh weight).

been in extremely short supply throughout China, not to mention the poor areas, before the 1980s.

Expanded Public Commercialization

Data on total marketing in poor counties are not available, but approximations suggest a degree of aggregate commercialization of about 30 percent in poor areas versus 55 percent in nonpoor areas. Detailed data on government foodgrain purchases and resales to rural residents also give indications. The national marketing rate for foodgrains reached 35 percent in the mid-1980s (table 7.1). But, despite record production levels, public foodgrain sales by farmers in poor counties averaged only 4 percent of production in 1983 and 1984, one-fourth of which was resold to rural residents within the country. This proportion rose to 6 percent of production in 1986, then ballooned to 19 percent in 1987 (with around one-third resold to local farmers), despite a decline in production per capita to a level 21 percent lower than that of 1984. Thus, state action for procurement expansion was part of the increased commercialization of agriculture in these years.

Declining stocks nationwide, combined with greater difficulty and higher prices in obtaining grain from outside these counties, forced local procurement authorities to obtain more grain locally in 1987 to meet the needs of local urban areas and county towns. For example, in Shanyang, a poor mountainous county in Shaanxi Province, large grain imports did

TABLE 7.1 Marketing rates for farm products, 1978, 1984, and 1987

Product	1978	1984 (percent)	1987
Staple food crops[a]	20.3	34.8	34.4
Cotton	94.3	98.9	95.9
Edible oils	55.9	67.4	74.6
Swine and piglets[b]	67.9	69.1	68.9
Aquatic products	67.7	58.2	44.7

SOURCE: SSB *Rural Statistical Yearbook of China* 1986, 1987, 1990; SSB *Statistical Abstract of China's Rural Economy by County* 1990.

NOTE: Marketing rate is defined as farmer sales to government and private markets as a percentage of the total production, in volume terms. Note that from 1978 to 1984, production grew by 34 percent for staple food crops, 189 percent for cotton, 202 percent for edible oils (through 1985), 37 percent for slaughtered swine, and 33 percent for aquatic products. From 1984 to 1987, production fell by 1 percent for staple food crops, 32 percent for cotton, and 3 percent for edible oils (from 1985), but grew by 19 percent for slaughtered swine and 54 percent for aquatic products.

[a] Rice, wheat, corn and other coarse grains, soybeans, broad beans, mung beans, peas and other bean crops, potatoes, and sweet potatoes are included.
[b] For meat as a proportion of the total numbers slaughtered.

not begin until after the 1983 and 1984 national bumper harvests. These deliveries were provided in 1985 and 1986 to the county government, which sold the grain, not to farmers, but to town residents and migrants from rural areas working in county towns. This facilitated rural migration to urban areas and provided a financing mechanism for development of housing, industry, and infrastructure, based on the labor of rural migrants. But after two years of disappointing national harvests, grain deliveries to poor counties were substantially reduced or eliminated in 1987. With rising market prices and high costs and difficulties associated with purchases from outside the county, local governments increased purchases from within the county in order to continue development.[13] The balance between use of "bureaucratic coercion" and the offer of higher prices cannot be completely verified. In principle, poor areas are exempt from quotas or contract responsibilities, so all sales should have been voluntary and at negotiated (near-market) prices. In practice, many poor county purchase organizations pay quota prices for some or all purchases.

Table 7.2 shows that state grain purchases on a rural per capita basis were generally higher among higher-income poor counties and that the large increase in procurement that occurred in 1987 was spread across all income strata (as well as all regions of China). But, between 1983 and 1987 in the very poorest counties, total resales to local residents (including any imports from outside) increased faster than procurement, so that net procurement from the poorest counties declined. This was not the case for any of the higher-income counties (above 150 RMB).

Growth in Incomes and Shifts in Income Distribution

Average incomes for rural residents in the 664 poor counties increased by 74 percent between 1983 and 1987, as measured in current prices. Over the same period, income growth in China's other ("nonpoor") counties increased by 45 percent. Still, due to differences in base-period values, absolute increments in rural per capita income during the 1983–87 period were greater in nonpoor counties than in poor counties (RMB 171 versus RMB 115), thus widening the absolute gap. Major difficulties arise in estimating real income growth due to the inadequacy of official price deflators.[14] However, regardless of the actual

13. Based on field notes and Shanyang County data collected by Li Jianguang, Rural Development Institute, Chinese Academy of Social Sciences, in 1990.

14. If the market price index for consumer goods is used, real income is estimated to grow by 2.5 percent per year in nonpoor counties, but by 7.5 percent per year in poor counties (1983–87). If the price index for state-distributed consumer goods is used, estimated real per capita incomes increased in poor counties by 5 percent but fell in other counties by 13 percent during 1983–87. As rural dwellers obtained consumer goods from a mix of market and state-controlled sources in unknown proportions, a rigorous estimate of real income changes probably lies between these two series.

TABLE 7.2 Foodgrain production, procurement, and resale, by income strata, in poor counties, 1983 and 1987

Year	All Levels	< 150	150–200	201–250	251–300	> 300
1983						
Counties in income strata (percent)	100.0	52.6	23.8	15.8	4.1	3.8
Grain output per capita (kilograms)	348.7	287.0	383.3	420.7	437.3	599.0
State grain purchased per capita (kilograms)	13.1	8.8	15.6	17.2	17.5	43.5
Resold share of state grain purchases (percent)	28.3	40.3	24.0	22.9	18.2	12.5
1987						
Counties in income strata (percent)	100.0	28.2	27.1	22.1	13.6	9.0
Grain output per capita (kilograms)	329.0	242.6	309.7	373.6	413.3	482.4
State grain purchased per capita (kilograms)	62.9	28.7	52.2	94.5	77.9	117.9
Resold share of state grain purchases (percent)	41.8	74.3	52.9	30.0	22.2	20.6

NOTE: Data are calculated according to real income strata deflated to 1983 constant RMB, using the state purchase price index for agriculture products. Use of this deflator implies somewhat less real income growth in poor counties than is probably warranted.

magnitudes, real incomes in poor counties grew at a faster rate than those in nonpoor counties in 1984, and appear to have lost less ground than the latter during the post-1984 period of declining real incomes.

The distribution of poor counties across real per capita income strata shifted to the right and appears to be more evenly spread between 1983 and 1987 (figure 7.3). Real incomes increased in most counties

FIGURE 7.3 Distribution of 664 poor counties by real per capita income, 1983 and 1987

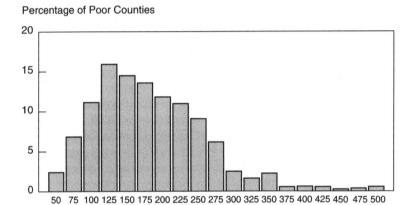

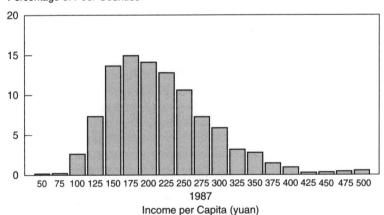

Note: Bars represent percentages of China's 664 poor counties with real per capita incomes in 25 yuan ranges.

within all income strata in 1984, although some counties experienced little growth. The number of counties that earned less than 75 RMB per capita continued to decline throughout the 1983–87 period. The reforms related to agricultural commercialization thus meant growth and more equality for poor areas.

Conclusions

China's success with rapid development of commercialization since the late 1970s rests upon a combination of preexisting conditions, liberal reforms, and public policy initiatives. The major growth in rural incomes that occurred during the 1978–84 period as a result of this process involved increased specialization and diversification in the rural economy, development of private trade, and increased factor market activity.

Rural income growth has also resulted from a deliberate public policy to promote productivity growth and adjust rural incomes in order to promote the commercialization process and achieve social purposes. This policy involved increased public investment in market-related infrastructure and agricultural inputs that eased market bottlenecks and increased productivity. It also operated via the still dominant systems for public procurement of farm products, distribution of industrial goods (especially agricultural inputs), and rural taxation and credit.

While China is still home to 100 million rural residents living around or below the poverty line, there has been considerable improvement in the welfare status of the poor since the initiation of the reforms, when the numbers of those living in absolute poverty were two to three times higher.

To a large extent, the rural poor benefited from the relaxation of past policies that were particularly damaging to their welfare—especially those policies that limited their ability to trade, choose crops, and engage in risk-diversifying and income-enhancing economic activities. But, they also benefited from a public input distribution system that greatly increased low-cost supplies to poor regions during the period, a public procurement system that provided high relative prices for incremental deliveries of major farm products, and public mechanisms that provided necessary credit. Even in poor regions, returns to incremental inputs were often high, owing to past public investments in water control and in agricultural research and seed delivery systems, coupled with high prices and reforms linking rewards to productive effort. Autonomous specialization and trade were selectively accelerated through agricultural regionalization plans backed by increased input supplies, public market guarantees for farmer-supplied and demanded goods at attractive relative prices, and extension and credit. While some farmers in successful poor areas may not have benefited directly through these public efforts, they were increasingly assisted by a permissive policy toward labor movement and rural industrial production of consumer and other goods whose demand had long been suppressed. This allowed profits and rapid growth in rural industry to occur in such areas and provided steady growth in off-farm labor income.

There is evidence that these same processes during the same rapid rural growth period even benefited most of the counties that were still poor in the mid 1980s, but to a lesser degree. Rural incomes in these poor counties grew at even faster rates than in the nonpoor counties (although by smaller absolute increments) at the end of China's rapid rural growth period.

Since 1985, China's government has formulated special policies for poverty alleviation and favored increased commercialization in poor areas, with an initial expansion of resources in 1986. Still, the impact of these expenditures relative to liberalization efforts, development of independent markets, and diversification, as well as other public initiatives, cannot yet be assessed at the household level. Active governmental efforts through the normal public systems for input distribution and farm procurement were responsible for the startling growth in public grain purchases and resales in poor counties in 1987.

The seeds of a more effective policy for targeting development of poor areas may be found in the new administrative mechanism for promoting and funding infrastructure development in poor counties. But, the government has not yet been able to deal effectively with two remaining constraints to commercialization and productivity growth in poor areas: first, lack of a marketing system for grain and other food crops that is capable of adequately supplying these areas and of permitting a higher degree of specialization; and, second, lack of a mechanism (involving more flexible land and labor markets) to permit emigration from poor areas that simply cannot support the existing population.

Finally, the hypothesis of sharp duality among China's rural areas must be rejected, although with qualifications. While diffusion of income growth and commercialization within poor counties is not addressed in this study, and while some entire counties may indeed have participated little in China's rapid commercialization and rural development process, most of the poorer counties seem to have benefited. Poor counties still experience major disadvantages of isolation and weak market integration with China's greater economy, and they chronically suffer from important shortcomings in hard and soft infrastructure. Yet there is evidence from the 1978–87 period of both autonomous economic development in poor counties, relating to liberal reforms, and of their responsiveness to public commercialization initiatives.

8 Investment in Rural Infrastructure: Critical for Commercialization in Bangladesh

RAISUDDIN AHMED

This chapter focuses on the role of rural infrastructure development on the pace of commercialization and the effects of such development on various income classes, especially the food-insecure poor, in the rural economy. Public expenditure and institutional policies for development of rural infrastructure are also considered. Although evidence is mostly drawn from Bangladesh, an effort is made to draw upon experiences elsewhere.

Infrastructure and Commercialization: Theoretical Links

Different people define infrastructure differently. However, a common definition is important for conceptual clarity and for establishing a common ground for tracing infrastructural effects. Development literature of the 1950s and 1960s included public utilities, ports, water supplies, electricity, transport and communication facilities, schools, public health, law and order, and flood control in the list of infrastructure. More systematically, Albert Hirschman (1958) set out four conditions for distinguishing infrastructure from other productive assets: (1) the services provided by the activity are necessary to facilitate or are, in some sense, basic to the carrying out of a wide variety of economic activities; (2) the services are provided in almost all countries by public agencies or by private agencies subject to public control, and they are provided free of charge or at publicly regulated rates; (3) the services cannot be imported; and (4) the investment needed to provide the services is characterized by "lumpiness." Consistent with Hirschman's emphasis on the first condition, the results presented in this chapter are based on the effects of rural transport and communication, electricity, markets, and financial institutions. These are considered to be the hard-core elements of rural infrastructure. Rural transport and communication are very closely associated with the development of electricity, markets, and financial institutions in rural areas.

The best way to explain the process and the logic of the link between infrastructure and commercialization is to trace the causal relation between infrastructure and economic change, and indicate the role of commercialization in this relation. Commercialization is both a cause and an effect in this relation; it strengthens the effect of infrastructure and, in turn, is strengthened by infrastructural effects.

Change in Attitude and Values

Perhaps the most profound effect of infrastructural development is on the attitude and values of rural households, even though such changes are least visible to casual observers. Development of transport and communication infrastructure enhances mobility of people and information through reduction in costs and time. The resulting increase in interaction with the outside world and the informal education process that such interactions involve contribute to changes in attitude and human capital development. The effects of these attitudinal changes are reflected in higher adoption of family planning practices, diminishing faith in superstitions, increasing preference for consumer goods produced outside (the so-called demonstration effect), greater appreciation for formal education, and a rising utility for income. The effect on motivation, combined with the effect on entrepreneurial development that increases the capacity to perceive and the ability to seize comparative advantages, bears immense implications for economic progress. These types of changes in household attitude and values render the rural structures congenial for diffusion of modern technology and ideas.

Prices and Markets

The positive impact of infrastructural development on reduction of marketing costs is well known and is the central driving force for agricultural commercialization. However, what is not fully perceived is the multifaceted effect of infrastructural development on expansion of markets, economies of scale, and improvement in market operation, particularly factor markets. Market operation is the primary precondition for commercialization. Therefore, the effect of infrastructure on market development is a critical link between infrastructure and commercialization.

The market expansionary effect of infrastructural development, through expansion of production of perishable and transport cost-intensive products that is facilitated by easy access to markets, is an important element in the commercialization process. The conversion of latent demand into effective commercial demand similarly expands market size. For example, if there is no road, people would walk from one place to another; if a road is developed and transport services operated, people would use these services. A latent demand for transport services is converted into an effective and commercial demand for transport services. These types of market expansion combined with the expansionary effect of specialization result in an increased scale of trade. Very often, this process leads to reduced per unit trading costs due to economies of scale.

Rural informal credit markets in developing countries are quite extensive and credit in kind is prevalent in such markets. Infrastructural development can reduce the extent of in-kind credits by transforming them to a monetary exchange at lower transactions costs. Facilitated by infrastructural development, credit markets tend to expand.

The effect of infrastructural development on labor markets is complex. Some features of related changes are the following. Availability of attractive consumer goods and income-earning opportunities increases the tendency to work longer hours. Participation of female labor in the work force increases as traditional taboos against female labor are overcome. The practice of bonded labor may diminish when alternative opportunities for labor are created by infrastructural development (Ahmed and Hossain 1990). Labor mobility improves and labor markets become less fragmented due to infrastructural development. These changes result in increased commercial transactions in labor markets and reduce the dualism between family and hired labor.

Diffusion of Technology and Agricultural Production

The effect of infrastructure on agricultural production and diffusion of modern agricultural technology is mediated through a number of forces. First, because of the attitudinal changes discussed above, farmers are more apt to accept a new technology in areas with developed infrastructure than in areas with underdeveloped infrastructure. Second, marketing of modern inputs such as fertilizers, pesticides, and irrigation equipment is logistically easier and cheaper in infrastructurally developed areas than underdeveloped areas. Agricultural extension workers find it convenient to work in places where they can move easily and live comfortably. Third, both factor and product markets operate more efficiently in infrastructurally developed than in underdeveloped areas. The combined effect of all these developments results in increased agricultural production.

The increased demand for modern inputs that are usually supplied from outside the agricultural sector directly creates a demand for cash and exchange. This, in turn, indirectly creates pressure for sales of farm products to finance the increased demand for inputs. The increase in production is generally associated with an increase in marketed as well as marketable surplus, thereby increasing the tempo of commercialization. In developing countries, where part of the agricultural production is meant for home consumption and part for sale, an increase in production leads to a more than proportionate increase in marketable surplus, partly because the additional production is not required for home consumption and, partly, because the demand for financing of purchased inputs and consumer goods increases.

Nonfarm Production and Employment

If infrastructural development is accompanied by policies permitting relatively free trade, then the effect on production of nonfarm goods and services in rural areas can be quite substantive. These effects are generally realized through an increased flow of consumer goods and services, increased processing of agricultural products in rural locations, increased flow of agricultural inputs and investment goods from outside the area, and dispersion of small-scale industries from urban to rural areas.

Among these four factors, the effect of household expenditures on generation of demand for nonfarm goods and services and the associated increase in employment is considered to be most profound. Households allocate an increasing share of their total consumption expenditures to nonfarm goods and services, partly due to easier availability and lower prices of such goods and services in infrastructurally developed, compared to underdeveloped, areas, and partly due to demonstration effects and the resulting change in taste. This change in expenditures is not limited to consumption items only. Investment patterns change toward better housing, sanitation and water supply facilities, and acquisition of durable goods. Obviously, an increase in consumption of nonfarm goods and services and in investment expenditures implies a simultaneous increase in commercialization.

Infrastructure and Commercialization: Empirical Evidence

A review of the literature reveals that empirical studies on the impact of infrastructure, particularly rural infrastructure, in developing countries are quite thin (Ahmed and Donovan 1992). The types of extensive effects hypothesized in the previous section are difficult to support adequately with empirical evidence from developing countries. Further research to test these hypotheses is essential. Therefore, a recent study (Ahmed and Hossain 1990) that examines many of these hypotheses in the context of Bangladesh provides the principal source of empirical evidence presented in this chapter. Empirical work conducted in other parts of Asia is also cited to supplement the evidence from Bangladesh.

Effect on Agricultural Production

The effect of infrastructure[1] on agricultural production can be approximated through a multivariate econometric model, where it is hypothesized that the effect is realized through the impact of infrastructure on prices, diffusion of technology, and efficient combination of inputs and outputs (Ahmed and Hossain 1990). The results show that improved rural infrastructure increases agricultural production by 32 percent in Bangladesh. The effect on the level of inputs that arises from differential input prices is small. The largest impact comes from the infrastructural effect on diffusion of technology, efficiency in use of inputs, and crop combination (more high-valued products). Differences in the diffusion of technology and prices between infrastructurally developed and underdeveloped areas in Bangladesh are shown in table 8.1.

The intensity of use of modern technology (that is, irrigation, high-yielding varieties [HYVs], and fertilizer) is substantially higher in villages with developed infastructure than in underdeveloped villages. Further analysis indicates that installation of private tubewells, tubewell electrification, and availability of fertilizers are extremely dependent on the presence of good roads, markets, and financial institutions.

TABLE 8.1 Differences in the use of modern technology and price levels between infrastructurally developed and underdeveloped villages, Bangladesh, 1982

Item	Underdeveloped Villages	Developed Villages	Difference[a] (percent)
Irrigation (percent of owned land)	20.5	42.1	105
Area under high yielding varieties (percent of cropped area)	24.5	42.0	71
Fertilizer (kilogram/hectare of cropped land)	78	150	92
Labor (man-days/hectare)	115	119	4
Paddy price (Tk/maund)	151.2	154.1	2
Rice price (Tk/maund)	276.7	293.8	6
Fertilizer price (Tk/maund)	138.6	119.3	−14
Labor wage rate (Tk/day)	18.1	20.2	12

SOURCE: Ahmed and Hossain (1990).
[a]Difference = [(Developed − Underdeveloped) / Underdeveloped] × 100.

1. A composite measure of infrastructure was used for this analysis. The index was computed by summing the individual costs of access to items of infrastructure (such as market, school, post office, bus stop, and so forth). The individual items were summed (aggregated) in a weighted form. The respective weights were the individual correlation coefficients with total cost of access (Ahmed and Hossain 1990).

Although comparable studies are quite rare in Asia, a recent study (Binswanger, Khandakar, and Rosenzweig 1989) in India shows that the combined effect of roads, markets, and financial institutions on agricultural production is greater than the individual effect of irrigation development. In India, heavy investments have been made in large-scale irrigation projects, but a major proportion of the total irrigated area comes from private tubewell irrigation that spread mainly in infrastructurally developed areas.

Another study, based upon district-level data for the 1960–79 period, found that adoption of HYVs and of private irrigation methods in the Indian Punjab was positively related to density of roads (McGuirk and Mundlak 1991). Long-run output was constrained by road deficiencies. As the constraint was relaxed through construction of roads, the output elasticity increased. A study in the Philippines (Evenson 1986) finds that output elasticity with respect to road density (measured in miles of road per thousand hectares of land) is 0.32, which implies that, if road density is increased by 100 percent, agricultural production will increase by 32 percent.

A simple comparative study of farmers in Malawi (Devres International 1980) found that a higher percentage of farmers near a road used oxen, fertilizers, and plows compared to farmers farther away from the road. Nearly 40 percent of farmers near a road had cash expenditures for farm production, whereas only 12 percent of farmers farther away had such expenditures; such expenditures of nearer farmers averaged two-and-a-half times the amount spent by farther farmers. A study from Central America (von Braun 1988a) highlights that proximity to a paved road is a major determinant of adoption of high-value export vegetable crops by smallholders.

What is the relation between production and marketed surplus—an indicator of commercialization? A separate study on Bangladesh estimated that the elasticity of marketed surplus of paddy with respect to production is 1.6 (Ahmed 1981), which means that when production goes up, marketed surplus goes up more than proportionately, and vice versa. A similar relation was also found in India by Narain (1981). Direct evidence, again from Bangladesh, on marketing of paddy due to infrastructural development is presented in table 8.2. Two observations stand out: first, as expected from theory, the extent of marketing is higher in infrastructurally developed than underdeveloped areas; and, second, and perhaps more significant, even small farmers in infrastructurally developed villages who are net buyers of rice sell paddy more than do small farmers in underdeveloped areas. The extent of commercialization is much larger in infrastructurally developed than underdeveloped villages.

TABLE 8.2 Marketing of paddy by landownership group in infrastructurally developed and underdeveloped villages, Bangladesh, 1982

Landownership Group	Developed Villages		Underdeveloped Villages	
	Farmers Marketing Paddy	Paddy Marketed	Farmers Marketing Paddy	Paddy Marketed
	(percent)			
< 0.5 acre	92	52	46	14
0.5–2.5 acres	81	44	74	21
2.5–5.0 acres	92	43	91	36
> 5.0 acres	100	53	100	47

SOURCE: Ahmed and Hossain (1990).

Effect on Household Income

Estimates of the infrastructural effect on income that are based on a recursive system of equations that sums up the direct and indirect effects of infrastructure indicate that infrastructural development in Bangladesh enhanced household income by about 33 percent (Ahmed and Hossain 1990). More interesting than this average increase in income is the 78 percent increase in income from labor-intensive products such as livestock and fisheries. Similarly, wage income increased by 92 percent. It was argued earlier that infrastructural development provides quick access to markets so that products such as livestock, fisheries, and vegetables would be produced in larger quantities for sale in markets. This argument seems to be supported by evidence from Bangladesh (Ahmed and Hossain 1990). Functionally landless households derive the highest rate of income increase (109 percent) from livestock and fisheries. These activities are labor intensive, and landless households have a comparative advantage in producing these commodities, if they can sell them in the market, which infrastructure enables these households to do. Although the results do not explicitly demonstrate that these increases in income also result in a higher degree of commercialization, such a conclusion would be logical.

The empirical evidence presented here is almost entirely from Bangladesh. However, studies that make indirect observations on the potential role of infrastructure are not uncommon. Thus Hayami et al. (1988a), studying the marketing of soybeans in Java and an outer island of Indonesia, observed that lack of development of rural infrastructure played a critical role in the lack of competition, high marketing margins, and low profits on the outer island compared to Java. Frederickson's study (1985) in the Philippines, based on aggregate data, established econometrically a robust and positive contribution of rural electrification to regional income. The scattered evidence indicates income effects in other countries similar to those observed in Bangladesh.

Effect on Employment

The effect of infrastructural development on employment of household labor in Bangladesh is shown in table 8.3. The employment effect was measured by probing into both supply and demand for labor. However, the demand for nonagricultural activities could not be estimated because of data problems. We assume that supply in nonagricultural occupations is synonymous with demand in the context of rural households. The total supply (first row in the table) can be compared with total demand thus derived. Following this logic, it appears that in infrastructurally developed areas, supply of labor (301 days worked) equals demand (177 + 124 = 301 days worked), implying full employment. In underdeveloped areas, labor supply (320 days worked) is greater than demand (163 + 108 = 271 days worked), implying unemployment/underemployment of 15 percent. However, from the commercialization point of view, we find an extremely high rate of increase (157 percent) in the hired labor intensity in employment, due to infrastructural development. This is primarily why the increase in wage income was so large (92 percent) in developed villages.

Shand (1986) documented empirically the experiences of generating off-farm employment in Asian countries. Infrastructural development came out sometimes explicitly, and at other times implicitly, as instrumental to the generation of off-farm employment.

TABLE 8.3 The effect of infrastructure on employment, Bangladesh, 1982

Variable	Underdeveloped Villages	Developed Villages	Difference (percent)	Statistical Significance
	(days worked/year)			
Total labor supply	320	301	−6	Not significant
Agriculture	222	187	−16	Weakly significant
Nonagriculture	108	124	15	Not significant
Self-employed	275	185	−33	Significant
Hired labor	45	116	157	Significant
Demand for farm labor	163	177	9	Not significant

SOURCE: Ahmed and Hossain (1990).

Effect on Consumption Pattern

The effect of infrastructure on commercialization runs from its effect on income through to the expenditure of that income. In this process, if expenditures increase on items produced outside the household, the rate of commercialization will increase. The consumption pattern, that is, the average and marginal budget shares and expenditure elasticities of the major groups of commodities and services, in infrastructurally developed and underdeveloped villages in Bangladesh is shown in table 8.4. The propensities to consume are estimated for the same income level in both infrastructurally developed and underdeveloped villages, so that the effects are not influenced by income differences between the two areas.

Table 8.4 shows that average and marginal budget shares for non-food goods and services are much higher in infrastructurally developed villages than in underdeveloped villages. The average budget share for services in infrastructurally developed villages is almost three times the share in underdeveloped villages. Among food items, the shares for milk and milk products and for fish, meat, and eggs are also higher in infrastructurally developed villages. Of more significance in this change of consumption patterns is the change regarding cereals: the average budget share of cereals in infrastructurally underdeveloped villages is about 20 percent higher than that in developed villages.

Effect on Social Attitude

Two pieces of evidence are presented on changes in social attitude: (1) female participation in the labor force, and (2) adoption rate of birth control contraceptives. While these are not directly related to commercialization, commercialization is the end product of a nexus of social changes as already discussed, and, therefore, this evidence is supportive, albeit indirectly, of the process of commercialization through the process of changes in social attitude.

Table 8.5 shows that the female participation share in income-earning activities, although generally low, is more than twice as high in infrastructurally developed villages as in underdeveloped villages in Bangladesh. This definitely indicates that development of infrastructure generates forces that bring about changes in traditional attitudes towards women's work outside the household. Similar evidence of the effect of infrastructure on traditional values in rural areas is shown in table 8.6: at similar income levels, 41 percent of women in electrified villages have adopted birth control contraceptives compared to only 18 percent in nonelectrified villages.

TABLE 8.4 Effect of infrastructural development on consumption pattern, Bangladesh, 1982

	Developed Villages			Underdeveloped Villages		
	(percent)			(percent)		
Consumption Item	Average Budget Share	Marginal Budget Share	Expenditure Elasticity	Average Budget Share	Marginal Budget Share	Expenditure Elasticity
Food						
Cereals	47.5	30.9	0.65	59.6	47.6	0.80
Potatoes and vegetables	4.0	3.2	0.80	5.4	4.9	0.91
Pulses	0.6	0.9	1.50	0.8	1.2	1.50
Spices	3.9	3.7	0.95	3.0	2.5	0.83
Narcotics	3.9	4.5	1.15	2.3	1.8	0.78
Sugar	1.5	2.8	1.87	1.4	2.0	1.48
Tea	1.1	1.3	1.18	0.0	0.0	0.40
Oils	2.7	3.1	1.15	2.3	2.4	1.02
Fruits	1.3	2.6	2.00	1.8	2.4	1.33
Milk and milk products	0.9	2.2	2.44	0.3	1.4	4.52
Fish, meat, and eggs	6.2	9.9	1.59	4.1	6.5	1.58
Total food	74.6	65.1	0.87	81.00	72.8	0.90
Nonfood						
Fuels/energy	8.3	7.2	0.87	8.7	5.5	0.63
Other consumer goods	1.9	2.4	1.29	1.4	2.2	1.55
Clothes	5.6	9.7	1.73	5.2	9.2	1.77
Services	9.3	15.7	1.70	3.3	9.9	3.30
Total nonfood	25.1	35.0	1.40	18.6	26.8	1.44
Total food and nonfood	100.0	100.0	—	100.0	100.0	—

SOURCE: Ahmed and Hossain (1990).

NOTE: Due to rounding, the average and marginal shares may not add up to exactly 100.

TABLE 8.5 Participation in income-earning activities in infrastructurally developed and underdeveloped villages, Bangladesh, 1982

	Female Workers as Percent of Total Workers		
Landownership Group	Underdeveloped Villages	Developed Villages	Difference[a]
< 0.5 acre	10.1	21.8	11.7
0.5–2.5 acres	8.5	14.4	5.9
2.5–5.0 acres	3.0	10.7	7.7
> 5.0 acres	2.3	10.1	7.8
All households	6.4	15.1	8.7

SOURCE: Ahmed and Hossain (1990).
[a]Difference = Developed − Underdeveloped.

Implications for Public Investment and Policies

It is evident that development of rural infrastructure plays a critical role in promoting rural development and commercialization. The task, then, is to examine the constraints that hold back development of rural infrastructure. These constraints generally lie in the arena of public expenditures and economic policies.

The considerations of governance of a country automatically force some development of transportation, communication, markets, schools, electricity, and so forth. During the 1950s and 1960s, the World Bank, other multilateral organizations, and bilateral donors placed very high priority on infrastructural development that led to a period of high growth in a number of countries that are now in either the developed or

TABLE 8.6 Percent of women practicing family planning in electrified and nonelectrified households and villages, Bangladesh, 1985

	Electrified Villages (Developed)			Nonelectrified Villages (Underdeveloped)
Women's Age	Households With Electricity (percent)	Households Without Electricity (percent)	All Households (percent)	All Households (percent)
10–19 years	50.0	40.0	45.5	14.3
20–29 years	65.3	48.0	59.3	26.8
30–39 years	48.1	31.3	41.7	25.4
40–50 years	29.7	7.7	22.9	3.4
Total	46.6	29.5	40.6	18.2

SOURCE: Hoque (1987).

developing middle-income categories. However, during the last two decades, this priority has diminished. It is also observed that once urban-oriented development has reached saturation, public allocation to infrastructure falls sharply before being extended to rural areas.

Infrastructure is expensive. A central question is whether the benefits of rural infrastructure are commensurate with the costs and whether the rural infrastructure benefit/cost ratios are superior to those of other alternatives. Unfortunately, the basis for estimating the benefits of rural infrastructure remains very weak, partly because the effects are indirect and difficult to measure, and partly because the established procedures for identifying benefits are erroneous. For example, the benefits of a road are traditionally measured only by the user cost savings in transportation, excluding the broader developmental effects. In a recent study of rural roads in Bangladesh, Chowdhury and Hossain (1985) found that the internal rate of return (IRR) of such roads, including their production effect, is 45 percent, compared to the general estimates of 14 to 24 percent for projects in the portfolio of the Planning Commission. Moreover, the benefit/cost criterion is hardly a realistic principle for guiding very long-term public resource allocation across sectors; it is, of course, a reliable criterion for allocation among projects. In reality, however, public resource allocation is influenced by the judgment of policymakers. Therefore, popular perceptions play critical roles. Studies on infrastructure have the potential to guide popular perceptions in the right direction.

For development of rural infrastructure, local-level political and administrative institutions are a necessity. This implies a decentralization of political power and sharing of public resources between the central and local governments. Most developing countries have local governments but they are inadequately staffed, particularly with technical personnel, and underfunded. Giving due priority to rural infrastructure may result in indirect pressure on political systems to decentralize power and share resources.

Improved infrastructure is a primary driving force and precondition of commercialization. The benefits of commercialization and specialization depend to a considerable extent upon infrastructure. However, much of the incentive for infrastructure provision results from the powerful effects of commercialization, once infrastructure is put in place. Thus, the two, infrastructure and commercialization, are in a favorable push-and-pull relationship. Public policy has a key role to play in infrastructure provision, whereas the commercialization effects may be much more a matter of the private sector and market forces.

9 Agricultural Processing Enterprises: Development Potentials and Links to the Smallholder

JOHN C. ABBOTT

Introduction

In countries where farming and fishing are major productive activities, processing enterprises can have a strategic developmental role. Infrastructural, institutional, and contractual issues arise around them. Whereas consumption and nutrition effects of agricultural commercialization linked to specific processing enterprises are traced in some of the detailed studies in Part V (for example, export vegetables, chapter 12; spices plantation, chapter 14; dairy, chapter 15; sugarcane factories, chapters 13 and 16), this chapter gives an overview of the broader experiences, potentials, and problems of agricultural processing enterprises.

Processing agricultural and fish products so that they may be stored, transported conveniently over distances, and presented in forms appealing to consumers greatly extends the markets in which these products can be sold, enables access to markets not otherwise accessible, and permits sales at higher prices and in larger quantities. Successful processing enterprises can generate foreign exchange, provide employment, contribute to food security by making food available at times when it would not otherwise be available, and be a stable and profitable source of income to farmers and fishermen.

The establishment of processing enterprises was an integral element of development plans of many Third World countries as well as a feature of international aid programs and recommendations of organizations such as the Commonwealth Development Corporation and the World Bank. The existence of effective processing and marketing enterprises was concomitant to the stance of the United Nations Conference on Trade and Development (UNCTAD) that climatic advantages and lower costs pointed to profitable export of various agricultural and fish products from tropical countries, provided that developed-country markets were open to them. Yet, processing plants that were established were not uniformly successful. The lessons from these past failures, based on a Food and Agriculture Organization of the United Nations (FAO) study of some 70 plants, will be discussed in the next section.

A spate of literature during the 1970s (Beckford 1972; George 1977; Lappé and Collins 1977) that attacked export processing for diverting resources from meeting domestic food requirements and for widening income gaps led to a reconsideration of the role of agricultural processing enterprises. During 1984–86 UNCTAD sponsored, in collaboration with FAO, an assessment of the arrangements by which agricultural and fisheries processing enterprises in developing countries acquired their raw materials and marketed their products, and the extent to which they transmitted more advanced technology to associated primary producers, assemblers, and suppliers of production inputs. Forty case studies of processing enterprises dealing with a wide range of products were commissioned (see examples of these listed in table 9.1). The review and analysis of these case studies, together with other relevant studies, form the remainder of this chapter.

Past Failures and Their Lessons

All of the plants studied by FAO that had not lived up to expectations appeared to be well designed from an engineering standpoint. Their main causes of failure or loss were inability to acquire sufficient suitable raw materials for processing, failure to market the processed products successfully, and weaknesses in management that exacerbated the first two problems. While problems in marketing the processed products and in management played an important role in these failures, analysis by Mittendorf (1968) showed that difficulty over the supply of raw materials for processing was the most frequent problem.

Problems of raw material supply included overestimation of potential supply, lack of suitable varieties for processing, insufficient incentives to suppliers, and lack of production support services. Access to a constant supply of suitable raw materials is vital for the efficiency of most processing operations; depending on scale and technology, processing can involve a major investment in fixed plant and equipment, and strategic personnel must be retained and paid through the year to ensure smooth plant operations. In turn, a steady flow of the processed product is essential for success in its marketing. Failure to maintain regular deliveries can have adverse consequences: buyers will be reluctant to enter into firm contracts and promote onward distribution;

prices obtained will be those for fillers around established products; and opportunities for discounts on transport and sales charges may be lost.

Many of the plants in the FAO study were set up in areas where the raw material was apparently in surplus, but in actuality, low yields set limits to increased production. Potential supply was overestimated, as, for example, in the case of the Bolgetanga tomato paste plant in northern Ghana, which was established in conjunction with an irrigation project. The farmers there were inexperienced with irrigation technology and the quantities grown during the succeeding years barely met local demand for fresh produce.

Apparent surpluses can also be bid away from the processor by competing marketing channels. For example, Liebig, a meat processing and marketing transnational, set up an abattoir at Kosti in Sudan, where the prices of live animals offered an attractive operating margin and the location was convenient for supply from large traditional herds and for transportation of the processed products to export markets. However, the plant was dismantled after a few years, as it never received a significant flow of stock; the arrival of the processor had stimulated the traditional buyers to offer better prices and the "surplus" was so thin that they were able to do so with the resources they could mobilize.

For some projects, no proper market outlet studies had been conducted before an investment decision was taken. In others, it was simply assumed—without investigation of production response to a new market outlet—that the establishment of a processing plant would attract a supply of raw materials.

Problems of market demand that included overestimation of prospective demand; misjudgment of tastes, preferences, and habits of consumers; underestimation of competition from other sources; and difficulties of entering foreign markets were also encountered. Management problems also contributed to failures of processing plants.

Errors in decision making of this nature have continued as the following cases tell. Of seven government-sponsored cassava processing plants built in Venezuela, only two were operating in 1980 (Abbott 1988). The Kenana sugar project in Sudan had to be rejuvenated by efforts of externally contracted management (Stauffer 1986). A tea processing plant was established in Rwanda under the assumption that small farmers nearby would collaborate in the provision of supplies, but the farmers were constrained by high staple food prices and concentrated on potato and grain production (von Braun, de Haen, and Blanken 1991). Eventually, a tract of land near the plant was expropriated for direct tea production to meet established plant capacities.

TABLE 9.1 Case studies of agricultural processing enterprises

Commodity	Country	Enterprise	Activity
Rice	Malaysia	Hanapi and Sons	Mill-polished rice, bran, very fine rice, broken rice, and husk from paddy
Maize	Kenya and Pakistan	Corn Products Corporation International	Convert maize to starch, sweeteners, oil-and-gluten feed and flour for livestock
Cassava	Thailand	Cassava chipping and pelleting factories	Transform cassava into chips and then pellets for export
Feed	Nepal	Ratna Feed Industries	Poultry feed processing and marketing of poultry products
Pineapple	Thailand	Siam Food Products	Can pineapples
Limes	Ghana	Rose's Lime Products	Process limes into lime juice, lime oil, and skins; process skins as raw material for pectin production
Fresh orange juice concentrate	Brazil	Juice plants	Process oranges into juice

Product	Country	Company	Description
Fruits and vegetables	Tanzania	Dabaaga Fruit and Vegetable Canning Company	Process products into jams, pickles, juices, and soups
Tomatoes	Taiwan	Taiwan Kagoma Food Company	Produce canned tomato products
Herbs	India	Chalam's Herbochem	Process a range of herbs, meat, and other rural products into extracts for use in drugs
Sugar	Swaziland	Mhlume Sugar Company	Produce and process sugarcane and market refined sugar
Sugar	Thailand	Kumphawapi Sugar Company	Japanese/Thai joint-venture processing sugarcane
Tea	Kenya	Kenya Tea Development Authority	Provide inputs to farmers to grow and pick tea; process and market tea
Tobacco	Thailand	Adams International	Joint venture producing Turkish tobacco
Cotton	Uganda	Bunyoro Cooperative Growers' Union	Gin cotton and sell cotton lint and seed separated by ginning
Livestock	Botswana	Botswana Meat Commission	Purchase and process livestock export products
Broilers	Jamaica	Jamaica Broilers	Produce and market broilers
Milk	India	Anand Dairy Cooperative Union	Process and market milk, butter, milk powder, and other dairy products
Milk	Bolivia	Dairy Industries Enterprise	Bring small farmers into the dairy economy; process milk and produce milk products

SOURCE: Abbott (1988).

Conditions for Success

Case studies of 40 enterprises that have been notably successful in the developing countries were assembled and reviewed (Abbott 1988). These case studies cover internationally known processing enterprises such as the Anand Dairy Cooperative Union in India and the Kenya Tea Development Authority as well as other smaller and less well known representative enterprises, and even the experience of individual persons. Nineteen commodity products are covered, as are an array of developing countries. The cases reflect the processing of raw materials from traditional modes of farming and other activities as well as of materials produced specifically for processing in a new, fully integrated system. These case materials are reviewed for significant leads on (1) effective relationships between processors and raw material suppliers, by commodity groups, and (2) implications for small-scale suppliers, including the scope for their participation in processor-supplier systems, transmission of technology to them, bargaining over terms of sale, and access to food supplies (see chapter 10).

Effective Links Between Processor and Raw Material Suppliers

In most of the successful cases, processors had to lay out money to be sure of their raw material supplies. An enterprise such as Siam Food Products in Thailand experienced poor performance when it was unable to receive its raw material, pineapple, in sufficient quantities from local farmers with whom it had contracts; in response, it had to expand its own direct production (Lee 1976). To ensure supplies, private firms made contracts with producers directly or through private agents. Cooperative unions used their local societies. Unilever Is in Turkey, for example, used a sunflower producers' cooperative as well as independent oilseed crushers (Abbott 1988). In the dried fish production-processing-marketing system in Indonesia, credit went back to the fishermen through an intricate channel of contributions and risk sharing (Abbott 1988). The state dairy enterprise in Bolivia made available in-kind credit (Abbott 1988). Haji Mansur's rice mill in Indonesia advanced funds to farmers to help meet their production costs, and, hence, ensured substantial direct supply of paddy (Harper and Kavura 1985). Straits Fish Meal in Malaysia only paid a market-clearing price for its raw material, but to be sure of supplies, it advanced funds to fishermen before they put out to sea (Abbott 1988).

Many processors also had to channel productive resources in appropriate proportions to their suppliers to obtain a suitable flow of raw materials. This was commonly done through producer/processor supply contracts, which have been found to be effective mechanisms for ensuring that supplies meet the requirements of processors, and, in turn, that producers are guaranteed a convenient outlet. Such contracts have provided an assured market for producers' output, access to the company's technological and associated services, and easier access to credit. Access to technological services involved field advice and training, supply of inputs such as seeds and fertilizer, and distribution of irrigation water. For instance, the contract offered to farmers by the Taiwan Kagoma Food Company included provision of seedlings to ensure uniform quality and harvesting time, provision of pesticides and fertilizers on credit, an agreed price per grade delivered with premiums as incentives, provision of field crates to keep supplies in line with plant capacity, among other features (Meneguay and Huang 1976). In terms of credit, processing enterprises often made loans to farmers at interest rates that were less than bank rates of interest and for longer periods; where enterprises did not provide loans, banks generally accepted a processor contract as collateral.

Producer/processor contracts reduce but do not eliminate risks for both parties. The risk remains that farmers can become dependent on a processor as their key market outlet. The degree to which the producer is captive to such a contract depends on its prices for inputs and outputs relative to those of competing enterprises. The newer the production line in a country, the more limited, often, are the alternative options. Problems can also arise when quality standards, where the contracts specify these as the sole domain of the buyer, are manipulated to reject supplies or accept them at substantially lower prices, to the detriment of the producer.

Diversion of contracted supplies onto local markets for fresh produce is the corresponding risk for the processor. This is an obvious temptation when shortages force up prices. In the case of the already-mentioned pineapple canning project in Thailand undertaken by Siam Food Products, of the 10,000 tons of fresh pineapple expected by the company, only 3,000 tons were actually received in 1972-73 and only 674 tons in 1974-75, one reason being that the local market price was substantially higher than the price negotiated with the cannery (Lee 1976).

The possibility of expected price relationships being upset by external events is best foreseen in processor supply price arrangements. Fixing a specific price in advance can impose strains on both the parties involved. An agreed relationship to a market indicator that reflects current conditions of production and of sale through alternative outlets—for example, a major national market for fresh produce or an international commodity exchange—may be more convenient for both parties. Further alternatives include offering a basic price plus bonuses that vary with changes in the price of the processed product on designated markets. In any event, the value of contracts with large numbers of suppliers hangs on the mutual interest of the parties involved. Legal enforcement is rarely practicable.

Different Experiences by Commodity Groups

Although the number of plants in each of the 19 commodity groups or subgroups covered in the 40 case studies is small, some observations are permitted, taking into account other relevant studies.

Milk producers always benefit from commitments by processors to purchase their supplies; the perishability of milk makes its prompt disposal onto a liquid market or conversion into a more durable processed product essential. Guaranteed access to a processor is particularly important where production conditions lead to large seasonal surpluses. The processor/distributor is equally concerned to maintain a steady supply. In India, both cooperative and private processors/distributors advance dairy farmers funds to pay for milk cattle, feed supplements, and veterinary services in return for a commitment to maintain regular supplies of raw material (Harper and Kavura 1985; and chapter 15).

Until relatively recently, the processing of poultry was undertaken only by individual producers and retailers. The development of the broiler industry is a classic example of the contract system. The contract system, by reducing both production and marketing risks, has opened the way to large-scale operations: sale at the optimum stage economizes on feed costs, and veterinary, processing, and marketing services can be provided more efficiently. There are various stages of development of the broiler industry. In Lebanon, the industry began as a private enterprise initiative characterized by individual and specialized processors (Reda 1970). Poultry production, formerly a side activity of general farmers, was transformed into a specialized activity as businessmen took advantage of the opportunity of applying imported broiler technology for exporting broilers to the Middle East market, where they already had commercial contacts. In Nepal, a feed processing firm, Ratna Feed Industries, has helped in the development of the poultry industry by distributing feed to dealers on credit as well as broiler chicks, providing technical and economic skills and veterinarian services, and assisting in the marketing of eggs (Mathema 1976). In some countries the broiler industry has developed to the point that large vertically integrated companies handle the production, processing, and marketing of chickens. Contracts are negotiated with farmers to raise chickens. Case studies from Thailand (Poapongsakorn et al. 1985) show that farmers under such contracts with the Charoen Popkhan Group received a fixed fee or wage, whereas in Jamaica (Freiwalds 1983) growers were paid under three categories of rental, performance, and price that rewarded efficiency.

Producer/processor contracts are particularly important where new crops or new methods of crop husbandry are to be introduced. Case studies of three vegetable processors—Taiwan Kagoma Food Company (tomatoes), Tabasco (peppers), and Basotho Fruit and Vegetable Canners (asparagus)—as well as processing of marigold heads in Ecuador indicate that all companies provided intensive assistance under their contracts to ensure supply in the desired quantities and over an extended supply season (Meneguay and Huang 1976; Truitt 1980; Abbott 1988). This assistance encouraged farmers to commercialize by reducing their risks though providing them with the technical know-how and a guaranteed market outlet. The contracts were quite detailed and the companies essentially had direct control over production. For instance, Basotho Fruit and Vegetable Canners, in order to establish asparagus, a new product in Lesotho, had to develop production from the ground up—it set up its own nursery; supplied farmers with clones, sprays, fertilizer, packing containers, trimming equipment, and shade huts; prepared the soil for them; and financed the cost of boreholes to provide water (Abbott 1988). To ensure that the supply season was extended beyond its normal period, the Taiwan Kagoma Food Company offered a price incentive and contracted with farmers operating in different microclimates (Meneguay and Huang 1976).

Contract processing of fruit for processing is less common. Establishment of fruit tree orchards is a long-term investment, and once they are established, the grower's need for finance is small. The risk is present, however, that the grower's investment may be lost for lack of a market outlet, should the processing plant within convenient reach go out of business or offer very low prices. Long-term comparative advantage in supplying the market is essential. The processing of limes in Ghana by Rose's Lime Products demonstrates how such production can survive radical changes in government policy and highly adverse exchange rates, if production conditions are favorable and plant overhead costs are low (Abbott 1988).

Processing of fruit on a large scale in developing countries is a domain of transnationals, which supply technology, management, and access to the export market. The growth of the frozen concentrated orange juice industry in Brazil came about when North American processors looked around for alternative cheaper and additional sources of supply and brought with them the latest technology (Strohl 1985). In some instances, transnationals even produce their own fruit to ensure uniform quality and state of maturation, which are critical for successful processing on a commercial scale. Del Monte in Kenya and Del Monte and Dolefil in the Philippines found direct production of pineapple preferable to purchasing from independent farmers, even under contract (Abbott 1988). Siam Food Products in Thailand learned this lesson the hard way, as discussed already (Lee 1976).

Commodities such as sugarcane and tea, where the time period between harvesting and processing is necessarily short to avoid substantial losses in quality, imply close integration between growers and processors. However, labor problems associated with harvesting sugarcane during the short period of maximum yield have led processors to prefer contracts with independent growers. The Mhlume Sugar Company in Swaziland, which had a sugar estate of 3,000 hectares in 1961, began in 1963 to clear and provide 30 to 40 plots of about 4 hectares each per year to Swazi farmers on long-term leases that regulated crop production, agricultural methods, construction, grazing, and other activities (Business International Corporation 1981). The company provided services such as distribution of irrigation water, seed cane, and fertilizers, as well as the cutting and transportation of cane, and partly recovered these costs through user fees. Contracts with smallholders are also advantageous in tea production; with a direct incentive for quality in the leaf price received, smallholders tend to pick tea more carefully than paid laborers. The Kenya Tea Development Authority's smallholder base has consistently produced higher quality tea than the commercial estates (Lamb and Muller 1982).

Transnational Versus National Enterprises

Developed-country enterprises or transnationals were believed at one time to have an advantage in organizing the processing and marketing of agricultural and fish products: they were familiar with consumer preferences and import controls, and they had established distribution channels and brand names. Sales techniques, perfected elsewhere, could give them an advantage in developing-country markets.

Export enterprises in developing countries, linked to a transnational, tend to do better than independent enterprises. Analyzing the Malaysian canned pineapple export business, Jabbar (1972) showed that independent enterprises, that is, enterprises without links to transnationals, sold their products on terms comparable with enterprises linked to transnationals when markets were favorable, but when markets turned down, the independent enterprises were the first to suffer.

In recent years, national enterprises have been gaining ground. By 1986, national enterprises including state trading organizations had a larger share of agricultural-based exports from developing countries than did transnationals. This has been achieved through a combination of sales through brokers, direct contacts with importers, and sales by state trading organizations, including barter and countertrade transactions.

In domestic markets, the traditional marketing channel is through established wholesalers with their own links to retail outlets. Processors who want to follow their products in distribution use a combination of branch outlets and retail sales. For example, Ratna Feed Industries in Nepal used its regular feed dealers to assemble eggs produced by its feed customers (Mathema 1976).

The failure of processing enterprises in domestic markets lies mainly in their overestimation of domestic demand; sponsors of processing enterprises have tended to assume wider prevalence of high-income consumer tastes, preferences, and purchasing power than was actually the case. Many plants were set up to process goods that would substitute for imports. Protected against competition, they operated with high costs and priced their products accordingly. The result was a spiral of declining markets, reduced sales, and an increasing burden of overhead costs. Problems are associated with external management or orientation. New plants tend to be locally owned and managed, and better adapted in terms of overhead commitments, product design, and sales organization to domestic markets.

Transmission of Technology and Managerial Capacity

The production/marketing contract with a processor has proved to be a highly effective vehicle for the transmission of technology. Farmers with no prior experience with a particular crop have been supplied with a tested package of inputs and husbandry procedures backed up by credit and a guaranteed market outlet, as was the case, for example, with asparagus in Lesotho and marigolds in Ecuador (Abbott 1988). Growers have been protected against risk by all-inclusive contracts, as, for example, the broiler contracts in Thailand and Jamaica (Poapongsakorn et al. 1985; Freiwalds 1983).

Standard technology can be acquired quickly if the price incentive is sufficiently attractive. Seeing their neighbors become rich is stimulus enough for others to learn and emulate. The spectacular growth of the broiler industry in Lebanon during the 1960s came about because of local businessmen taking advantage of opportunities arising from American and European imported broilers (Reda 1970).

A producer cooperative processing enterprise can be an effective mode of transmission of technology to members and staff through group pressure and individual participation in decision making. An example of this mode is the Anand Dairy Cooperative Union in India.

Government advisory and veterinary services, among others, can contribute to the extension and dissemination of new technology, in a more general sense. For example, in Lebanon, vaccines were distributed free of charge, which greatly reduced the risk of epidemic disease (Reda 1970). However, in many developing countries, the research and development undertaken by governments and their extension messages tend to be too general. They are insufficiently adapted to the needs of particular sets of producers and to the conditions under which they operate. The prevailing government-sponsored system of supplying credit for inputs through a financial intermediary, extension through a government agency, and buying through a marketing board also encounters problems of input timing, coordination, payment, and collection (Mittendorf 1968). In many areas of production, precise recommendations, provision of inputs without cash outlays, and a guaranteed outlet and price by a processor can be decisive in bringing farmers and fishermen out of a static situation.

Processors have found it much easier to introduce new technologies as a complete package associated with a new product than to improve the handling and management of traditional agricultural products. Here, government intervention may be needed to promote new processing technology.

Bargaining Power over Terms of Sale

Once a farmer has oriented his operations towards meeting the requirements of a particular processor, the farmer becomes, to a considerable extent, a captive of that processor (see chapter 10 on this issue). As discussed already, the farmer may also be subject to the manipulation of quality standards, where contracts specify that these are the sole domain of the buyer. Tomato growers for the Taiwan Kagoma Food Company complained that their tomatoes received a lower grade than warranted because of delays in handling their produce after it had arrived at the plant (Meneguay and Huang 1976). They also complained that delays in the provision of crates by the factory, because of a lack of capacity at that time for processing tomatoes, resulted in tomatoes rotting in the fields and the farmers' not being able to profit from them. Of course, there is the risk to the processor that inadequately remunerated suppliers will shift eventually to other activities, but there can be a phase of exploitation until this occurs.

The government can intervene to regulate the contract process. For instance, in Brazil the government requires that the basic producer-to-processor price for juice oranges be negotiated annually (Strohl 1985). In Taiwan, mediation over prices has become accepted practice, after discussions arose over asparagus and mushrooms.

Conclusions

Processing enterprises have various advantages attached to them: they can be major earners of foreign exchange; they can facilitate the capturing of value added by developing countries; they can create employment; they can raise the long-run productivity and incomes of raw material producers; they can transmit more advanced production and handling technology; and so forth. For example, the 138,000 smallholders growing tea for the Kenya Tea Development Authority have cash incomes well above the average for their area, as do the 28,000 growers and associated workers for the Mumias Sugar Company in Kenya. Continuing access to the favorable market provided by the Botswana Meat Commission has stimulated a livestock off-take rate that is much higher than on similar range grazing conditions in other parts of Africa and has been of broad and immediate benefit to the low-income livestock-dependent rural population.

In many countries, the establishment of processing enterprises was left to private initiative and risk or, where initiated under government programs, received only superficial attention with respect to the conditions of economic viability. Now, the scope for development through processing is becoming more widely recognized in developing countries. While a free-trade environment is needed to offer advantages in securing and maintaining export markets, the future of processing in developing countries lies in national participation, appropriate institutional arrangements, and national market and enterprise growth.

10 Contract Farming and Commercialization of Agriculture in Developing Countries

DAVID GLOVER

Introduction

The distributional benefits of commercialization of agriculture, access to commercialization opportunities, and sharing of commercialization risks are functions of institutional arrangements. Obviously, the indirect food security and nutritional effects are, thereby, partly a function of such institutional arrangements. This chapter explores the relevance to food security of one form of contractual relationship in agriculture: formal contracts between producers and buyers (generally processors or exporters), a production and marketing system known as contract farming. The chapter does not refer to the extensive literature on informal contractual relations, such as sharecropping, or on traditional systems of contract farming, such as the extensive "strange farmer" system in West Africa's groundnut sector. The chapter draws generalizations and conclusions from studies done by the author and by other researchers. The latter include two research networks initiated by the author. One network surveyed the experience with contract farming in seven East and Southern African countries (*Eastern Africa Economic Review* 1989); the second examined the experience in Thailand, Malaysia, and the Philippines (Glover and Lim, forthcoming).[1]

The first section provides a basic definition and description of contract farming.[2] The second section describes several aspects of contract farming that are relevant to food security and provides evidence about the effects where it is available and plausible estimates where evidence is not available. The third section presents some corresponding policy recommendations.

1. The research was financed by the International Development Research Centre.
2. Further details can be found in Glover (1984) or Minot (1986).

Theory and Practice of Contract Farming

In contract farming, a central processing or exporting unit purchases the harvests of independent farmers. These purchases can supplement or substitute for company production. The terms of the purchase are arranged in advance through contracts, the exact nature of which can vary considerably from case to case. Contracts are generally signed at planting time and specify how much produce the company will buy and at what price. Often the company provides credit, inputs, farm machinery rentals, and technical advice. It always retains the right to reject substandard produce.

Contracting is most commonly practiced by food processing companies. Since their processing plants have high fixed costs, these companies have an interest in keeping raw material inflows at a steady level, close to plant capacity. Reliance on open market purchases is unlikely to achieve this interest. Contracts, on the other hand, can specify planting dates (and thus delivery dates) as well as total quantities to be delivered. Contracting reduces much of the uncertainty that would exist if the company simply bought crops on the open market, and gives the company some control over the production process (for example, over the variety grown). There is no reason, of course, for a company not to use more than one method of obtaining its supplies, and some companies use a combination of company farms, contract growers, and open market purchases.

Many contract-farming schemes in less developed countries are multipartite arrangements involving private firms (usually foreign, but occasionally local), the host-country government, and international aid or lending agencies, such as the U.S. Agency for International Development (USAID), the World Bank, or the Commonwealth Development Corporation (CDC). The CDC has been particularly active in this type of scheme. In one common variation, a national development bank provides growers with credit for the purchase of fertilizer, seeds, and other inputs. At harvest time, the company pays growers the contract price, but takes off a sum that goes to the bank to repay its loan to the grower. In this system, private companies avoid the problems of assessing creditworthiness and prosecuting defaulters. In some cases, government agencies provide inputs or technical assistance.

Contract-farming schemes tend to be of two types. One type produces traditional tropical commodities, such as sugar, rubber, or oil palm, which tend to be produced at lowest cost on large tracts of land. Contract-farming schemes in such commodities usually involve a large number of growers, tight central control, and provision of numerous services by the central processing unit (for example, irrigation, harvesting, and aerial spraying). There is usually heavy involvement of external donors in these schemes. Many such schemes originated as resettlement schemes. In extreme cases, some schemes are, in effect, disguised plantations, for example, the Papua New Guinea cardamom scheme (see chapter 14). These large projects, often referred to as outgrower schemes, are particularly common in Indonesia and Malaysia (rubber and oil palm) and in Africa (oil palm, sugar, and tea). The case studies in Kenya, Malawi, and Sierra Leone provide insights at the scheme and household levels for these types of schemes (see chapters 16, 20, and 21).

Another type of contract-farming scheme, usually on a smaller scale and with more private-sector involvement and less tight centralized control, is frequently used in fruit and vegetable production, particularly in Central America and Thailand. Most frequently, this type of scheme entails the export of high-value items, such as asparagus, cucumbers, melons, or strawberries, with the company providing quality control, brand names, and marketing channels. Business-oriented growers, cooperatives, and individual small farmers are all involved. Fresh vegetable exports from Guatemala are an example of such schemes (see chapter 12). Total developing-country employment in contract farming of these nontraditional crops is much less than in traditional crops, such as bananas and sugar. However, there is some evidence that contract farming of nontraditional crops is expanding at a faster rate and that these labor-intensive products are more promising outlets for small farmers.

Contract Farming, Commercialization, and the Food Question

Several aspects of contract farming impinge on food production and consumption by growers, their families and employees, and other segments of the population. This section attempts to summarize some of the evidence about the relevant effects of contract farming. The sample of cases studied in East and Southern Africa is presented in table 10.1. An inventory of schemes in the region from which the sample was selected indicates a wide range in size of outgrower schemes. It is noteworthy that contract farming need not necessarily entail large units (Glover and Kusterer 1990).

Income Generation via Market Access

One of the key features of contract farming is that it provides farmers with access to markets that would not otherwise have been available to them. Without the quality control and tight coordination offered by contract farming, it is frequently unlikely that smallholders would be able to sell perishable goods overseas through open market sales. The most significant income increases have been generated in

TABLE 10.1 Network of case studies in East and Southern Africa, by crop

Country	Sugar	Tea	Cotton	Nontraditional[a] Crop
Kenya	X[b]	X[b]		X
Tanzania	X	X		
Malawi	X	X		
Zambia	X		X	
Zimbabwe	X		X	X
Swaziland			X	X
Lesotho				X

[a]For example, fruits, vegetables, and oilseeds.
[b]From secondary data.

those schemes in which smallholders gain access to lucrative export markets for labor-intensive luxury crops (Glover 1986). Important income increases have resulted from traditional crops as well, however, particularly tea. The Kenya Tea Development Authority's success in this field is often cited (Lamb and Muller 1982).

Data on income generation are difficult to compile, partly because farmers themselves have poor data. In many schemes, numerous deductions from crop payments made it difficult for farmers to assess their profit position. In most cases studied in the two research networks mentioned earlier, farmers expressed satisfaction with their returns and many schemes had long queues for entry, both clear indicators of the net benefits available from such schemes.

We basically find that substantial income increases can and do result from contract farming. Furthermore, contract farming in cash crops often brings significant changes with respect to both size and frequency of payments to the recipients. It is difficult to generalize about the direction in which contract farming will shift the payment system—this is highly case specific. A contract-farming crop with a fairly continuous harvest and payments may replace one harvested and sold once a year in the local market, or the reverse may be true. It is also possible for contract-farming schemes to separate the payment system from the harvest cycle by providing weekly or monthly advances that resemble a wage, as often occurs in outgrower schemes. This is a possibility in contract-farming schemes; in open market sales it is not, at least not without recourse to the capital market.

Displacement of Alternative Crops

Contract farming could reduce food production if a new contract cash crop displaced a previously grown food crop. Evidence suggests that such displacement does not frequently occur when farmers are allowed to make their own decisions. In such cases, the contract crop tends to displace a less profitable cash crop rather than a food crop.

In centrally managed schemes, these decisions are often made by the scheme authority. Many schemes require growers to maintain a certain acreage of food crops, though a few have gone in the other direction and required growers to specialize in the contract crop. A series of African case studies found that when land was fairly abundant and farmers had the freedom to diversify their crops and income sources, food supply was adequate and no special measures were needed from the scheme authorities (*Eastern Africa Economic Review* 1989). In Swaziland, for example, households produced much of their food on family plots outside the scheme, allowing them to use their scheme plots for sugar. Reduction of food supply was most likely to be a problem in areas where availability of land was a constraint. Schemes that did restrict production of noncontract crops tended to encounter serious problems, not only in food production, but also in maintaining adequate levels of recruitment and participation in the schemes. For example, some schemes in Zimbabwe restricted farmers by allowing them to grow only the contract crop on their plot or by admitting only full-time farmers, thus forcing them to forgo off-farm income. In each case, farmers either evaded the restrictions (sugar and cotton schemes) or the scheme became unviable (tea schemes).

Local agroclimatic and market conditions are sometimes more suitable for cash or export crops than food crops. In such cases, there can be significant welfare gains from trade. For example, a cooperative of former banana workers was formed in Honduras in the early 1960s to farm land formerly given over to bananas. The cooperative experimented with many food crops but found that they did not grow well in their soil and drainage conditions. On reverting to bananas (the export crop), the cooperative's income increased to levels far above those of food crop producers.

A key feature of contract farming, which bears on production response, is risk sharing and risk reduction. In fact, contracting is fundamentally a way of allocating risk between the company and its growers. The latter assume most of the risks associated with production, while the former assumes the risks of marketing the final product. Total risk is reduced relative to a noncontract situation of that crop. However, it is unclear whether total risk for the farmer is reduced, since nontraditional crops are inherently risky; when the addition of the contract crop to the farm's product increases total risk for the farm, it may induce a farmer to increase the percentage of subsistence requirements.

Multiplier Effects for Employment, Infrastructure, and Market Development

Local governments often favor contract farming in the belief that it will produce greater spillover or linkage effects with the local economy than would plantation production. Studies have found significant variations in this respect. For labor-intensive fruit and vegetable crops, a multiplier is clearly present in the great expansion of daily farm labor employment made necessary by the new contracted crops. This was found to be the case with cauliflower and broccoli cultivation in Guatemala (Glover and Kusterer 1990). There are also significant new employment opportunities in the transport and processing sectors. Again, the effects of this employment and income generation on food supply will depend on income elasticities of demand by laborers and on domestic supply response to that demand.

For traditional export crops, and in cases where a highly mechanized and centralized production system is transferred to large outgrowers, the situation is different. Here, the nominal transfer of legal responsibilities via a contract does little to change the economic imperatives of the production system. In the Central American banana industry, for example, the linkage effects of local production are not much greater than those provided by transnational corporation production; nearly all of the inputs used by associate producers are provided by exporting companies. Moreover, as Ellis (1977) shows, the linkage effects of any form of banana production are very slight.

In addition to direct employment effects, some large contract-farming schemes in remote areas have had broader rural development impacts. To some extent, these schemes have acted as growth poles. This has occurred in sugar schemes in western Kenya, tea schemes in Tanzania, and in a frontier asparagus scheme in Peru: all schemes have performed well in terms of opening up underdeveloped areas of the countries in which they are located. Construction of roads and other infrastructure and expansion of interregional trade have been some of the direct consequences of the establishment of contract-farming schemes. Thus, while infrastructure is a precondition for many contract-farming schemes and a driving force of agricultural commercialization, as pointed out in chapter 8, it is also sometimes the case that the growth potentials of commercialization push infrastructure development, which then may result in further second-round benefits from the infrastructure. There is some evidence that these rural development effects could have been greater if more deliberate planning had been carried out and local resources had been used more frequently. The chief relevance of contract farming as an organizational form in this respect is that the market, on its own, was not opening up these regions.

Market development also results from the production process itself in some environments: the reject rate in fruit and vegetable export operations, which usually rely on contract farming, often reaches 50 percent. These rejects—often set aside merely for reasons of ripeness or size—can be sold in local markets for a fraction of the price paid by industrial-country consumers. The produce has nutritional value and can complement the traditional starchy diet of low-income consumers. In some cases, exotic vegetables are not well accepted initially but catch on as their characteristics and preparation methods become better known. In Guatemala, for example, reject cauliflower and broccoli are so widely and cheaply available that they have become a nutritious staple of the poorest people; the leaves and stalks are also used as animal feed and as an organic fertilizer for the noncontracted (mainly subsistence) crop fields (Glover and Kusterer 1990).

Effects on Extension, Inputs, and Services

The contractual relationship between growers and a processing company in contract farming provides the latter with the assurance that it can appropriate a share of the benefits from the investments it makes in production at the farm level. This is most apparent in the areas of extension and input provision. The company has a direct interest in providing effective extension services because it wants high-quality, low-cost produce. Public extension services have no such incentive and regulate their performance in accordance with bureaucratically defined criteria (for example, number of farmers served, quantity of inputs distributed). These criteria are much less effective in assessing performance and providing incentives than the profit-related criteria used by contracting companies.

A pirori, then, one would expect the quality of extension provided in contract farming to be superior to that found in purely public or market-oriented systems. In most private contract-farming schemes, extension tends to deal only with the contract crop, although some of the multipartite schemes in Africa use a multicrop approach.

Some of the production techniques learned in contract-farming schemes are highly crop specific and are not transferable to other commodities. Management skills learned through participation in an agribusiness scheme are more widely applicable, however, and include accounting practices, negotiating skills, and awareness of the importance of quality, characteristics of export markets, and contract provisions. Generally speaking, there tends to be some transfer of contract-farming–

induced production and management skills to other cash crops and to the farm enterprise as a whole.

The situation is similar with respect to input provision. The volume and timeliness of delivery of agrochemicals should be close to optimal in a contract-farming situation, since the company has a secure means of ensuring repayment of in-kind credit through deductions from crop payments. Chemicals applied can have residual effects on adjacent or rotated food crops. In addition, inputs delivered for use on the contract crop are occasionally (and illicitly) diverted to food crops, depending on the farmer's estimate of relative marginal returns. Irrigation water provided through contract-farming schemes has also been used for secondary crops, as in some African sugar schemes that support the production of vegetables for local markets.

On most schemes, basic agronomic research is fairly limited; the emphasis is on disseminating existing technologies rather than on developing new ones. There is, however, unexplored potential for contract-farming schemes as sites for the testing and introduction of new food crop technologies. At present, much basic and adaptive research is carried out on experiment stations and tested in study villages under fairly tight control. From there to nationwide dissemination is a big leap; contract-farming schemes could provide an intermediate step.

Policy Implications

A number of policy implications flow from the effects described above. The chapter's basic hypothesis is that the largest and most direct effect of contract farming on food security is likely to result from income changes and that this effect will be positive. Other aspects of the contracting relationship may have moderate to weak effects in either a positive or negative direction (see table 10.2). The priorities, then, are first, to maximize the income-generating effects of contracting schemes; second, to moderate those secondary aspects that have potential negative effects; and third, to increase the secondary positive effects.

Measures to improve the financial viability of schemes are of particular importance. These measures include setting appropriate pricing policies, rewarding risk taking by private companies in new crops or regions, and improving the autonomy and accountability of parastatals. Since much income is generated as a result of access to export markets, identification of potentials of nontraditional exports, studies of marketing channels, and market promotion efforts will be useful.

Payment systems can be modified to provide smaller, more frequent payments. Where feasible, project authorities should sign contracts and

TABLE 10.2 Contract farming linkages to food security and policy implications

Contract Farming Variable	Likely Effect on Household Food Security	Policy Recommendation
Income generation via market access	Strong +	Measures to improve financial viability; market promotion
Payment system		
Size/frequency	Indeterminate	Provide small, frequent payments and disbursements to women
Recipient	Strong −	
Displacement of alternative crops	Weak −	Leave cropping decisions to farmers
Multiplier effects	Moderately + on decreased supply	Greater use of local resources
Extension	Moderate +	Multicrop extension; emphasis on management skills
Input and service provision	Moderate +	None
Research	Potentially +	Testing and trials of food crops using contract-farming infrastructure
Risk	Indeterminate	None
Other qualifications		
Import substitution	Moderate +	Attention to price trends in tradables
Comparative advantage	Moderate +	Attention to price trends in tradables
Use of rejects	Moderate +	Marketing and information programs
Pricing policy	Moderate −	Realistic price levels

disburse payments to household members who actually carry out the work. Where these are women, there will likely be a positive effect on food expenditures.

Farmers should be given as much freedom as feasible in managing their enterprises, particularly with respect to choice of crop mix and off-farm activities. Restrictions on noncontract crop activities should be avoided.

The multiplier effects of contract farming can be maximized by encouraging project authorities to plan for development of investment opportunities into which growers can channel their new income. Greater use of local resources in transport, maintenance, and manufactured goods provision can also make a contribution.

Extension services should be designed to provide learning effects that go beyond production of the contract crop. It is unlikely that the debate over single-crop versus multicrop extension systems will ever be resolved: it is difficult to assess the trade-off between the technical superiority that comes from specialization and the efficiency in delivery that comes from multicrop extension. Contract farming studies, however, tend to give more support to the latter. Farmers seem to prefer the farm management approach provided by multicrop extensionists, and it could be argued that specialized extension services are often not feasible in very poor countries.

As noted previously, many of the benefits from contract-farming extension lie not in production but in management. These skills are more readily transferable to food crops than are production techniques. Management skills are most likely to be developed in schemes where producer prices closely reflect quality and final market prices; where farmers receive detailed accounts of the company's payments for crops and deductions for inputs; and where farmers are given substantial responsibility for managing their operations, rather than operating within a scheme where control is highly centralized.

The use of rejects and by-products from cash crops, such as fruits and vegetables, can often be increased through greater attention to marketing and information. The establishment of cooperatives to market rejected produce locally has been very successful (for example, in Honduras), and programs in local markets to inform consumers about the nutritional value and methods of storing and preparing nontraditional foods can be useful.

11 Income and Employment Generation from Agricultural Processing and Marketing at the Village Level: A Study in Upland Java, Indonesia

TOSHIHIKO KAWAGOE

Introduction

In the rapidly growing economies of Southeast Asia, besides the emergence of processing enterprises (as discussed in chapter 9) catering to the diversifying and growing domestic and international demand, partly combined with contract-farming systems (as discussed in chapter 10), sizable informal village-level processing and marketing activities are a feature of the rural growth process. The employment and income effects for the poor can be a major factor in the spreading of rural growth benefits of commercialization for development and nutritional improvement, as was so successfully achieved in the 1970s and 1980s in Indonesia (Java). The informal sector is known to have a high labor absorptive capacity as well as the potential to contribute to alleviation of poverty and unemployment in developing countries (ILO 1972, 1974; Oshima 1984). The informal sector's potential to generate local income and employment through farm product processing and marketing activities has been demonstrated with a case study of soybeans in Indonesia (Hayami et al. 1987, 1988a). Those results suggest a critically important role of such activities in generating income in local economies, as well as in improving income distribution by increasing employment.

The purpose of this chapter is to illustrate that significant additional income and employment can accrue to farm producers from agricultural marketing and processing activities at the village level. The policy question that arises is how to foster and enhance these favorable effects. The principal insights from the studies on market policy reform (China) and

---

I am greatly indebted to Yujiro Hayami, Yoshinori Morooka, and Masdjidin Siregar for allowing me to incorporate the results of our joint research into this chapter, and for helpful comments. This study is based on a research project commissioned by the UN/ESCAP Regional Coordination Centre for Research and Development of Coarse Grains, Pulses, Roots, and Tuber Crops in the Humid Tropics of Asia and the Pacific (CGPRT Centre). The CGPRT Centre is not responsible for the opinions expressed in this chapter.

Income and Employment Generation in Upland Java, Indonesia 177

on the critical role of infrastructure (Bangladesh) in chapters 7 and 8 are a relevant backdrop to this village-level study. Moreover, the case studies in Part V largely have a household focus in the context of commercialization projects. This village-level study of dynamic indigenous commercialization in a growing economy serves as a complementary case to these case studies.

This chapter uses case studies of village-level processing and marketing activities of soybean, cassava, and tobacco in Indonesia. These crops are grown in typical upland areas under rain-fed conditions.

Study Site and Farming System

The study site is located in Garut District, West Java, in a typical village of a rain-fed upland area. Farming is the main occupation of the majority of the village households; more than 80 percent of the population is engaged in farming. Sixty percent of the farm households are full-time farmers and another 34 percent are part-time farmers. Only 6 percent of all households are agricultural laborers. Operational farmland area per farm household is very small, an average of 0.5 hectare. Only 8 percent of the farmers cultivate more than 1.0 hectare.

Farmers traditionally intercrop to use their limited land more efficiently. Eight main crops—soybeans, corn, tobacco, cassava, peanuts, upland rice, fruits, and vegetables—are cultivated in various combinations. The wet season normally begins in September and continues into May. There are many variations in the farming system, but a typical cropping pattern is to plant soybean with corn from September to January, followed by soybean and tobacco from February to June. Cassava is usually grown at the edges of the fields.

Only a small share of the crops produced in this village are consumed by the producing households. For instance, more than 80 percent of the total soybean output is sold, while the remainder is held back by farmers for use as seeds for the next season's production; home consumption is negligible. Tobacco leaf is processed by farm households into dried-cut tobacco materials, and almost all of the product is sold. Sixty percent of cassava is processed into chips by farm households, and most of it is sold outside the village. The remaining 40 percent of the cassava root is sold directly to middlemen at the field as standing crops. Roughly speaking, more than 70 percent of total farm output is sold in this village.

Processing Activities in the Study Site

In the study area, many farm products harvested in a village are processed within the village. Some crops are processed by farm producers themselves at home; these activities are referred to here as "home processing." There are also part-time or full-time processors who purchase farm commodities from neighboring farmers and process them into various products. These processors usually operate on a small scale and depend mainly on family labor; they are referred to here as "cottage industry." Cottage industries and home processing are classified into two distinct categories, since the former use purchased materials and the latter process their own farm products. In reality, however, these two categories are not mutually exclusive.

Tobacco and cassava are processed by farmers themselves.[1] Most of the tobacco leaf is processed in farm households by slicing and drying, using a combination of male and female labor. Slicing with a large blade is usually done by males, and sorting and arranging the sliced tobacco for drying in the sun is usually done by females.

Farmers process cassava root, which they have harvested themselves from their fields, into several products for sale. In the study village, 60 percent of the cassava root is processed into chips, called *opak*. The process of making *opak* is very simple,[2] and the product is storable and can be fried as a snack. Dried *opak* is sold to consumers in urban areas through peddlers, as well as directly to neighboring villagers and grocery stores (*warung*) for retail. Farm home consumption of *opak* is less than 10 percent of total production. *Opak* is also processed by landless laborers, who purchase cassava root from neighboring farmers.

Several types of cottage industries also operate in the study area by purchasing farm produce from farmers in the village as well as from other nearby villages. Typical outputs are fermented soybean cake (*tempe*) and crude tapioca flour. *Tempe* is processed in rural areas on a very small scale.[3] In the study village, about 20 *tempe* processors operate, handling, at most, 10 kilograms of soybean per day. Most operators are housewives of farmers and use simple facilities found in most ordinary farm kitchens.

Besides the home processing of *opak* by farmers, cassava roots are

---

1. The highly storable nature of dried tobacco material enables it to be held for a long time until attractive marketing opportunities arise.

2. This is usually done by a housewife with the aid of her daughters. Ground fresh roots are mixed with several ingredients, such as oil, salt, and spices, and are formed by hand into flat elliptical shapes for steaming. Then they are spread out on a bamboo tray for drying under the sun for a few days.

3. The process of making *tempe* is simple and does not require special equipment. *Tempe* is the fermented product of soybean, using the mold *Rhizopus oligosporus* (Winarno, Haryadi, and Satiawiharja 1985). The soybean is soaked, boiled, and hulled. Then it is inoculated with mold, wrapped in plastic or banana leaves, and kept one or two days for fermentation in a warm place. Finally, the soybean is knitted together tightly by fibrous mold.

sold through village-based collectors to processors to make crude tapioca. Many local factories of differing operational size process fresh cassava root into crude tapioca flour in cassava-producing areas.[4] Some cottage-industry–type factories process less than 500 kilograms of cassava roots at one time, but other local factories process more than 40 tons of roots every day. The cottage industry processors are mostly farmers who engage in processing only when fresh cassava root is available in the village.

Marketing Channels

Marketing of farm products is carried out mainly by informal agents who do business on a small scale and have few permanent employees. These agents are linked to each other through informal contracts and tacit understanding to form an intricate marketing network. Typical marketing and processing channels for soybean, cassava, and tobacco as well as their products are illustrated in figure 11.1. Typical of the informal sector, there exists an almost infinite variety of agents and channels for marketing agricultural commodities and their products. Hence, figure 11.1 is limited to showing typical marketing channels in a village.

Most marketing tasks at the village level are done by farmers as a side job or by housewives of farmers. We may classify these marketing agents into farmers themselves and middlemen. Farm producers play an important role in marketing their products. It is not uncommon for farmers' wives and daughters to peddle home produce outside the village. It is also not uncommon for farmers themselves to carry their products to town for sale to traders or for direct supply to processors.

Village-based middlemen are peddlers, grocery stores, and collectors. Peddlers, who are mostly women, collect a basketful of products from farmers for sale directly to consumers by door-to-door visits in nearby towns and villages or to stall vendors in town marketplaces. Peddlers handle mainly ready-to-eat or ready-to-cook perishable commodities, such as *opak*, fruits, and vegetables.

Grocery stores, which are usually attached to villagers' houses, are mostly managed by housewives. Such stores carry urban commodities for sale to villagers, as well as merchandise produced in the village, such as *tempe* and *opak*.

While peddlers and grocery store keepers are essentially retailers,

---

4. Cassava roots, brought to a factory by collectors, are first soaked in water to remove dirt and then crushed by a simple grinding machine. Then crushed cassava roots are separated into residue, and water that contains starch. The water is stored in a big wooden barrel and is allowed to settle for one night. After removing the water, crude tapioca starch is obtained. It is spread out on bamboo trays and dried in the sun for two or three days. Usually, crude starch is sold to refineries that process it into refined starch, and it is also sold to nearby towns through collectors.

FIGURE 11.1 Typical marketing and processing channels for farm products at the village level

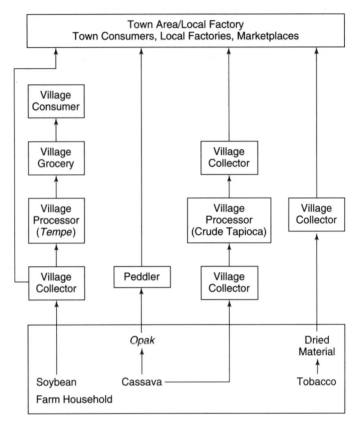

collectors are wholesalers. They collect farm products for delivery to processors in villages and towns and also to traders in towns. Most collectors are middle-class farmers whose farm operations are based mainly on family labor. They often use assistants, mainly landless farm laborers and marginal farmers, to collect directly from farmers. The wives and daughters of these landless farm households and marginal farmers are peddlers. Small returns from petty trades are important supplements to the income of people in the poorest class in the village community. The distinction between farmers and village-based middlemen is, therefore, not complete.

Production Structure of Agricultural Commodities and Activities

Table 11.1 presents input and output data as well as farm production costs and returns for individual crops in the study area. Total value

TABLE 11.1 Value added and employment indicators for selected crops per hectare of intercropped harvested area, Garut, Indonesia, 1986 (average of 18 sample farms)

Item	Soybean[a]	Tobacco	Cassava[a]	Others[b]	Total
Value-added ratio (percent)[c]	69	80	84	90	82
Labor income share (percent)[d]	60	37	56	49	47
Labor employment (days/hectare)[e]	73	99	11	170	352

[a]Sum of the first and second crop seasons.
[b]Corn, upland rice, vegetables, and fruits.
[c]Value added/output. Value added = output − current inputs.
[d]Labor income/value added.
[e]Assuming six hours of work per day.

added from one hectare of upland farmland per year was Rupiah (Rp) 783,000 (US$695 based on the exchange rate of Rp 1,126 per US$1 in 1986). About half of the total value added is derived from soybean, tobacco, and cassava production, and the rest comes from production of other crops, such as corn, upland rice, vegetables, and fruits, which are not processed in the village. Total labor income, including both family and hired labor, is estimated to be about half of total value added and is equivalent to 352 working days per hectare per year.

Home Processing

Since home processing is done by farmers themselves, using their own-farm produce, they acquire the income and employment generated by this activity. In the case of tobacco processing, the value of the dried material produced from one kilogram of leaf is estimated at Rp 420. The gross value added from home processing of tobacco amounts to Rp 170 per kilogram of leaf. The value-added ratio is about 40 percent, which implies that 40 percent of the market value of dried tobacco material is acquired by farmers as income from home processing. Labor's share of this income, which is measured by comparing the estimated cost of family labor with the total income from the processing, is as high as 97 percent, reflecting the highly labor-intensive nature of home processing of tobacco.[5]

The gross value added of cassava processing amounts to Rp 43 per kilogram and the value-added ratio is 55 percent. Assuming that 16 kilograms of cassava root are processed per day, the total income from processing is Rp 688, which is slightly higher than the typical female

---

5. In order to estimate the cost of family labor, typical daily wage rates for hired farm work for upland areas (Rp 1,200 for males and Rp 600 for females, assuming six hours of work per day) have been used. In the case of tobacco processing, an average wage rate of Rp 150 per hour is assumed.

wage rate for farm work of Rp 600 per day. However, if we take into account the capital depreciation of processing equipment, the net income would be almost equal the wage rate. The high labor-income share of 88 percent for *opak* processing also reflects the highly labor-intensive nature of home processing of cassava.

Cottage Industry

Compared with the home processing of tobacco and *opak, tempe* and tapioca manufacturing—typical examples of cottage industry—have very low value-added ratios, but similar shares of income for labor. This seems to reflect the fact that cottage industries and home processing are both highly labor intensive. It appears that both home processing and cottage industries supplement farm household income by increasing the use of family labor. Family members engaged in these activities receive incomes of about equal or slightly better than for those engaged in farm work.

Marketing

The prices of soybean, cassava, tobacco, and their products at various points in the marketing chain are summarized in table 11.2. For example, village collectors purchase soybean from farmers at the price of Rp 580 per kilogram and sell it to *tempe* processors at Rp 590, implying a unit marketing margin of Rp 10. Then, *tempe* processors make soybean into *tempe* and sell it to village groceries at Rp 40 per piece. Village groceries retail it to consumers at Rp 50, which also implies a unit marketing margin of Rp 10.

Total value added from the marketing can be estimated after deducting transportation costs paid to public transporting agents or private transporters for chartering trucks. This deduction may result in an underestimation of the value added from marketing activities, since value is also added from transportation activities themselves. However, most of the transporting agents are town based and full time in their operation. Since the analysis in this study limits income and employment generation to the village level, the decision to exclude transportation activities is realistic.

Judging from the labor-intensive nature of village-level marketing, income accruing to local labor is assumed to be the same as the marketing margins.[6]

Since the middlemen's working hours are difficult to measure di-

---

6. This calculation neglects the interest on working capital. Turnaround of working capital is usually very short in the case of village collectors, however, and thus it seems to be negligible in the middlemen's profit.

TABLE 11.2 Prices and marketing margins of selected agricultural commodities at various points in the marketing chain

Crop/Product	Seller	Buyer	Sale Location	Price (Rp/kilogram)	Marketing Margins (percent)
Soybean					
Soybean	Farmer	Village collector	Farmgate	580	
	Village collector	Village processor	Factory	590	2
Tempe	Village processor	Village grocery	Store	40[a]	
	Village grocery	Village consumer	Store	50[a]	25
Tobacco					
Dried material	Farmer	Village collector	Farmgate	3,000	
	Village collector	Local factory	Factory	3,500	17
Cassava					
Fresh roots	Farmer	Village collector	Field	30	
	Village collector	Village processor	Factory	40	33
Crude tapioca	Village processor	Village collector	Factory	240	
	Village collector	Local factory	Factory	250	4
Opak	Farmer	Peddler	Farmgate	250	
	Peddler	Town consumer	Consumer's house	400	67

[a]Price per piece of 90 grams.

rectly, they are estimated indirectly by dividing the income of middlemen by the standard wage rates of hired farm workers. This calculation assumes that village collectors and village grocery store keepers are earning an average income per hour used for marketing activities in the village. This is not an unrealistic assumption given the fact that those marketing agents are themselves farmers or farm laborers. It has already been identified (Hayami et al. 1987) that the local marketing system can be approximated by perfect competition. The price spread between different locations corresponds almost exactly to the cost of transportation, suggesting a competitive market in this area. This implies that collectors have little room to gain additional profit beyond an average income from the marketing activities.

184 *Toshihiko Kawagoe*

For the purpose of illustration, table 11.3 presents estimates of income and employment generated for the most typical marketing and processing channels in the village. Two marketing and processing channels are considered for both soybean and cassava. More than 35 percent of soybeans are processed into *tempe* in Indonesia (Winarno, Haryadi, and Satiawiharja 1985). We apply that share to the soybean produced in the village, processed into *tempe* by village processors and sold to villagers through village groceries (case 1), and assume that the remainder, 65 percent, is shipped to town areas by village collectors (case 2). In the case of cassava, 60 percent of fresh cassava roots harvested in the village is processed into *opak* by farmers, and the product is sold by peddlers to town consumers (case 1), and another 40 percent is processed into crude tapioca flour and sold in town areas by village collectors (case 2). All the harvested tobacco leaf is assumed to be processed into dried material by farmers and sold to local processors by village collectors. These marketing and processing channels are shown in figure 11.1.

Marketing margins in the case of *tempe*, for instance, were 25 percent of the farmgate price; for *opak*, they were 67 percent; and for dried tobacco, 17 percent. Marketing margins for unprocessed or less processed commodities were, of course, much less: 2 percent for soybeans and 4 percent for crude tapioca (table 11.2).

Income and Employment Generation

Significant income and employment were generated from processing and marketing of soybeans, cassava, and tobacco (table 11.3).[7] In the case of *tempe* processing and marketing, total value added per year was about Rp 77,000 of which about 60 percent was produced on the farm and the rest was added in processing and marketing. If the soybean crop was shipped directly to other sectors, little income was added to the village community (case 2). Thus, a high contribution of marketing and processing to income is expected when farm produce is processed in the village and then marketed.

In the case of cassava processing, although both *opak* and tapioca are processed in the village, *opak* is sold as a final product to consumers by peddlers, while crude tapioca is shipped to local factories for further processing. Processing and marketing of *opak* contributed more than 70 percent of total income compared with 35 percent for crude tapioca. This finding implies that more income is generated if the crop is pro-

---

7. The calculations are considered to represent lower-bound estimates, since only the direct contributions of marketing and processing are counted; the indirect multiplier effects through forward and backward linkages with other activities in the village communities are not counted.

TABLE 11.3 Income and employment generation from production, processing, and marketing of selected upland crops, per hectare of harvested area

Item[a]	Soybean		Cassava		Tobacco (Dried Material)	Total
	Case 1[b] (*Tempe*)	Case 2[c] (Soybean)	Case 1[d] (*Opak*)	Case 2[e] (Tapioca)		
Farm production (kilogram/hectare)	321		821		1,433	
Crop allocated to the marketing channel (kilogram/hectare)	112	208	492	328		
Value added (Rp 1,000/hectare)						
Farm production	44 (57)	83 (99)	13 (24)	8 (65)	275 (44)	423 (49)
Processing	11 (14)	0 (0)	21 (40)	3 (20)	257 (41)	291 (34)
Marketing	226 (28)	1 (1)	19 (36)	26 (15)	97 (15)	142 (17)
Total	77 (100)	84 (100)	53 (100)	13 (100)	629 (100)	856 (100)
Labor income (Rp 1,000/hectare)						
Farm production	27 (46)	49 (98)	7 (16)	5 (51)	103 (23)	191 (31)
Processing	9 (16)	0 (0)	19 (41)	3 (28)	245 (55)	275 (45)
Marketing	22 (38)	1 (2)	19 (43)	2 (21)	97 (22)	142 (23)
Total	57 (100)	50 (100)	45 (100)	9 (100)	445 (100)	608 (100)
(Labor's share, percent)[f]	(74)	(60)	(85)	(72)	(71)	(71)
Labor employment (days/hectare)						
Farm production	26 (33)	47 (96)	7 (10)	4 (50)	99 (22)	183 (28)
Processing	15 (19)	0 (0)	31 (44)	3 (31)	273 (60)	321 (49)
Marketing	37 (47)	2 (4)	32 (46)	2 (19)	81 (18)	153 (23)
Total	77 (100)	49 (100)	69 (100)	9 (100)	453 (100)	658 (100)

[a] Numbers in parentheses are percentages.
[b] Soybean produced in the village that is processed into *tempe* by village processors and sold to villagers through village groceries.
[c] Soybean that is shipped to town areas by village collectors.
[d] Fresh cassava roots harvested in the village that are processed into *opak* by farmers and sold by peddlers to town consumers.
[e] Fresh cassava roots processed into crude tapioca flour and sold in town areas by village collectors.
[f] Total labor income divided by total value added.

cessed into final products and marketed to consumers by village-based middlemen.

Total value added from home processing of tobacco nearly equaled income from tobacco production. This result shows that farmers could double their income from tobacco by processing it themselves. Since the product, dried tobacco material, was sold to local factories, not much income was added from marketing activities.

Almost half of the total income generated from these three products was added from marketing and processing activities (column 6, table 11.3). This suggests that about half as much income would have accrued to villagers if marketing and processing activities had not been developed. The relative contribution of marketing and processing to labor income and employment were approximately 70 percent, even higher than the relative contributions to total value added. In the case of *opak*, it is remarkable to see that the contributions of marketing and processing were almost 90 percent of total income and employment generation. It should be noted that most of this income and employment accrued to women whose opportunity cost is lower.

Conclusion

These findings suggest that the processing and marketing of farm products play a critically important role in generating income in village communities. Income from marketing and processing is almost equal to income from farm production, and it is especially prominent when crops are processed into final products and marketed to consumers by members of a village community.

Another important role of marketing and processing is in equalizing income distribution by increasing employment and the share of income accruing to labor. Such activities provide plenty of employment opportunities to farmers and their family members, especially to the poorest class of people in the village community, such as farm laborers and marginal farmers. It is clear that promotion of village-level processing activities may be used as a means of alleviating poverty and inequality.

The estimations here show only part of the total income and employment that is being generated from processing and marketing activities, since the analysis is limited to the village level. The crops shipped to local towns are processed widely by local factories, and if these activities are taken into account, the income and employment generation in local communities would be even greater. This should be resolved in future studies.

PART V

The Diverse Experience

12 Nontraditional Vegetable Crops and Food Security among Smallholder Farmers in Guatemala

JOACHIM VON BRAUN

MAARTEN D.C. IMMINK

Introduction

Export-oriented agricultural production has a long tradition in Guatemala that has resulted in a highly dualistic agricultural sector. Traditional exports, which include coffee, cotton, sugar, bananas, and beef, accounted for approximately 70 percent of total agricultural exports in the late 1980s. Most of the economic gains from agricultural development have been confined to the large-scale modern sector. Increasing concerns are being expressed about the food security of the low-income rural and urban populations. National self-sufficiency in basic food grains (maize) is a stated goal of the Guatemalan government. At the same time, crop diversification and expansion in nontraditional agricultural exports are perceived as necessary to counteract deterioration in market prices of traditional exports, a growing foreign debt, and dwindling foreign exchange reserves.

Vegetable production appears to be a promising option for meeting the two goals of crop diversification and expansion of nontraditional agricultural exports because of rising foreign demand. Targeted at the smallholder sector, such production creates opportunities for adoption of labor-absorbing production techniques, even as gross farm margins are shown to be considerably higher than for staple food production. At the same time, the increased dependence on market conditions for farm outputs and inputs and the larger farm production outlays imply significant risks, which, without appropriate financing, may be difficult for smallholder farmers to assume. Unstable market prices for export vegetables and farm inputs, crop failures, breakdown of marketing institutions, and collapse of export markets all represent significant risk factors that can profoundly affect incomes and food security of farm households growing export crops. In addition, macrolevel foreign trade, monetary, and fiscal policies can positively or negatively affect the economic and social outcomes of crop diversification and commercialization programs for smallholder farmers.

A Cooperative for Commercialization

The study was conducted in the cooperative of Cuatro Pinos, which is located approximately 35 kilometers west of Guatemala City, in the Western Highlands, at an altitude of approximately 2,000 meters (a more detailed description may be found in von Braun, Hotchkiss, and Immink 1989). The area is close to the capital, and good infrastructure facilitates market integration of fresh vegetables. Average farm sizes are below the average for the Western Highlands. A survey conducted in the 1970s in the area found that 90 percent of cultivated land was covered by maize and beans and the remainder by vegetables (Hintermeister 1986).

The area is characterized by high rates of population growth—above 3 percent per year. The incidence of diseases such as diarrhea and respiratory infections in children is high, particularly in the rainy season (May–October). The prevalence rates of growth retardation and weight deficiency are similar to those reported for other highland areas and affect more than half of the preschool children.

In 1989 the cooperative was active in eight villages. The storage and processing facilities for the export vegetables as well as the central offices of the cooperative are located in the village of Santiago, with access to Guatemala City via paved roads. The cooperative was formally organized in 1979 with 177 farmers, after a period of precooperative productive activities undertaken with foreign investment funds and technical assistance. By 1989, cooperative membership had grown to 1,600. In 1985, cooperative farmers were cultivating nearly 300 hectares of export vegetables, more than half under snow peas, with the remainder under cauliflower, broccoli, and parsley. The area under export vegetable production quadrupled between 1980–81 and 1985. At the cooperative's central facilities, employment for approximately 150 persons—more than half of them women—has been created in screening and packing tasks.

Export vegetables were introduced to Guatemala in the mid-1970s with the assistance of substantial foreign financing. The cooperative farmers of Cuatro Pinos entered collectively into a contract-growing arrangement with a single processing company in the early 1980s. It had become apparent to the company that there were significant economic advantages associated with contract-growing on smallholdings in the Western Highlands. Since then, the cooperative has diversified its sales contracts and actively promotes its own export markets in the United States and Europe.

The price variability of the new crops, especially snow peas, is extreme. For instance, within one year, 1985, prices fluctuated between 0.10 and 2.00 quetzales per pound. Farmers partially cope with this price variability by spreading the growing seasons and having a long harvest period.

Export channels for fresh commodities from developing countries are risk prone. Exchange rate manipulation is an important risk factor. This study shows that the exchange rate policy introduced in 1985 in Guatemala implicitly taxed the export vegetables of small producers by about 25 percent. This implicit tax had an adverse effect on employment in the small-farm export sector.

Recently, besides a single multinational company and the cooperative, many other traders have begun to engage in the export of these crops. Also, local processing and freezing of fresh produce has been initiated. These developments reduce the risk of a sudden collapse of the export marketing channel.

Nontraditional export crops are substantially more profitable than traditional crops to farmers. Net returns (gross margins) per unit of land of snow peas, the most important new crop, are, on average, 15 times higher than those of maize, the most important traditional crop. Returns of the new crops per unit of family labor were about twice as high as those from maize and 60 percent higher than from traditional vegetables produced for local markets in 1985. Input costs for snow peas, however, were, on average, about 1,600 quetzales per hectare, while they are about 430 quetzales per hectare for traditional vegetables and 120 quetzales per hectare for maize. Short-term financing of inputs poses a problem to small farmers and indicates the importance of rural credit.

Study Design

The results of this study are based upon household-level surveys conducted in 1983 and 1985 in the six villages in which the Cooperative Cuatro Pinos was active. The two rounds of surveys were conducted in the same households, which were visited during the same period of the year (November–January) in order to eliminate any seasonality effects that would otherwise have been introduced.[1]

---

1. The household-level data include information on demographic and socioeconomic characteristics of the household; agricultural production (land size and quality, crop and animal production, labor inputs by crop, input purchases, and crop sales); nonfarm income; food and nonfood expenditures, consumption of self-produced foods; height and weight measurements of children under ten years of age; and morbidity and child-feeding practices. Depending on their nature, the various types of data cover different recall periods; for example, the agricultural production data cover the crop years of 1982–83 and 1984–85, and off-farm income data cover the twelve months preceding the surveys. Food and nonfood expenditures are monthly, with an annual recall for lumpy expenditures (durable goods) in the 1985 survey.

A roughly equal number of cooperative members (195) and nonmembers (204) was selected at random by village, based upon 1983 census information. In order to ensure representative coverage of the situation in the smaller villages, the sample was purposefully biased toward the four smaller communities among the six villages. This brought the total sample closer to prevailing village patterns in the Western Highlands. The village samples covered from 38 to 75 percent of cooperative members and from 8 to 17 percent of nonmembers in each community. The sample included early and late members: 42 percent of the sample cooperative members had entered the cooperative five to seven years before 1985, and 28 percent had entered less than three years before the survey.

Export Vegetable Adoption by Smallholder Farmers

An in-depth study of 21 households who were members of the cooperative revealed that in all cases the decision to apply for cooperative membership was made by the male head of household, who in 16 cases had previously asked the opinion of his spouse (Nieves 1987). Members pay a one-time membership fee, which in 1985 corresponded to 12 days' wages for a typical farm worker.

A PROBIT model was formulated to better understand what determines whether smallholder farmers shift production to export vegetable crops. Results from this PROBIT analysis show that farmers with larger farms or farmers who were located close to a highway were more likely to adopt export vegetable production. The fact that a larger share of cropland was allocated to export vegetable production on smaller farms, as will be seen below, possibly indicates an economy-of-scale effect in the lower ranges. Higher incomes from relatively secure off-farm employment slightly decrease the likelihood of growing export vegetables, as do higher shares of female labor in household labor, but in a more robust way, reflecting the male-dominated decision-making process of adopting cash crops. As may be expected, farmers with traditional views toward growing maize and those who are located in remoter areas are less likely to adopt export-vegetable production. Land quality, total household labor force size, and human capital endowment of the male heads of household (age and educational level) do not seem to play a role in the adoption decision-making process. In general, then, less traditional farmers with larger farms and better access to rural infrastructure, but without access to secure off-farm employment, are most likely to adopt export vegetable production.

Production and Employment Effects of Export Vegetable Production

Noncooperative farm households grew traditional subsistence crops (maize and beans) on 78 percent of their land, compared to 52 percent of the land for cooperative farm households.[2] The smallest cooperative farmers allocated the largest shares of their land to the production of export vegetables. Cooperative farms of less than 0.25 hectare allocated 45 percent of their land to the new crops and only 38 percent to maize and beans. In contrast, noncooperative farm households of similar size planted 81 percent of their land with subsistence crops. Cooperative members with the largest farms (more than one hectare) cultivated export vegetables on 39 percent of their land, with 52 percent of the land allocated to maize and beans. Noncooperative households in the largest landholding category grew maize and beans on 67 percent of their land.

The new export vegetables were grown on plots of better land quality. Increasingly, more land was purchased and rented by farmers who grew the new crops, which may have contributed to the increase in land prices and rental values in the area. Population pressures have aggravated this trend. Higher land values represent indirect benefits to landowners who do not grow export vegetables, whereas farm units that rent land and do not grow export vegetables incur additional economic costs.

The new export crops have not only replaced traditional vegetables (cabbage, carrots, and radish) but have also reduced the area allotted to traditional subsistence crop production. Given that the study region was already a net importer of maize, the displacement of maize production by the new vegetable crops is not likely to have a significant effect on maize prices in this region.

Smallholder farm households in the Western Highlands have a strong preference for food security based upon self-produced maize. Unstable food markets, insecure off-farm employment, and limited institutionalized food security programs to cope with food crises are probably factors that condition this preference (Bossen 1984). Ninety-four percent of the export crop producers grew some maize. The great majority of export crop producers tended to have larger amounts of maize available on a per capita basis for household consumption when compared to traditional farmers of similar farm size, except for farmers with less than 0.25 hectare of land (table 12.1). Maize and bean yields were, on average, 30 percent higher for export crop producers because of their more labor-intensive cropping practices and higher levels of fertilizer application.

---

2. "Cooperative farm households" and "export crop–producing households" will be used here interchangeably, as will "noncooperative farm households" and "traditional farmers."

TABLE 12.1 Production of subsistence maize and use for household consumption in farm households, by farm size, Guatemala, 1985

Farm Size (hectares)	Export-Vegetable Growers		Traditional Farmers	
	Production	Use for Own Consumption[a]	Production	Use for Own Consumption[a]
	(kilograms per adult-equivalent)[b]			
<0.25	48	41	59	49
0.25–0.50	103	88	95	82
0.50–1.00	131	113	127	97
>1.00	182	137	180	138
Average	142	115	109	87

SOURCE: Data from the household survey made by the Institute of Nutrition of Central America and Panama and the International Food Policy Research Institute, 1985.
[a] Calculated from a complete production and disappearance balance (production minus sales, losses, animal feed, gifts, seed).
[b] Adult-equivalents are based on age- and sex-specific energy requirements of household members.

The analysis led to the conclusion that this difference was not the result of a self-selection bias of more efficient farmers becoming export crop producers.

The opportunity cost of maize produced for household consumption increased significantly for export crop–producing households. The shadow price of maize for a typical farm increased 137 percent, from 0.49 quetzal per kilogram to 1.16 quetzales. The difference between the shadow price and the actual market price (market price was 0.29 quetzal per kilogram in 1985) can be considered an insurance premium that farmers were willing to pay to maintain a degree of food self-sufficiency based upon maize and beans.

The export crop production created local employment directly on farms and indirectly elsewhere through forward and backward linkages and multiplier effects resulting from increased income spent locally. Labor input increased on the export crop–producing farms by 45 percent, approximately half of which was supplied by household labor and half by hired labor.

There was a shift of labor from traditional to new crops and a partial substitution of labor between household labor and hired labor. Total labor input on maize was cut back by 13 percent, on beans by 43 percent, and on traditional vegetables by 29 percent, but hired labor input on maize increased somewhat as household labor was partly displaced by hired labor. Women provided a substantial share of the increased household labor—44 percent of the labor increase on farms of less than 0.5

hectare and 32 percent on farms of more than 1 hectare. While women's share in total household labor decreased with increasing farm size, men's share remained stable and children's share increased. Women's labor input to maize was low (9 percent) compared to their input to traditional vegetables (25 percent) and the new export vegetables (31 percent). Most of the household field labor was provided by men (in maize, 83 percent; in snow peas, 59 percent; and in traditional vegetables, 61 percent).

Combining farm-level employment with estimated employment created through input supply and output marketing yields an overall 21 percent increase in agricultural employment in the six communities. This increase in agricultural employment reduced off-farm work and interregional migration of farm household members: 0.7 persons per household worked away for an average of 2.3 months among export-vegetable producers' households, while among other farm households, 0.9 persons per household worked away for an average of 4.2 months.

Income and Expenditure Effects

The export crop production led to increased household income, with the increase between 1983 and 1985 being most pronounced among recent participants. Total expenditures, a proxy for income, increased among recent cooperative farmers by 38 percent above the average nominal increase in the study population. Off-farm earnings were reduced substantially with increases in farm income and adoption of export crop production. Income gains from export crop production were highest among growers with the smallest farms, almost 60 percent higher than among nonparticipants. These income gains moved the poorest households upward on the income scale relatively more than the other income groups and thereby reduced the disparity in income distribution in the study area. Only 38 percent of the poorest export crop producers remained in the lowest tercile of the income distribution between 1983 and 1985, as compared to 55 percent of the poorest traditional farm households.

Export crop–producing households spent, on average, a slightly lower share of their total expenditures on food: 64 percent versus 67 percent for other households. However, in absolute terms, per capita food expenditures were, on average, 18 percent higher in export crop–producing households because of higher incomes. The lower budget shares to food among the export crop–producing households were mostly due to relatively lower expenditures on all foods, except for meat, fish, and eggs.

A model to estimate Engel curves was specified in which sources of income (their relative shares) as well as levels of income were included as explanatory variables. Household size is controlled to account for potential scale effects. Among sources of income, income earned from the new export crops, as well as male and female income earned from off-farm sources, can be distinguished. A testable hypothesis was that income earned under men's control (from off the farm and from the new export vegetables) is spent relatively less on food than farm income in general and that female-controlled off-farm income is spent relatively more on food than general household income.

Three different models were specified to evaluate hypotheses. The first model included cooperative membership to assess whether there was an effect over and above the relative income share effect from the new crops, which could be so if the institutional arrangements under which the export crops are grown and managed (such as cash payment schemes and savings opportunities) differ from cash cropping outside the cooperative scheme. The model was estimated separately for members and nonmembers to test additionally for differences in spending behavior between the two groups. Total expenditure is used as a proxy for expected permanent income in this analysis, which is not entirely satisfactory because of simplistic assumptions about savings. The results of the analysis are presented in table 12.2.

Food expenditures as a proportion of total expenditures decrease significantly but not rapidly with increased income. At the total sample mean, a 10 percent increase in total expenditure decreases the budget share to food by only 0.2 percent. Once income levels and sources are controlled for, cooperative membership (MIEM) does not seem to have a distinct effect on shifting the budget share for food in one direction or the other. At the margin, however, the budget share to food decreases more rapidly with rising income in cooperative households than in noncooperative households (see LTEXPCSQ).

There are indications that male-controlled income is spent more according to men's preferences; more food may not rank high among these preferences (Nieves 1987). Income-source–related variables indicate that an increased share of income from new export vegetables (RCASH) marginally decreases the budget share to food over and above the total income effect in cooperative households. The parameter is only weakly significant. If, for instance, the income share from new cash crops increases from 0 to 50 percent, the food budget share is reduced by 1.2 percentage points, holding income constant. The net effect for the food budget share of an increased income share of new export crops is very similar in order of magnitude to the net effect of an increased share of male nonagricultural income (see the parameters for RCASH in the member model and for RMNAGINC in the nonmember model). Women's share in total off-farm income (RFNAGINC) does not appear

TABLE 12.2 Determinants of budget shares to food in total sample, cooperative nonmember and member households, Guatemala, 1985

Explanatory Variable[a]	Total Sample		Nonmembers		Members	
	Parameter	t-Value	Parameter	t-Value	Parameter	t-Value
HHSZ	0.003765	1.424	0.005195	1.462	0.002795	0.704
RCASH	−0.023010	−1.691	−0.018140	−1.047	−0.038070	−1.655
RFNAGINC	0.009191	0.261	0.018070	−0.411	−0.007464	−0.104
RMNAGINC	−0.023100	−1.706	−0.033500	−2.263	0.020290	0.578
MIEM	0.010550	0.797	—	—	—	—
LTEXPC	0.279710	1.861	0.156420	0.913	0.579280	1.821
LTEXPCSQ	−0.035270	−2.809	−0.024610	−1.694	−0.060220	−2.314
Constant	0.248630	0.557	0.596770	1.193	−0.630700	−0.650
R^2	0.34		0.37		0.33	
Degrees of freedom	342		178		158	
F-value	25.6		17.7		12.8	

NOTE: The dependent variable is budget share to food.
[a]Definitions of variables:
HHSZ = household size (number of persons).
RCASH = ratio of income from new export vegetables over total income (total expenditure is used as proxy for expected total income).
RFNAGINC = ratio of female nonagricultural off-farm income over total income (total expenditure is used as proxy for expected total income).
RMNAGINC = ratio of male nonagricultural off-farm income over total income (total expenditure is used as proxy for expected total (income).
MIEM = membership in cooperative (members = 1, else = 0).
LTEXPC = log of per capita total expenditures serving as a proxy for expected (permanent) income.
LTEXPCSQ = (log of per capita total expenditures)^2.

to have an effect on the budget share to food over and above that of total income.

Food Availability and Nutrition Effects

Calories available for household consumption were derived from food expenditures. In all households, daily energy availability increased significantly with income. Half of the available dietary energy came from maize, but the share of maize tended to decrease at higher income levels. The calorie share of beans also decreased slightly, while that of meat, eggs, and other foods increased with higher incomes. Cooperative households acquired, on average, 7 percent fewer calories per capita, but spent about 10 percent more per available calorie than did other households. There were no significant differences, on average, in the food composition of the daily diet between cooperative and noncooperative households.

In order to explore further the income effect on food availability (including food from own production) among export crop–producing households, a regression analysis of the 1985 data was undertaken. The model is estimated for the total sample and separately for the bottom 50 percent of the income group among members and nonmembers. Results are presented in table 12.3.

The estimated response of calorie availability to changes in income levels is highly significant: additional income increases calorie consumption (LTEXPC) but at decreasing rates at the margin (LTEXPCSQ). The elasticity of calorie consumption with respect to income is 0.31 at the total sample mean. As the new export crops increased household per capita income substantially, a positive effect on calorie consumption is to be expected in cooperative households.

Most purchased food items in Guatemalan households are actually acquired by women. Only in the case of maize is a significant share purchased by men. One would expect that women's income shares would have a positive effect on food acquisition, over and above the total income of the household. The results, however, do not provide statistical support for this hypothesis. Furthermore, increased shares of total income from new export crops do not have a significant effect on calorie acquisition over and above the income-level effect (see variable RCASH in table 12.3). The results also indicate that—controlling for income level, income composition, household size, and demographics— cooperative members acquire 6 percent less calories than nonmembers (with a low level of statistical significance). This tendency observed for the total sample is further supported by separate estimates of the calorie consumption functions for members and nonmembers in the bottom

TABLE 12.3 Determinants of availability of calories to households in total sample and the lowest two quartiles of cooperative members and nonmembers, Guatemala, 1985

Explanatory Variable[a]	Total Sample		Lowest Two Quartiles			
			Members		Nonmembers	
	Parameter	t-Value	Parameter	t-Value	Parameter	t-Value
HHSZ	1,747.343	13.021	1,244.150	5.075	1,510.410	6.629
RCASH	57.254	0.082	553.030	0.438	173.124	0.194
RFNAGINC	−1,685.120	−0.971	1,096.170	−0.437	−1,675.740	−0.606
MIEM	−829.710	−1.218	—	—	—	—
LTEXPC	20,446.410	2.663	3,928.210	1.218	6,289.810	3.794
LTEXPCSQ	−1,328.330	−2.072	—	—	—	—
RCHILD	−6,627.550	−3.106	−4,957.520	−1.088	−4,923.010	−1.230
MPRICE	−82,424.880	−5.774	−46,732.990	−1.821	−54,545.190	−1.648
BFPRICE	505.477	1.020	949.183	0.826	1,139.320	1.168
Constant	−47,623.400	−2.060	−6,841.310	−0.394	−18,424.900	−1.417
R^2	0.41		0.33		0.64	
Degrees of freedom	340		60		98	
F-value	26.6		4.20		9.60	

NOTE: The dependent variable is calories available for consumption per day in the household (from purchases and consumption from own production, 1985).

[a] Definitions of variables:
HHSZ = household size (number of persons).
RCASH = ratio of income from new export vegetables over total income (total expenditure is used as proxy for permanent income in the ratio).
RFNAGINC = ratio of female nonagricultural off-farm income over total income (total expenditure is used as proxy for permanent income in the ratio).
MIEM = membership in cooperative (members = 1, else = 0).
LTEXPC = log of per capita total expenditures per year serving as a proxy for expected (permanent) income.
LTEXPCSQ = (log of per capita total expenditures)^2.
RCHILD = ratio of number of children under five over number of persons in the household.
MPRICE = price of maize.
BFPRICE = price of beef.

half of the joint income distribution: nonmembers show a highly significant positive response in calorie acquisition with rising income but members do not. Calculated at the identical income level at the mean of the joint sample, members increase their calorie acquisition by 2.8 percent with 10 percent more income, but nonmembers increase theirs by 4.4 percent.

Nutrition problems are poverty related. Health and environmental sanitation conditions along with household-level food availability and mothers' employment are important factors that determine the nutritional status of young children (Baldeston et al. 1981). New export crop production directly or indirectly affects all these factors. In general, cooperative households had better housing and water supply conditions, made more use of higher quality health services, and tended to receive food aid more often. Members of cooperative households also tended to have higher education levels.

The prevalence of weight deficiency and stunting among preschool children in the Western Highlands generally increased during the 1980s (von Braun, Hotchkiss, and Immink 1989). This may indicate worsening economic and social conditions in these areas. Displacement of large segments of the rural population due to political violence and repression is a significant contributing factor. The nutritional status of young children in the study population is similar to that reported for other highland populations during the same period (von Braun, Hotchkiss, and Immink 1989). Between 1983 and 1985, the prevalence of weight deficiency and stunting increased among children from households in the bottom and middle income terciles, but decreased slightly among children in the upper income tercile.

Children from export crop–producing households were somewhat less likely to be weight deficient or stunted than children from traditional farm households. Between 1983 and 1985 the prevalence of weight deficiency and growth retardation increased in both groups, but slightly more so among children from traditional farming households.

Regression analysis of the 1983–85 data for cohorts of children aged 6 to 60 months showed that household-level energy availability has a positive effect, although decreasing at the margin, on the likelihood of reduced acute or chronic malnutrition (low weight-for-height) in young children. Longer duration of cooperative membership was associated with reduced weight deficiency and stunting, which may reflect a long-term income effect as well as extended participation in the food and nutrition education programs of the cooperative.

In order to explore further the nutrition effects of changes in household income and in income control and composition—all associated with new export crop production—a second regression model was formulated and tested with the 1985 data (table 12.4). Increased income decreases stunting and weight deficiency, but at higher levels the marginal effect is reduced. Reduction of stunting, however, will require major income increases. Higher shares of off-farm nonagricultural income are associated with reduced weight deficiency in young children, independent of whether that income is earned by men or women, although the effect is greater in the case when it is controlled by women. Higher shares of nonagricultural income earned by men are also associated with reduced stunting. Higher shares of income from the new export crops are not adversely related to the nutritional status of the children.

Conclusions

Production of export crops by smallholder farms in Guatemala appears to have had a positive effect on household income and food security. Production of export vegetables increased on-farm employment, reducing the need to rely on uncertain off-farm employment in local and far-off labor markets for additional income. New employment opportunities were also created for nonmember households in the study communities. There were substantial increases in total household income, particularly for the smallest farms, which more than offset reduced off-farm earnings. Disparities in income levels were also reduced in the process. Export cropping was found to be associated with higher yields of staple foods (maize and beans) and, thus, export crop producers maintained own production of these foods for consumption in the context of a risky food-market environment.

At the same income levels, export crop–producing households spent less on food out of marginal income than did other households, perhaps because the income from export crop production is usually male controlled. On average, export crop–producing households spent more on food but also purchased a food basket that contained more expensive foods. The nutritional quality of the daily diet was also likely to improve.

Income was positively associated with the nutritional status of young children. To the extent that new export crop production was positively associated with higher levels of household food availability and better sanitary and environmental conditions, children were less likely to suffer from weight deficiency and stunting. Under conditions of expanding economic opportunities, educational efforts and improved access to better health, infrastructure, and services can reinforce the income effect on child nutrition. The cooperative's actions in these areas became feasible because of the economic success of the export vegetables, be-

TABLE 12.4 Multivariate analysis on effects of income and income source and composition on the level of nutritional status of children aged 6–120 months

Explanatory Variable	Weight-for-Age Z-Scores		Height-for-Age Z-Scores		Weight-for-Height Z-Scores	
	Parameter	t-Value	Parameter	t-Value	Parameter	t-Value
Age in months	−3.489E-03	−0.85	−0.0156	−3.03	−9.485E-03	−2.28
(Age in months)^2	3.497E-05	1.14	1.397E-04	3.64	1.053E-04	3.39
Birth order[a]	1.706E-04	0.01	−0.0240	−1.37	0.0220	1.55
Sex[b]	7.083E-03	0.12	0.0379	0.52	−0.0562	−0.96
Duration of breast-feeding[c]	9.663E-05	2.03	1.894E-04	3.17	−3.063E-05	−0.63
Income per capita[d]	8.231E-04	3.20	7.807E-04	2.41	5.162E-04	1.98
(Income per capita[d])^2	−3.930E-07	−2.74	−3.489E-07	−1.93	−2.625E-07	−1.80
Share of male nonagricultural income	0.1613	2.00	0.3291	3.24	−0.0531	−0.67
Share of female nonagricultural income	0.4953	2.20	0.4895	1.73	0.3140	1.37
Share of income from export crops	0.1569	1.88	0.1958	1.81	0.0640	0.73
Constant	−2.1165	−12.89	−3.110	−15.08	0.0549	0.33
R^2	0.032		0.073		0.028	
Degrees of freedom	785		785		785	
F-value	3.590		7.280		3.290	

[a]First-born child = 1, second = 2, and so on.
[b]Male = 1; female = 2.
[c]Variable only for children above 24 months of age (else = 0).
[d]Expenditure per capita per year is used as proxy for income.

cause this permitted spending of some of the co-op's profits on social services.

The economic, food security, and nutritional outcomes in this case study appear to be influenced by market conditions (both national and international) macrolevel policies, and complementary social investments. Production of sustained income growth from small-scale crop diversification will require certain conditions: (1) improvements in marketing efficiency, both domestically and internationally, through investment in infrastructure, marketing organizations, and transportation facilities; (2) rural credit programs to finance higher input costs and on-farm infrastructure (for example, mini-irrigation), particularly targeted to the smallest farmers whose relative income gains are the highest, thus reducing income disparities; (3) vertical integration of production, processing, and marketing, which will capture economies of scale and increase economic returns to diversified farmers; (4) foreign trade policies that stimulate nontraditional agricultural exports and simplification of export licensing; and (5) development of rural financial institutions that are highly accessible to smallholder farmers, to capture and make available the savings potential generated by increased incomes to finance community development projects, so that economic benefits from diversification are spread among the rural poor.

Reduction of food security risks derived from lowered own food production requires appropriate technological changes that result in significant yield increases in maize production, coupled with effective technical assistance, particularly focused on the smallest farm households that are most prone to reduce consumption of own-produced food.

Large and sustained income increases are required to produce improvements in child nutrition. In the short run, household income increases seem to have no health effects, which may reflect inadequate access to health care. Reinforcement of the positive economic effects on the quality of life of smallholder farmers and their households requires complementary social investments that result in better access to adequate preventive and curative health care, as well as improved sanitary practices and environmental conditions.

13 The Nutrition Effects of Sugarcane Cropping in a Southern Philippine Province

HOWARTH BOUIS

LAWRENCE J. HADDAD

Introduction

This chapter summarizes findings on the nutrition effects of commercialization of agriculture in Bukidnon Province on the southern island of Mindanao in the Philippines. This region was engaged primarily in semisubsistence corn production until the establishment of a sugar mill led to a rapid expansion of sugarcane production in the area. The Bukidnon Sugar Company (BUSCO), which began operations in 1977, was established in response to high world sugar prices. Cane production was sufficiently profitable to generate a high demand for contracts by nearby farmers, so the mill's capacity was expanded in 1981.

Research Design

This study compares landowners, tenants, and landless laborers. Bias due to sugar adopter self-selection was a consideration in selecting a sample. Therefore, in the hope of obtaining roughly comparable adopting and nonadopting groups, the survey area was extended beyond the vicinity of the mill to include households that did not have the opportunity to adopt sugar (due to prohibitive costs of transporting the sugarcane to the mill) but shared a common growing environment and cultural heritage with sugar-adopting households.

Survey work began in August 1984 and included four rounds of data collection that ended in August 1985; 510 households were included in the study.[1] Households were divided into different crop-tenancy groups for both corn and sugar: owners, owners/share tenants, share tenants, and laborers. Furthermore, the sample includes corn growers who rent

---

1. For more detail on research design and sampling methods, see Bouis and Haddad (1990).

land on fixed-rate or other types of arrangements and households with other occupations.

A Comparison of Corn and Sugar Production Systems

From the post-World War II period until the early 1970s, Bukidnon experienced heavy immigration. New settlers typically homesteaded recently cleared rain forest. During the past two decades, corn yields have fallen dramatically, averaging declines of 50 percent over a 13-year period, because of reduced soil fertility in the former rain forest areas.

The main corn harvests occur in July and December. The average growing cycle of corn from plowing to harvest is 3.3 months, so there is ample time for three crops per year. However, production of a third crop depends on adequate rainfall at the onset of the relatively dry months after December. Thus, most households produce two corn crops a year, with a few households producing three. Average yields are highest for the first crop (0.9 tons of shelled corn per hectare) and fall by 25 percent for the second crop.

Of the 32.5 days of family labor inputs per hectare per corn crop observed for the total sample, about 15 percent is provided by women (typically the wife of the household head) and 25 percent by children. Participation of children in the hired labor market is very low. Women provide a quarter of hired labor inputs. Allocation of labor is task specific. Men do almost all of the work for tasks needing a water buffalo. The remaining tasks are shared by the husband, wife, and, to the extent applicable, several children.

All households that adopted sugar production (sugar households) continued to produce some corn, a situation quite similar to the Guatemalan case just reported (chapter 12). Per capita consumption of own-produced corn by sugar households, about 1.3 kilograms per capita per week, was much lower than the 2.0 kilograms per capita consumed by corn households, despite the fact that sugar households produced sufficient corn to have consumption levels equal to those of corn households. Sugar households preferred to purchase rice in place of corn. For corn households, per capita consumption of home-produced corn on small farms was only marginally less than on large farms, suggesting that households kept what they needed for home consumption and sold the remainder.

Average net returns to corn production, computed on a variable cost basis, were 1,023 pesos per hectare per crop (about US$51), partly in cash and partly in the form of own-produced corn consumption. Share tenants have lower returns than this average since the share paid to their landlords is subtracted.

The sugar milling season begins in late October and finishes in late July. The average length of the growing period for sugar is, on average, twelve months. Sugar production in the area generally involves a plant crop and ratoons.[2] There are substantially higher input costs for the plant crop than for the successive ratoons. The average sugarcane yields are nearly identical across tenure groups. In addition, yields do not decline with successive ratoons because fertilizer inputs increase with higher-numbered ratoons. Average net returns for sugarcane of 4,570 pesos per hectare are well above those for corn, 1,023 pesos. Economic returns to corn and sugar production are examined more closely in the following subsection.

Profits and Labor Allocation Patterns

Net profits for sugar production (which reflect a value for family labor inputs) as compared with corn production are even higher than are net returns expressed on a variable cost basis, since family inputs are greater for corn. Negative profits, on average, for smallholder corn households indicate that these households would have done better to have rented out their land (and entered the labor market) rather than to have undertaken corn production. This is not to say that farmer behavior is inconsistent with utility maximization. Yields and prices obviously cannot be predicted with complete accuracy and some value may be attached to working for oneself rather than for someone else. The low average value (a net profit of only 93 pesos per hectare) for the total sample indicates that a marginal activity corn production has become over time for smallholders.

Corn and sugar production use almost identical amounts of total labor per hectare per year (approximately 109 days). However, the mix of family and hired labor, and of men's, women's, and children's labor are quite different. Family labor accounts for two-thirds of total labor inputs for corn production while hired labor accounts for one-third. These fractions are reversed in the case of sugar production. Women's participation in own-farm production declines dramatically with a switch from corn to sugar production, from 12.4 days per hectare for corn to only 2.7 days per hectare for sugar.

Food Expenditures and Calorie Intakes

Food budget share declines and household calorie consumption increases with incomes (table 13.1). Incomes of sugar households are

---

2. Ratoon crop, in the case of sugarcane, grows without replanting in the season after a harvest.

TABLE 13.1 Income, expenditures, and calorie availability and intake, by income and expenditure quintile and crop-tenancy group, Bukidnon, the Philippines, 1984–85

Group	Per Capita Income	Total Expenditures (pesos/week)[d]	Family Food Expenditures	Food Budget Share[a] (percent)	Calorie Availability[b] (per day)	Calorie Intake[c]
Income quintile[a]						
1	13.1	30.0	24.0	80	2,170	2,266
2	21.9	36.6	26.6	73	2,237	2,313
3	29.8	39.7	28.5	72	2,321	2,336
4	41.4	48.1	33.3	69	2,639	2,433
5	101.7	76.2	43.2	57	2,826	2,443
All	41.7	46.2	31.1	67	2,439	2,358
Expenditure quintile[e]						
1	21.9	21.8	17.2	79	1,790	2,108
2	25.4	29.8	23.4	79	2,143	2,288
3	28.5	38.0	28.8	76	2,411	2,384
4	45.8	50.0	34.8	70	2,666	2,439
5	87.6	91.9	51.7	56	3,193	2,575
All	41.7	46.2	31.1	67	2,429	2,358

(*continued*)

TABLE 13.1 Income, expenditures, and calorie availability and intake, by income and expenditure quintile and crop-tenancy group, Bukidnon, the Philippines, 1984–85 (Continued)

Group	Per Capita Income	Total Expenditures (pesos/week)[d]	Family Food Expenditures	Food Budget Share[a] (percent)	Calorie Availability[b] (per day)	Calorie Intake[c]
Crop-tenancy group						
Corn						
Owners	35.2	41.4	29.3	71	2,375	2,372
Owners/tenants	47.7	49.2	32.7	66	2,445	2,387
Share tenants	46.4	46.6	30.4	65	2,368	2,329
Laborers	28.4	40.0	29.6	74	2,405	2,412
	26.6	32.3	24.9	77	2,266	2,326
Sugar						
Owners	52.2	53.5	33.7	63	2,534	2,343
Owners/renters	70.1	64.4	39.3	61	2,655	2,386
Renters	83.3	89.9	47.5	53	3,148	2,447
Laborers	43.0	43.5	30.0	69	2,350	2,371
	26.5	30.8	24.0	78	2,208	2,237

SOURCE: Bouis and Haddad (1990).

[a] Food budget share = family food expenditures/total expenditures.
[b] Per adult-equivalent, derived from food expenditures.
[c] Derived from 24-hour recall of foods consumed.
[d] 1984 pesos.
[e] Quintile 1 is the lowest rank and 5, the highest.

higher than of corn households when similar tenancy groups are compared. Calories purchased per peso decline with increasing income; higher income households seek more variety in what they eat, and thus staples occupy a smaller proportion of food expenditures. The food expenditure data indicate that the highest expenditure quintile households spend 60 percent more than the lowest expenditure quintile households to purchase an equal amount of calories. Overwhelmingly, it is meat consumption that increases dramatically with increased income and which accounts for the increased calorie costs at higher income levels.

Regression analysis for the relationship between calorie intakes and food expenditures shows that household calorie intakes increase significantly as food expenditures increase. At mean food expenditure levels, each extra peso spent on food increased household calorie intake (per adult-equivalent) by about 90 calories at the margin for the sample as a whole as compared with more than 400 calories purchased on average by a peso spent on food. Household calorie intake elasticities with respect to food expenditures are estimated to be 0.17 for the sample as a whole, and 0.15 and 0.21 for corn and sugar households, respectively.

When the caloric intake of preschoolers who are not breast-fed was assessed, it was found that they consumed an average of 75 percent of their caloric requirements compared to adults, who were consuming in excess of 100 percent of energy requirements. Preschoolers in corn households consumed more calories (76 percent of requirements) than preschoolers in sugar households (72 percent of requirements). While the difference between the two groups is not large, it is surprising, given the higher incomes in sugar households.

Table 13.2 presents the regression estimates for preschooler calorie intakes as a function of household calorie intakes. For the total sample and the two subsamples, household calorie intake was a positive and significant determinant of preschooler caloric intake, as shown by regression analysis. At the margin, calories were distributed more or less equally among household members, increasing the share of household calories going to preschoolers who consume below the average household levels. For the total sample as well as for the corn and sugar household subsamples, the estimated preschooler calorie intake elasticities with respect to household calorie intakes are 1.18, 1.20, and 1.17, respectively.

Nutritional Status of Preschool Children

Table 13.3 presents Z-scores for height-for-age (ZHA), weight-for-age (ZWA), and weight-for-height (ZWH) for all preschool children

TABLE 13.2 Regression results for preschooler calorie intakes as a function of household calorie intakes

Explanatory Variable[a]	All Households		Corn Households		Sugar Households	
	Parameter	t-Value	Parameter	t-Value	Parameter	t-Value
HCALAEQ	0.82028	30.86*	0.84577	25.52*	0.77578	16.68*
RATIOPAR	−1,483.48814	−14.01*	−1,588.46312	−11.22*	−1,320.44308	−7.57*
MOTHED	16.79655	2.26*	17.66845	1.93	5.11738	0.35
MOTHAGE	1.03590	3.37*	1.02134	2.68*	0.79280	1.46
NUTRSC1	−3.42859	−0.55	−13.81724	−1.86	25.15519	2.11*
CHILDCRE	0.32042	1.63	0.51105	2.04*	0.05842	0.18
SICK	−81.70871	−1.90	−137.87299	−2.57*	38.38262	0.51
SEX	65.98609	1.73	45.64234	0.96	72.80922	1.07
ACCAGE	43.84393	4.84*	39.98655	3.27*	35.74011	2.45*
AGESQ	−0.54548	−4.56*	−0.52775	−3.32*	−0.35900	−1.81
ADEQVHH	49.24556	3.61*	60.88718	3.35*	34.94330	1.62
RD1	−2.93130	−0.06	−47.44410	−0.71	60.91730	0.65
RD2	−65.62428	−1.23	−94.31894	−1.39	−13.65748	−0.15
RD3	−110.98226	−2.08*	−110.95572	−1.65	−106.97079	−1.17
Constant	263.42399	1.12	530.87080	1.69	98.99475	0.26
R^2	0.769		0.792		0.567	

F-value	99.08	68.91	30.90
N	975	587	345
Preschooler calorie intake elasticity with respect to household calorie intake	1.18	1.20	1.17

SOURCE: Bouis and Haddad (1990).

NOTE: The dependent variable is preschooler calorie intake per adult-equivalent per day.

^aDefinitions of variables:

HCALAEQ = household calorie intake per adult-equivalent per day.
RATIOPAR = ratio of average of father's and mother's calorie intake per adult-equivalent over the household calorie intake per adult-equivalent.
MOTHED = years of formal education of the mother.
MOTHAGE = mother's age in months.
NUTRSC1 = measure of nutritional knowledge of the mother.
CHILDCRE = minutes spent by mother in child care in previous 24 hours.
SICK = zero-one dummy for reporting sickness by mother in previous two weeks.
SEX = 0 = female, 1 = male.
ACCAGE = age of preschooler in months.
AGESQ = (age of preschooler in months)^2.
ADEQVHH = number of household members expressed in adult-equivalents.
RD1, RD2, RD3 = zero-one dummy variables for round.

*Significant at 5 percent level.

TABLE 13.3 Z-scores for height-for-age, weight-for-age, and weight-for-height for preschool children, by age and expenditure quintile, Bukidnon, the Philippines, 1984–85

Expenditure Quintile[a]	Age of Preschool Children in Years					
	0	1	2	3	4	0–4
	Height-for-Age Z-Scores					
1	−2.08	−2.75	−2.62	−2.44	−2.69	−2.57
2	−1.24	−2.37	−2.26	−2.30	−2.46	−2.22
3	−1.20	−2.03	−2.04	−2.13	−2.17	−2.02
4	−0.91	−1.97	−1.86	−2.28	−2.30	−2.02
5	−0.82	−1.88	−1.76	−1.94	−1.91	−1.80
All	−1.31	−2.24	−2.15	−2.24	−2.34	−2.16
	Weight-for-Age Z-Scores					
1	−1.82	−2.15	−1.77	−1.53	−1.61	−1.75
2	−0.90	−2.06	−1.69	−1.62	−1.52	−1.62
3	−1.24	−1.76	−1.45	−1.42	−1.39	−1.47
4	−1.44	−1.71	−1.47	−1.45	−1.41	−1.49
5	−0.86	−1.60	−1.39	−1.30	−1.33	−1.35
All	−1.25	−1.88	−1.57	−1.48	−1.46	−1.55
	Weight-for-Height Z-Scores					
1	−0.47	−0.81	−0.73	−0.48	−0.50	−0.60
2	−0.07	−1.06	−0.82	−0.61	−0.40	−0.64
3	−0.46	−0.83	−0.62	−0.45	−0.40	−0.55
4	−0.92	−0.81	−0.77	−0.46	−0.51	−0.65
5	−0.47	−0.80	−0.66	−0.38	−0.42	−0.54
All	−0.43	−0.87	−0.72	−0.48	−0.45	−0.60

SOURCE: Bouis and Haddad (1990).

NOTE: The heights and weights of preschoolers were measured in each round so that Z-scores for any one preschooler are typically included in the mean calculations for two columns. U.S. National Center for Health Statistics (NCHS) standards were used. The Food and Nutrition Research Institute (FNRI) in the Philippines has recently issued a set of reference values based on a national sample of apparently healthy Filipino children. Healthy Filipino children are close to the NCHS standard during the first half of infancy, gradually deviating from it as age advances. It may then be expected that Z-scores based on the NCHS standards gradually decline with age.

[a] Quintile 1 is the lowest rank and 5, the highest.

disaggregated by expenditure quintile and by age. The ZHA scores for preschoolers less than one year in age indicated a very strong association between height and income. This pattern is probably a reflection of better maternal nutrition in high-income groups during pregnancy and breast-feeding (see Bouis and Kennedy 1989).

As age increases and children are weaned, ZHA scores for all expenditure quintiles decline. However, ZHA scores decline more rapidly for higher-income quintiles so that by the age of four, heights of higher-income children are only marginally better than those of lower-income children. There appears to be little association between income and weight-for-height. ZWA scores show a pattern that is a mix of the patterns for the ZHA and ZWH scores.

Higher-income households (sugar owners and sugar owner/renter households) have significantly taller preschoolers at the outset (age tercile 1). However, after starting out significantly taller, these children are shorter on average—although not significantly—than their corn household counterparts by the time they reach the third age tercile. Children of corn laborers and sugar laborers in the oldest age tercile are significantly more stunted than children in any of the other six crop-tenancy groups; this result is not surprising given the low incomes of these households.

The total length of time that preschoolers are ill does not vary significantly by income group or by crop tenancy group—averaging about 32 percent. However, diarrhea and fever occur 25 percent more frequently in children from sugar households than from corn households.

A growth model using weight-for-height to measure children's nutritional status is presented in table 13.4. Preschooler caloric intakes are a positive and significant determinant of weight-for-height for the whole sample as well as for the corn and sugar household subsamples. While the magnitudes of Z-score elasticities are sensitive to population means, which can approach zero, the estimated elasticities calculated from the coefficients on preschooler calorie intakes are 0.39, 0.34, and 0.57 for the whole sample and the corn and the sugar household subsamples, respectively. Greater calorie intakes mean better nutrition in the short run, and presumably in the long run also if these intakes can be sustained over longer periods. Morbidity is negatively and significantly associated with short-run nutritional status for the whole sample and for sugar households.

Conclusions

Production Policies

Corn productivity in the study area is low. In many instances corn tenants would realize higher incomes by working as agricultural laborers

TABLE 13.4 Regression results for relationships between preschoolers' calorie intake and weight-for-height

Explanatory Variable[a]	All Households		Corn Households		Sugar Households	
	Parameter	t-Value	Parameter	t-Value	Parameter	t-Value
PCALAEQ	1.0775×10^{-04}	3.90*	9.6935×10^{-05}	2.62*	1.5494×10^{-04}	3.05*
DIARR	−0.52031	−3.62*	−0.17917	−0.90	−0.79902	−3.59*
FEVER	−0.22818	−3.44*	−0.06216	−0.68	−0.33448	−3.04*
MNTHBFED	2.4400×10^{-03}	0.77	-5.844×10^{-03}	−1.28	0.01162	2.28*
BRTHSP1	-4.406×10^{-03}	−3.62*	-2.851×10^{-03}	−1.89	-7.761×10^{-03}	−2.84*
MOTHED	0.02909	2.50*	0.02247	1.42	0.05561	2.77*
FATHED	−0.01673	−1.66	−0.03022	−2.02*	-2.062×10^{-03}	−0.13
MOTHAGE	1.4876×10^{-03}	4.45*	1.1664×10^{-03}	2.73*	1.6346×10^{-03}	2.64*
NUTRSC1	-5.475×10^{-03}	−0.65	-4.256×10^{-03}	−0.40	−0.01784	−1.16
CHILDCRE	-4.263×10^{-04}	−1.72	-3.577×10^{-04}	−1.10	-7.576×10^{-04}	−1.77
SEX	−0.15879	−3.13*	−0.11764	−1.73	−0.26390	−2.96*
HTFATH	-2.294×10^{-03}	−0.53	2.8451×10^{-03}	0.45	-8.935×10^{-03}	−1.28
HTMOTH	0.01455	3.97*	0.01453	3.60*	8.4126×10^{-03}	0.90
RD1	−0.09839	−1.40	−0.07415	−0.80	−0.04898	−0.40
RD2	−0.14613	−2.06*	−0.16582	−1.75	−0.07918	−0.67
RD3	−0.06586	−0.91	−0.09738	−1.02	0.08283	0.68
Constant	−2.80493	−3.17	−3.41129	−2.82	−1.06934	−0.65

R^2	0.115	0.089	0.231
F-value	7.91	3.07	6.20
N	995	522	347

SOURCE: Bouis and Haddad (1990).

NOTE: The dependent variable is weight-for-height.

[a] Definitions of variables:

PCALAEQ = preschooler calorie intake per adult-equivalent per day.
DIARR = zero-one dummy for diarrhea being reported in past two weeks.
FEVER = zero-one dummy for fever being reported in past two weeks.
MNTHBFED = months that child was breast-fed before stopping.
BRTHSP1 = months between births of present and previous child.
MOTHED = years of formal education of the mother.
FATHED = years of formal education of the father.
MOTHAGE = age of the mother in months.
NUTSCR1 = measure of the nutritional knowledge of the mother.
CHILDCRE = minutes spent in child care in previous 24 hours.
SEX = 0 = female, 1 = male.
HTFATH = height of the father in centimeters.
HTMOTH = height of the mother in centimeters.
RD1, RD2, RD3 = zero-one dummy variables for round.

*Significant at 5 percent level.

(assuming employment were available) rather than on their farms. Declining soil fertility is a major reason for low corn yields. There would appear to be high returns to developing low-cost technologies for improving soil fertility and to investing in extension programs for disseminating these technologies to farmers.

Sugar production in Bukidnon is more profitable than corn production. Smallholders who kept their land were able to raise their incomes by switching from corn to sugar production. However, average returns of 4,500 pesos per hectare per year (US$225 on a variable cost basis without valuing family labor and without subtracting interest on loans) are not high in an absolute sense, considering the risks involved and the amount of capital invested.

Nutrition Policy for Agricultural Households

Raising household incomes appears to be a necessary, but not a sufficient, condition for substantially improving preschooler nutrition. Calorie intakes of preschoolers are positively and significantly related to their nutritional status. Yet higher-income households choose to purchase nonfood items and higher-priced calories, while preschoolers continue to consume well below recommended intakes. Improvement of preschooler calorie intakes, however, is not a sufficient condition for substantially improving their nutritional status, because of the high prevalence of sickness, even among high-income groups. Reducing illness may involve both education and improvement of community-level health and sanitary conditions.

Winners and Losers

There are a substantial number of households that are "losers" in the process of agricultural commercialization in Mindanao. Households that lost access to land, as corn tenant farms were consolidated into larger operational units for the production of sugar, have suffered severe declines in income. In addition, preschoolers in these households are significantly more stunted by the age of four than children from households that maintained their access to land and continued to grow corn. Winners in this agricultural commercialization process are the landowners who switched to sugar.

What were the fundamental factors that caused the apparent deterioration in land tenure patterns and in income distribution? First, declining corn productivity constituted a force, in a sense "pushing" landowners and tenants off the land. Returns to household labor and other inputs were quite low, making the decision to leave the land much easier in times of difficulty. Second, the greater ability of the larger farm households to bear risk, their better access to credit facilities, and their generally higher education, know-how, and access to important political and social institutions put them in a much better position to take advantage of new agricultural production technologies when they became available.

Development Strategy for Mindanao Agriculture

The introduction of sugar in southern Bukidnon provides some important lessons for how not to pursue development of export cropping in Mindanao in the future. From the outset, most of the milling capacity of BUSCO was met by sugarcane produced on large-scale haciendas. Small-scale producers entered the scheme relatively late, and in a marginal way in terms of the total mill output. About half of the small-scale sugar producers surveyed had no grower contracts with the mill and so had to negotiate individual arrangements with those who had contracts, reducing their income from sugar.

What should be done instead? To prevent further deterioration in land distributed patterns, first, smallholder corn productivity needs to be improved. Both open-pollinated and hybrid varieties are available, but typically only larger landowners in Bukidnon are experimenting with the new corn technologies. Second, the government needs to make a conscious effort to develop the inevitable expansion of export cropping in Mindanao on a smallholder basis. This involves reducing the barriers to entry by providing smallholders with credit and information through extension and by actively promoting their access to processing and marketing facilities where necessary. Although such characteristics unfortunately do not always coincide with high output prices and low production costs, *ceteris paribus*, the government should seek to promote export crops that are labor intensive, have diseconomies of scale in production, have low transportation costs in marketing, and can be stored for relatively long periods after harvesting.

14 The Effect of a Short-Lived Plantation on Income, Consumption, and Nutrition: An Example from Papua New Guinea

JOHN R. McCOMB
M. P. FINLAYSON
J. BRIAN HARDAKER
PETER F. HEYWOOD

Agricultural Commercialization and Nutrition Problems in the Study Setting

This case study captures the story of a short-lived, temporarily successful, yet unsustainable commercialization project. Many such commercialization projects have surfaced and disappeared soon after in low-income countries (see chapter 9), and little is known about their effects on people's welfare, including children's nutrition.

Papua New Guinea is well endowed with natural resources relative to its population. In 1980, the population was 3.0 million and was projected to reach 4.0 million by 1991. Agriculture is the most important sector of the economy, supporting, directly or indirectly, about 85 percent of the population and, according to the 1993 World Bank *World Development Report*, accounting for about 26 percent of the GDP and 35 percent of exports. At the time of the 1980 census, over 90 percent of rural households and over 80 percent of all households were dependent on small-scale farming systems, which combine subsistence food production with variable amounts of cash cropping (DPI 1984).

The main food staples are root crops (sweet potato, taro, and yams) and sago, with bananas being an important co-staple in some areas and a secondary staple in others. The value of subsistence production is not well documented, but recent estimates indicate that subsistence food cultivation accounts for approximately half the total value of agricultural production (World Bank 1988b).

The extent and type of involvement of the rural population in cash

---

Based on work done by the authors for a joint project by the Papua New Guinea Institute of Medical Research, Madang, and the Department of Agricultural Economics and Business Management, University of New England, Armidale, Australia. The project was originally funded by the Australian Centre for International Agricultural Research (ACIAR). ACIAR also contributed funds for the preparation of this chapter.

cropping varies with the region of the country, itself a determinant of the nature and length of contact with Europeans, the cash crops produced, and the mode of production. The most important cash crops are coffee, cocoa, coconuts, oil palm, tea, and rubber. Cardamom is also produced now. These cash crops are produced mainly for export; together, they earned 80 percent of Papua New Guinea's nonmining export income and 33 percent of all export income in 1986 (World Bank 1988b). In the early 1970s, plantations provided three-quarters of total production, but by the mid-1980s smallholders accounted for 70 percent of total production. The majority of the smallholders are based in villages, but there are increasing numbers located on land settlement schemes, group-owned blocks, and subdivided plantations.

Malnutrition is a significant health problem in Papua New Guinea. Estimates from the 1982–83 National Nutrition Survey indicate that approximately 38 percent of children under five years of age were below 80 percent of the Harvard standard for weight-for-age (Heywood, Singleton, and Ross 1988). The general growth pattern is of a rapid decline in weight-for-age during the first year of life and a continuing but much slower decline over the next four years, at the end of which approximately half the children are below 80 percent of the standard (Heywood 1983).

Debate has been vigorous, but inconclusive, about the nutrition effects of increased participation by rural households in Papua New Guinea in the cash economy through shifts from subsistence crops to cash crops (Lambert 1979; Hide 1980; Harvey and Heywood 1983). Cash cropping has been associated with increased food imports, leading to concerns about food dependency and national and local food security. Increases in cash incomes have led to increased supplementation of traditional diets with imported foods, chiefly canned fish and rice.

The Cardamom Plantation at Karimui

The Karimui census division is in the southern part of the mountainous Simbu Province. The area is remote, accessible only by small aircraft or on foot. About 5,000 people live in dispersed hamlets on the plateau of an extinct volcano, about 1,100 meters above sea level. Almost all households grow subsistence crops, chiefly sweet potatoes, in a slash-and-burn farming system. Most households also own small plots of cash crops. The main smallholder export cash crops are coffee, cardamom, and chillies. Peanuts and other crops grown in excess of family needs may also be sold locally.

Smallholder cardamom production in Karimui commenced in the early 1970s. Production remained at a low and fluctuating level through the 1970s, due to management and marketing difficulties. In 1978, local smallholders formed the Karimui Cardamom Growers Association (KCGA), with the immediate aim of improving marketing and a long-term aim of establishing a nucleus plantation to dry and market smallholder production. In 1981, a new company, Karimui Spice Company (KSC), 70 percent of whose equity was owned by the KCGA and 30 percent by the provincial government, was granted lease over 230 hectares (later increased to 388 hectares) to establish a plantation. By 1985, 300 hectares had been planted. The even distribution of rainfall, combined with high expenditure on agricultural inputs, resulted in very high yields (over 300 kilograms of dry cardamom per hectare per year), compared to the major producer, India, where average yields are approximately 120 kilograms per hectare per year.[1]

The plantation provided an important source of paid employment to men for clearing forest, planting, and maintenance; and to men, women, and children for harvesting. By mid-1987, the KSC employed about 260 male laborers for maintenance and grading and about 300 female pickers. Picking was on a casual employment basis, with a high turnover of pickers from one picking fortnight to the next. Since labor was abundant and people (especially women) were eager to work, the KSC was usually assured of sufficient pickers to harvest the crop. The flexible picking program allowed women to work on the days convenient to them.

Despite increasing cardamom production (figure 14.1), high costs resulted in capital shortages for the KSC. The capital constraint that had prevailed throughout the development stage of the project was exacerbated by a drought that reduced production in 1987. The situation reached a crisis point in the latter part of that year with banks and agricultural produce suppliers demanding to be paid. In November and December 1987, laborers' wages were reduced (figure 14.2), most laborers were laid off, and maintenance was reduced to a minimum for several months while graders processed the existing stock. Low prices and poor quality of the cardamom, combined with destruction of the new cardamom-curing facility by fire, prevented the company from solving its financial problems. A second consecutive dry year (1988) resulted in low production, making financial management of the plantation difficult. By October 1989, the plantation and facilities were closed.

1. The major producers of cardamom are India and Guatemala, each contributing about one-third of the annual world production of 10,000 tons. Cardamom is largely consumed in the Middle East, South Asia, and, to a lesser extent, Europe. Cardamom price is variable and determined by production levels and quality of cardamom in the two main producer countries, which, in turn, are affected by seasonal growing conditions. Lower grades of cardamom are severely penalized on the world market.

FIGURE 14.1 Cardamom production and wages at the Karimui Spice Company, July 1981 to June 1988

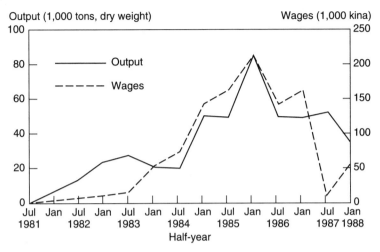

Source: Finlayson et al. (1991).

FIGURE 14.2 Total wages paid and daily wage rates at Karimui Spice Company, August 1987 to July 1988

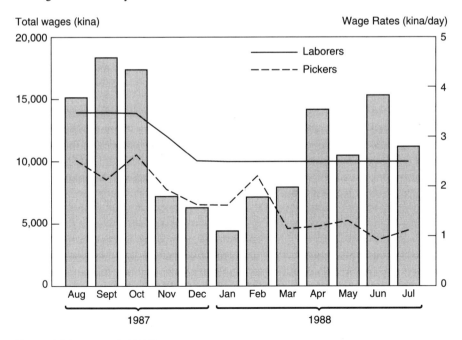

Source: Finlayson et al. (1991).

This chapter evaluates the effects of wage incomes at Karimui on the nutritional status of young children on the basis of a detailed one-year study conducted between July 1987 and August 1988, when the KSC was already generally in decline. Thus, this case study captures commercialization-nutrition linkages under adverse circumstances.

Research Design and Analytical Framework

The settlement patterns in the area, the distances between villages, the lack of vehicular access, and the necessarily high intensity of data collection dictated selection of households in clusters. Three clusters were chosen to represent a range of distances from the cardamom plantation and, hence, varying degrees of involvement in the cash economy.[2]

Initial selection and stratification of households were based partly on household wage employment status. The most significant factor affecting cash incomes during the course of the study was KSC's changing economic situation. By January 1988, almost all KSC laborers had been laid off, and picking had ceased. Alternative opportunities for wage employment were limited. When the company recommenced operations, wage rates for "permanent" (male) laborers were reduced from K3.50 per day, which had been paid during the early part of the study, to K2.50 per day. Rates for pickers (mainly women) were maintained at K0.25 per kilogram, but lower cardamom yields, which resulted from lower rainfall and also, perhaps, from changes in management practices, meant generally lower rates of pay per day by the end of the study period.

In much of the analysis below, the results are presented separately for two consecutive periods:

- The first period, from July 1987 to January 1988, which corresponded to the time when employment and wage income was falling;
- The second period, from February 1988 to August 1988, which corresponded to the time when employment increased temporarily and then fell once more.

Sample Household Characteristics

For the purpose of this analysis, households were grouped into wage-earning and non-wage-earning categories in each survey period. Households earning more than K6.00 per fortnight in wages, or spending more than 10 percent of surveyed days in wage employment, were

---

2. Details of survey and study design are in Finlayson et al. (1991).

classified as wage households; all others were classified as nonwage households. There were 34 nonwage and 53 wage households in the first survey period and 38 nonwage and 49 wage households in the second survey period. Fourteen wage households from the first period became nonwage households in the second period, and 10 nonwage households from the first period became wage households in the second period.

The size of surveyed households varied from 3 to 21 persons. To control for variation in household size and composition, the analysis was carried out on an adult-equivalent basis (Finlayson et al. 1991). There were small differences between nonwage and wage groups in the age of household head, size of household (adult-equivalents), and educational standard of the household head (grade), but they were not found to be significant.

Ownership of land is traditionally by clans. The right of a household to use of land is established by clearing virgin forest, and subsequent rights to use that land are usually passed on patrilineally. Land was not a scarce resource in the sample villages, although utilization of available land appeared to be more difficult as distance from villages to new land increased.

Households obtained food from both cultivated and uncultivated land. The size of cultivated food gardens varied considerably between households, as did the type of crops grown. Food trees were grown in abandoned food gardens or forest. Wage households had lower average total garden areas than nonwage households in both periods, but the only differences between wage and nonwage households that were statistically significant were those for coffee garden areas ($p < 0.05$) and total garden areas ($p < 0.01$), in the second period (table 14.1).

Coffee, cardamom, and chillies are perennial crops grown for sale. Eighty-six percent of households had coffee gardens, and 31 percent had cardamom gardens. The average areas were small—for both these cash crops, the average area across all surveyed households was 0.27 hectares.

The division of labor in domestic agricultural activities remains pronounced at Karimui. Men perform tasks such as cutting forest and making fences and houses, while women (including teenagers and young girls) are responsible for burning and clearing the undergrowth and planting, maintaining, and harvesting food gardens. Females cook, take care of children, and perform other domestic activities more often than males. Females are involved in wage employment as frequently as males and, overall, perform production and income-generating activities more frequently than males (Finlayson et al. 1991).

The sample households lacked material possessions, with few capital assets other than houses, land, and crops, indicating the limited extent to which the economy has changed from subsistence to an emergent cash

TABLE 14.1 Selected demographic features and crop areas of nonwage and wage households, Papua New Guinea, 1987–88

Feature	Nonwage Households		Wage Households	
	Period 1[a]	Period 2[b]	Period 1[a]	Period 2[b]
Number of households	34	38	53	49
Household wage (kina/fortnight)[c]	1.71	1.53	15.50	16.48
Age of household head (years)	39.3	37.9	38.0	38.9
Education of household head (years)	1.4	1.5	1.9	1.8
Household size (adult-equivalent)	4.8	4.6	5.2	5.0
Number of food trees	13.1	12.7	12.2	12.5
Food garden area (square meters/adult-equivalent)	803	769	671	646
Coffee area (square meters/adult-equivalent)	650	674*	317	271*
Cardamom area (square meters/adult-equivalent)	239	318	124	53
Total area (square meters/adult-equivalent)	1,692	1.761**	1,112	969**

SOURCE: Surveys undertaken by the authors.
[a]July 1987–January 1988.
[b]February 1988–August 1988.
[c]K1.00 = US$1.1256, averaged over the duration of the study.
*Significant difference at the 0.05 level.
**Significant difference at the 0.01 level.

economy. Almost all households owned their own house (built from local materials) and, on average, had two axes, two to three bush knives, and a spade—tools important for food production. Only a few households owned tools for maintenance of cash crops such as pruning saws, coffee pulpers, or secateurs.

Household Income Sources and Role of Wages

Average net income levels[3] per adult-equivalent per fortnight for wage and nonwage households were similar (table 14.2). Subsistence income for nonwage households was significantly higher than for wage households in the second survey period, with an increase in subsistence food production being the main contributor to the difference. Food gifts received by households significantly increased in the second period for

---

3. Net income is defined as follows: (subsistence food production + food gifts received + animal sales + cash crop sales + market sales + wages + trade store receipts + cash gifts/traditional exchange received) − (food gifts given + farm inputs + animal costs + trade store inventory purchases + cash gifts/traditional exchange given).

TABLE 14.2 Household income and expenditure of nonwage and wage households, Papua New Guinea, 1987–88

	Nonwage Households		Wage Households	
Income and Expenditure	Period 1[a]	Period 2[b]	Period 1[a]	Period 2[b]
	(kina per adult-equivalent per fortnight)			
Income				
Subsistence income	6.69[c]	9.81	6.12	7.20[c]
Subsistence production	5.91	6.92	5.53	4.86[d]
Food gifts received	0.78[c]	2.89	0.59	2.34
Farm production	1.04	0.81	1.32	0.88
Farm (animals/cash crops)	0.55	0.33	0.73	0.50
Market (subsistence surplus)	0.49	0.47	0.59	0.38
Off-farm income	4.22[c]	2.07	7.33[c]	5.73[c]
Wages	0.49	0.39	3.80[c]	3.67[e]
Karimui Spice Company wages	0.31	0.31	2.46	2.36
Other wages	0.18	0.08	1.32	1.31
Cash gifts received	3.02[d]	1.17	2.93	2.02
Store receipts	0.71	0.51	0.60	0.04
Total income	10.91	11.88	13.45	12.93
Expenses				
Inputs	0.05	0.17	0.27	0.39
Animals	0.10	0.03	0.12	0.03
Store inventory purchases	0.90	0.48	0.85	0.14
Cash given away	1.78	0.67	3.01[c]	2.14[d]
Food gifts given	0.24	0.50	1.01[c]	0.29
Total expenses	3.07	1.85	5.26	2.99
Net income	7.84	10.03	8.19	9.94

SOURCE: Surveys undertaken by the authors.
[a] July 1987–January 1988.
[b] February 1988–August 1988.
[c] Significant difference at the 0.05 level, same group.
[d] Significant difference at the 0.01 level, between groups.
[e] Significant difference at the 0.001 level, between groups.

both wage and nonwage households. The value of farm production for both wage and nonwage households fell during the latter part of the study, but the difference between the two survey periods and between the two groups of households was not significant. Farm inputs for wage households were larger, on average, than for nonwage households, while expenditures for the purchase of animals and store inventory were similar for both. In the second period, there was a reduction in the average amount spent on animals and store inventory.

As expected, wage households received significantly more off-farm income than nonwage households in both periods. Nonwage households earned significantly less off-farm income in the second period than in the first, mainly due to a reduction in receipts of cash gifts, reflecting the contraction of the cash economy. Although fortnightly wage earnings from KSC fell sharply during the first period, then recovered somewhat and fell again late in the second period, average wages received were not significantly different between the two periods for either the wage or nonwage groups. Thus, the averages hide to a large extent the considerable fluctuation in wage incomes that many households suffered over the full period of the study, and they particularly hide the fall in wage incomes during the first six months.

The fall in wage rates from KSC coincided with low household sales of coffee and cardamom. Coffee sales were reduced, in part, because of the discovery of coffee rust in the region. Yields were reduced by the rust itself, but also by coffee renovation practices that included stumping of existing plants to combat rust and permit new growth to occur.

The relatively high values of food and monetary transfers shown in table 14.2 indicate that gifts and exchanges are important components of the Karimui socioeconomic network. The differences between food gifts received and food gifts given may be due, in part, to some underreporting of food given away before returning to the house, where most of the recording was done. Consequently, subsistence production may be undervalued, but the estimates of subsistence consumption should not be affected.

Differences in the ability or willingness to be employed contributed to considerable inequality of cash incomes between sample households. Comparing the distributions of wage and nonwage households by income tercile and survey period shows that there was a higher percentage of wage households than nonwage households in the upper income tercile, but the difference was not significant for either period, or overall, using a chi-square test at $p < 0.10$ level of significance.

Females contributed 86 percent of the total value of subsistence food and accounted for 53 percent of total income (cash and subsistence) (table 14.3). These percentages exclude the contribution males make in establishing food gardens by felling trees and fencing the area.

Household Expenditure and Energy Availability Effects

Household expenditure is defined net of the costs and transfers already deducted in the calculation of net income (see table 14.2). On

TABLE 14.3 Mean cash and subsistence income received by males and females over ten years of age, by source of income, Papua New Guinea, 1987–88

Income Source	Males[a]	Females[b]
	(kina/capita/fortnight)	
Wages	3.82	1.60
Farm sales	1.55	0.95
Cash gifts	5.32	0.73
Store receipts	1.20	0.01
Total cash receipts	11.89	3.29
Value of subsistence food	2.07	12.49
Total	13.96	15.78

SOURCE: Adapted from Finlayson et al. 1991.
NOTE: Average across all sample households for the full recording period (26 fortnights)
[a] 1.64 males over ten years of age per household.
[b] 2.37 females over ten years of age per household.

average, about 70 percent or more of household expenditure was allocated to food by both wage and nonwage households (table 14.4). Wage households spent significantly more on store and market foods than did nonwage households in both periods. Store foods were more expensive in terms of energy or protein than local foods when the latter were valued at prevailing average market prices. For example, sweet potatoes provided approximately 83 kilocalories per toea, while rice supplied approximately 37 kilocalories per toea (1 toea = 0.01 kina).

Subsistence sources, including forests and gifts, supplied most of the average energy consumed in both nonwage and wage households. Subsistence gardens and other "tree" food plants (such as sago, breadfruit, and pandanus) contributed, on average, about 80 percent and 70 percent of available energy for nonwage and wage households, respectively. Overall, while nonwage households apparently had more available food energy (2,800–3,100 kilocalories per adult-equivalent per day), on average, than wage households (2,400–2,600 kilocalories), the difference was significant only in the second survey period. Nonwage households obtained significantly more energy from subsistence food gardens in both periods than did wage households. Conversely, as expected, wage households obtained significantly more energy from purchased food than nonwage households for both periods.[4]

---

4. It should be noted, however, that the methods of data collection for food from purchased and nonpurchased sources were different, and the possibility of overestimation of subsistence consumption for nonwage households cannot be discounted. If it were present, it would explain at least part of the apparent difference in energy availability between the wage and nonwage households.

TABLE 14.4 Expenditure shares of foods and nonfoods of nonwage and wage households, Papua New Guinea, 1987–88

Expenditure Category	Nonwage Households		Wage Households	
	Period 1[a]	Period 2[b]	Period 1[a]	Period 2[b]
	(percent)			
Food	69.8	83.5	61.4	72.0
Subsistence food	56.7[c]	73.1	38.0[c]	54.4[d]
Store food	9.5[e]	5.3	13.0[e]	11.8[e]
Food gifts given	2.1	3.9	7.5[d,e]	2.3
Market food	1.5	1.2	2.8[e]	3.5[f]
Cash gifts given	15.6	7.6	22.5[d]	16.4[d]
Store inventory expense	7.9	3.7	6.3	1.1
Animals	1.0	0.2	0.9	0.2
Farm inputs	0.4	1.3	2.0	3.1
Other expenditure	5.4	3.6	6.9	6.8[d]
	(kina/adult-equivalent/fortnight)			
Total expenditure	11.38	12.74	13.41	12.69

SOURCE: Surveys undertaken by the authors.
[a]July 1987–January 1988.
[b]February 1988–August 1988.
[c]Significant difference at the 0.05 level, between periods, same group.
[d]Significant difference at the 0.05 level, between groups, same period.
[e]Significant difference at the 0.01 level, between periods, same group.
[f]Significant difference at the 0.001 level, between groups, same period.

What are the effects of changed sources and levels of income on household food energy availability? In general, calorie availability may be expected to increase with increases in net income. In addition, subsistence foods will be replaced by purchased foods as the proportion of income earned as cash rises. Unless there is perfect substitution, the proportion of income earned as cash will affect the amount of food energy available. Hypotheses about these possible impacts on energy availability are tested using the combined cross section and two-period time series of observations from the sample households. In interpreting the results, the reservations already noted about the reliability of the subsistence data should be kept in mind.

Females at Karimui were involved in income-producing activities on more surveyed occasions than were males (Finlayson et al. 1991). Males earned more cash per capita than females, but females contributed a greater value of subsistence food. It is hypothesized that the proportion of cash income received by females had a significant and positive effect on the amount of food energy available for consumption at the household level.

Certain household characteristics may also have affected the availability of energy. In particular, the effect of household size on available food energy is explored in the model. Finally, the market prices of the main food staples—sweet potatoes, taro, and cooking bananas—are expected to have affected the availability of energy.

Regression analysis suggests that energy availability increased with net income. The elasticity of energy availability with respect to income, evaluated at the mean of the observations, is 0.34 (table 14.5). The proportion of total income earned as wages was found to have a negative, but insignificant, effect on energy availability. The share of cash income to females in the household had a significant and positive effect on the level of food energy available. Household size appeared to have a significant and negative effect on energy availability, suggesting that, in larger households, there was less food per adult-equivalent. The price of staples was found to have a negative, but insignificant, effect on food energy available for consumption.

Morbidity and Nutritional Status Effects

Nutritional status in the Karimui census division is poor in comparison with many other parts of the country. Cross-sectional anthropometric surveys in 1981 (Harvey and Heywood 1983) and 1985 (Groos 1988), respectively, showed 47 and 41 percent of children under five years of age to be less than the 80 percent weight-for-age standard, using the

TABLE 14.5 Regression analysis of energy availability

Explanatory Variable[a]	Parameter	t-Value	Mean	Standard Deviation
TOTINC	100.7005	3.459	9.9757	6.0974
TINCSQ	−.0163	−.014	136.4787	155.4409
PROPWAGE	−251.7340	−1.321	.2348	.3773
PROPFEM	5.9633	2.040	36.5527	24.9242
STAPRICE	−42.4235	−.891	10.1696	1.4597
AVADEQ	−82.1084	−2.772	4.9189	2.4227
Constant	2,355.8876	4.496	—	—
R^2	0.3852			

NOTE: The dependent variable is kilocalories per adult-equivalent per day.
[a]Definitions of variables:
  TOTINC = total net disposable income in kina per adult-equivalent per fortnight.
  TINCSQ = (TOTINC)^2.
  PROPWAGE = proportion of total income from wages.
  PROPFEM = percentage of cash income obtained by females.
  STAPRICE = staple price index.
  AVADEQ = household size in adult-equivalent.

Harvard growth standards. While there appeared to be a general improvement in nutritional status, Groos (1988) found that children of mothers who had been working at the KSC did not show the same level of improvement on all indices of nutritional status as did other children in South Simbu. By 1987, the generally favorable trend appeared to have been reversed, since 43 percent of children surveyed were less than 80 percent of the standard weight-for-age (Groos and Hide 1989).

Since morbidity and nutritional status appear to be interrelated in many situations (Lehmann, Howard, and Heywood 1988), fortnightly recall surveys of morbidity were carried out by selected symptoms for mothers, and children under five years of age. Six broad categories of symptoms were considered in the analysis of factors affecting growth: diarrhea, breathlessness/pneumonia, fever, coughs, nasal/upper respiratory tract infections, and all others. In general, the frequency of sickness increased with age for young children, reaching a peak at about 12 months of age and decreasing thereafter. Children aged between 51 and 56 months in the first period, and between 57 and 62 months in the second period, were sick almost as often as those in the 15- to 20-month age group.

Diarrhea was found to be important in explaining growth and nutritional status of children. Diarrhea generally occurred more frequently in the first period than in the second, with children between 9 and 20 months having the highest incidence for both periods. Comparison of the frequency of episodes of diarrhea by net income tercile and household wage status indicates that in nonwage households the frequency of diarrhea declined with increasing levels of income, while, conversely, in wage households, the frequency of diarrhea increased with increasing levels of income (table 14.6). The causes of these relationships are not clear at this stage.

TABLE 14.6 Frequency of diarrhea in children of nonwage and wage households, by income tercile, Papua New Guinea, 1987–88

Income Tercile	Percentage Frequency of Diarrhea[a]	
	Nonwage Households	Wage Households
1	7.5 (49)	7.4 (56)
2	6.9 (51)	9.7 (54)
3	5.6 (28)	10.5 (64)
Total	6.8 (128)	9.2 (175)

SOURCE: Surveys undertaken by the authors.
NOTE: Number of observations are in parentheses.
[a] Every two weeks, mothers were asked if their under-five-year-old children had diarrhea in the previous two weeks.

Anthropometric measurements of sample children were made three times during the study, at approximately six-month intervals. Nutritional indices were calculated using U.S. National Center for Health Statistics standards. Averages for each index and the percentage of individuals below commonly accepted cut-off points for children from wage and nonwage households are presented in table 14.7. Children who were less than three months of age at the end of either the first or the second interval were left out of the analysis for that interval. On average, the nutritional status of children from nonwage households and wage households was not significantly different. However, in some cases, a greater percentage of children from wage households were below cut-off points for the indices of nutritional status.

Overall, rates of change between survey rounds for each nutritional index were similar for children from wage and nonwage households. However, children from nonwage households in the lowest income tercile appear to be worse off than children from wage households, while, in

TABLE 14.7 Means of three nutritional indices and percentages of children below cut-off points for each index for Karimui children, by nonwage and wage households, Papua New Guinea, August 1987, February 1988, and August 1988

Nutritional Index	Nonwage Households			Wage Households		
	Mean of Standard (percent)	N	Below Cut-off Point[a] (percent)	Mean of Standard (percent)	N	Below Cut-off Point[a] (percent)
Weight-for-age						
August 1987	83.7	117	39*	81.6	171	47*
February 1988	83.6	128	38*	80.4	174	44*
August 1988	81.8	127	48*	79.3	172	52*
Height-for-age						
August 1987	92.5	113	30	92.5	168	30
February 1988	92.0	128	32	91.2	174	36
August 1988	91.2	127	48	90.9	172	36
Weight-for-height						
August 1987	96.0	112	26	93.6	164	37
February 1988	97.0	128	22	94.6	174	31
August 1988	96.6	127	14	94.0	172	30
Average age (months)						
August 1987	23.3	117		24.0	171	
February 1988	28.9	128		27.5	174	
August 1988	34.0	127		32.1	172	

SOURCE: Surveys undertaken by the authors.

[a]Unless otherwise indicated, the cutoff point is 90 percent. An asterisk indicates a cut-off point of 80 percent.

the upper-income tercile, children from nonwage households appear to have made greater improvements in their nutritional status than children from wage households.

Children's nutritional status is influenced by age, sex, genetics, and factors such as morbidity and food intake. Chronically low levels of available food energy can be expected to result in low levels of attained nutritional status. Two sets of models are specified to examine the effects of household energy availability and other factors on nutritional status. In one set of models, the dependent variables are changes in the percentage points of the three nutritional indices per month in the two intervals between anthropometric measurements. The corresponding independent variables relate to measurements within the corresponding periods. The growth of young children is very sensitive to changes in their health, and these models are designed to reveal any short-term impacts of health on growth.

Results indicate that measures of initial nutritional status and age were generally the most powerful explanators of change in the nutritional status of young children (table 14.8). Models for all three indicators show that children who started the growth interval with poor nutritional status had greater rates of improvement in nutritional status than those who were better off initially. The effects of age were more complex, as reflected in the coefficients on the variances of MIDAGE and AGESQ.

Availability of food energy at the household level had an insignificant effect on change in the nutritional status of children. The share of available energy from purchased sources was also found to have an insignificant effect on weight and height rates of growth. The frequency of diarrhea was found to be significant and negative in its effect on change in percentage of standard height-for-age. The effect of being weaned was negative on the rate of change in percentage standard for all nutritional indices, and, to a significant degree, in the case of weight-for-height growth. The sex of child and size of household were not found to be significant predictors of rate of change of nutritional status.

In the set of models in table 14.9, on the other hand, factors affecting the attained level of nutritional status are examined, using, as dependent variables, percentages of standard of three nutritional indices at both the midpoint and the end of the study. For the midpoint observations, the independent variables were measured over the first period only, whereas those for the observations at the end of the study are averages over both periods. In interpreting the results in table 14.9, therefore, the data limitations should be recognized. That is, although attained nutritional status reflects the longer-term impacts of nutrition and other factors on growth of children, energy consumption, frequency of diarrhea, and weaning status are based on one year of observations at most, and, in about half the cases, on only six months of observations. Effects on the attained level of nutritional status of events that occurred before the time of data collection obviously could not be included in the models.

Age had significant negative impacts on nutritional status in all three models (table 14.9). Boys generally had lower scores than girls, and children being weaned did less well than others in terms of weight-for-height. As expected, diarrhea had a negative effect overall on the nutritional status of children and significantly so for weight-for-age and height-for-age. Frequency of diarrhea over the survey period appears to have had a significant negative effect on height-for-age, while a child's recent history of diarrhea (in the previous fortnight) was shown to have a significant and negative effect on weight-for-age, but an insignificant impact on weight-for-height. Energy availability had a significant and positive effect on height-for-age. Negative coefficients on KCALSQ indicate diminishing returns to increased availability of energy. The coefficient for the share of available food coming from purchased sources was not significant.

Conclusions

The experience of the KSC underscores that long-term viability of small- and medium-scale export-oriented agricultural projects depends upon adequate capital and sound financial management. Management and administrative skills are often lacking, particularly in marketing and in cash flow and risk management. Technical assistance is therefore an essential ingredient in these areas as well as in the more generally recognized areas of agronomy and postharvest processing.

The failure of the KSC demonstrates the riskiness of export crop production. Planners in Papua New Guinea need to give more thought to ways of reducing these risks. Some spreading of risk will usually be possible by diversification. For example, cardamom production could have been combined at KSC with production of other cash crops, such as coffee. Or, smallholder production of cash crops could have been encouraged in parallel with the development of the plantation. In many areas of the country there is scope for smallholder growers to expand food production for local sale, often with lower production risks than production of cash crops for export. To exploit this scope more fully, major improvements may be required in the food marketing system, including transportation services, handling, market facilities, and market information.

The experience at Karimui demonstrates that there are important

TABLE 14.8 Regression models for change in percentage of standard for weight-for-age, height-for-age, and weight-for-height per month of children in sample households

Explanatory Variable[a]	Mean	Model 1 Change in Percentage Weight-for-Age/Month (mean = −.2489)		Model 2 Change in Percentage Height-for-Age/Month (mean = −.1608)		Model 3 Change in Percentage Weight-for-Height/Month (mean = .0835)	
		Parameter	t-Value	Parameter	t-Value	Parameter	t-Value
MIDAGE	28.12	.0312	1.971	.0153	2.731	−.0535	−2.939
AGESQ	1,074.76	−5.4E-04	−2.017	−2.5E-04	−2.420	1.1E-04	1.975
SEX	.51	−.1531	−1.184	.0127	.277	−.2665	−1.767
AVADEQ	5.50	.0038	.134	−.0049	.488	.0096	.289
WEAN	.63	−.6232	−3.774	−.0866	−1.490	−.96318	−4.945
WAMED	82.34	−.0530	−8.841	—	—	—	—
HAMED	92.12	—	—	−.0316	−5.722	—	—
WHMED	95.15	—	—	—	—	−.1011	−11.503
DIAR	8.21	.0056	.799	−.0061	−2.474	.0138	1.662
KCAL	2,652.26	2.0E-04	.676	6.7E-05	.663	1.1E-04	.315
KCALSQ	8.2E+06	−2.6E-08	−.550	−1.1E-08	−.669	−1.2E-09	−.022

	Coef.	t	Coef.	t	Coef.	t
EXPCAL	.0047	.608	.0029	1.053	−.0032	−.352
Constant	3.8424	4.900	2.5152	4.586	10.8810	9.839
R^2	.29393		.22031		.39000	
$\bar{R}^2$	.26966		.19352		.36904	
F-value	12.11377		8.22249		18.60513	
Degrees of freedom	291		291		291	

12.34

[a] Definitions of variables:

MIDAGE = mid-interval age in months.
AGESQ = (age)^2
SEX = male = 1, female = 0.
AVADEQ = household size in adult equivalent.
WEAN = pre/post = 0 weaning = 1.
WAMED = initial percent standard weight-for-age.
HAMED = initial percent standard height-for-age.
WHMED = initial percent standard weight-for-height.
DIAR = percent of occasions of diarrhea in the fortnight preceding anthropometric measurement.
KCAL = available energy in kilocalories per adult-equivalent per day.
KCALSQ = (available energy)^2.
EXPCAL = percent of available energy that was purchased.

TABLE 14.9 Regression model for attained levels of anthropometric standards of children

Explanatory Variable[a]	Mean	Model 1 Weight-for-Age (mean = 80.92)		Model 2 Height-for-Age (mean = 91.26)		Model 3 Weight-for-Height (mean = 95.40)	
		Parameter	t-Value	Parameter	t-Value	Parameter	t-Value
AGE	30.87	−.3761	−2.809	−.2045	−3.788	−.4183	−4.068
AGESQ	1,238.42	.0050	2.373	.0023	2.745	.0056	3.483
SEX	.51	−2.4920	−2.346	−1.1149	−2.604	−1.2632	−1.549
BIRTHNUM	4.04	.0643	.260	.1716	1.718	−.3078	−1.617
HEIGHT	150.10	.6328	5.313	.2953	6.151	.1066	1.166
WEAN	.63	−.6453	−.472	.4760	.864	−3.0483	−2.905
DIAR	8.21	−.1173	−2.019	−.0630	−2.691	−.0319	−.714
KCAL	2,652.26	.0030	1.156	.0021	2.067	−2.9E-04	−.147
KCALSQ	8.2E+06	−3.1E-07	−.779	−2.8E-07	−1.731	1.3E-07	.411
EXPCAL	12.34	.0045	−.068	.0324	1.217	−.0683	−1.346
Constant		−11.4508	−.616	46.7110	6.238	89.9921	6.307
R^2		.19593		.25248		.17106	
$\bar{R}^2$		.16830		.22679		.14258	
F-value		7.09098		9.82883		6.00518	
Degrees of freedom		291		291		291	

[a]Definitions of variables:
AGE = age in months.
AGESQ = (age)².
SEX = male = 1, female = 0.
BIRTHNUM = birth order (first = 1, etc.).
HEIGHT = mother's height in centimeters.
WEAN = pre/post = 0, weaning = 1.
DIAR = percent of occasions of diarrhea.
KCAL = available energy in kilocalories per adult-equivalent per day.
KCALSQ = (available energy)².
EXPCAL = percent of available energy that was purchased.

advantages where commercial plantation crop production can be introduced in a manner that is compatible with the preservation, to a large degree, of the integrated smallholder subsistence/cash cropping system. Not only does the low-investment smallholder system provide a platform from which efforts to attain economic growth in the rural sector can be launched, but it can also serve as a fallback should commercial cropping fail. Planners need to be alert to the dangers of those types of rural development that prematurely cut people off from their subsistence base and, therefore, place at risk the sustainability of more intensive, commercially oriented production systems.

An evaluation of the relationship between agricultural commercialization and nutritional status of children at Karimui indicates that long-term household food security is affected by the level of involvement in the wage labor economy, which can affect the nutritional status of young children. In general, however, these food intake effects are small compared with the nutritional impacts of morbidity and weaning practices. There is an urgent need to improve the long-term health of the community through improvements in hygiene and sanitation, maintenance of health services, and improvements in education about nutrition and weaning practices.

Analyses show that, at the time of this study, households that engaged in wage labor employment had approximately the same income level as households that remained, more or less, out of the wage labor market. However, wage households apparently had less available energy, on average, and significantly less in the second half of the study. Wage households that were more reliant on purchased foods appeared to be at a disadvantage in terms of access to food energy when employment opportunities declined. Nonwage households generally had larger garden areas than wage households, which seemed to provide greater food security. The fact that there were not more severe food shortages may be attributed, in part, to the flexible employment system at KSC—which permitted villagers to largely maintain their food gardens while intermittently engaging in employment—but also to the traditional social support system that involved considerable giving of gifts of food and money.

Energy availability was higher in households with greater proportions of cash income received by females. The opportunities offered by the KSC facilitated women's access to cash—a rather unusual situation compared to other rural areas of Papua New Guinea. The results emphasize the important role of women in the commercialization of the rural economy. Not only are women involved in earning cash to buy food and other necessities through wage labor or through marketing of surplus subsistence agriculture produce, but they also have the main responsibility for maintaining food supplies for the household for subsistence food gardens.

Neither the level of available food energy nor the share from purchased food appears to have had a significant overall effect on changes in the nutritional status of young children. However, when cumulative effects of availability and source of food energy on attained nutritional status were examined, some response to household calorie availability was found for the height-for-age of children.

The prevalence of diarrhea had a very negative effect on children's weight-for-age and height-for-age. Results presented here thus suggest the need for complementary inputs from the health sector. Children who had diarrhea more often had slower growth rates in height-for-age than other children. Curiously, rates of growth in weight-for-age and weight-for-height were not affected significantly by the prevalence of diarrhea. Furthermore, children from wage households with higher income levels showed a trend toward higher prevalence of diarrhea. Although it is not clear at this stage what factors are most important in determining diarrheal morbidity at Karimui, experience in other areas suggest that weaning patterns, water supply, sanitation, access to health services, and quality of child care are likely to contribute to the prevalence of diarrhea.

Risk from fluctuating export cash crop prices, unreliable marketing systems, and crop disease have encouraged Karimui farmers to seek alternatives to smallholder cash cropping as a means of earning money. The cardamom spice plantation provided alternative opportunities in the form of wage labor. However, poor management and unseasonable conditions undermined the viability of the plantation. Karimui people are now forced back on to their traditional resources in their efforts to support themselves and also to find opportunities to participate in the cash economy. In view of their recent experiences, the people may now view other commercial agricultural ventures with healthy skepticism. Benefits may flow if the lessons of the KSC collapse are learned and if people exercise more caution before cutting back too sharply on their subsistence production activities.

15 Why Should It Matter What Commodity Is the Source of Agricultural Profits? Dairy Development in India

HAROLD ALDERMAN

Introduction

In 1896, an obscure out-of-office politician rose to prominence and the Democratic presidential nomination on the basis of a rousing speech condemning the gold standard. Working up to the conclusion of that oratory, William Jennings Bryan claimed:

> ... the great cities rest upon our broad and fertile prairies. Burn down your cities and leave our farms, and your cities will spring up again as if by magic; but destroy our farms and the grass will grow in the streets of every city in the country (Bryan 1967).

This speech is not a statement about economic growth linkages or a political statement about the need to reverse urban biases; it is an expression of philosophy of agricultural fundamentalism in which the family farm is viewed as a moral bastion, in contrast to urban business interests. In its extreme, such a position presents trade from rural to urban areas as inevitable exploitation of the former if not actually a step in the path to moral degradation. In the Nebraska of Bryan's day, as in the Nebraska of Willa Cather's fiction a few decades later, farmers who brought cream to the market would be castigated for ruining their children's health.

Similar sentiments have been observed in contemporary India. For example, Alvarez (1983) helped to prompt a high-level governmental evaluation of the world's largest dairy development program, Operation Flood, by an attack in a leading Indian weekly, in which he lamented that all the milk in his village had gone to towns. Alvarez does not express a similar concern about the manufactured goods that have left the cities, purchased with the cash earned by farm households. Agricultural fundamentalism is rarely as explicit in policy discussion, although it is often latent. This is particularly true in debates on the nutritional

239

implications of cash cropping, in general, and of commercialization of dairying, in particular.

Both supporters and critics of Operation Flood have used the general public's interest in nutrition to gain adherents to their broader views on the project and in so doing have shifted the criteria of evaluation of the project (Alderman and Mergos 1987). Lipton (1985) is correct in claiming that the "irrelevance of consumed Operation Flood outputs to poor people's nutrition is confirmed by the EC evaluation [of Operation Flood]";[1] the poorest sectors of both urban and rural communities consume negligible quantities of milk. Lipton's further claim that this view of the poors' negligible consumption of milk is not seriously disputed, however, is only partially true. Increases (decreases) of milk consumption by population subgroups are often presented by proponents (opponents) of the strategy of dairy marketing cooperatives that underlie Operation Flood as evidence of Operation Flood's nutritional impacts. To a large degree, however, the debate has shifted to a discussion of whether producer incomes have increased and, if so, whether the new income is used to purchase nutritious foods.

It should, then, be relatively straightforward to evaluate dairy development, or any other agricultural development scheme, in terms of production and marketing strategies that maximize income, or incomes, of a particular population group. Under such a criterion, output mix is important insofar as there is an optimal output mix for a set of resource constraints, but changes in household welfare would be unambiguously measured in terms of real income gains. This book, however, is predicated on the proposition that all sources of profits are not equal, at least in terms of their welfare impacts. Agricultural fundamentalism aside, there are some economic explanations of why the source as well as level of profits may influence household welfare. This chapter discusses some of those possible explanations using the example of dairy development in South Asia, principally cooperative marketing in Karnataka, India, to illustrate some of the general points.

The Separability Paradigm

Agricultural households are both producing and consuming units: as producers, they maximize profits subject to prices, assets, and available technology, and as consumers, they maximize utility subject to prices and income.[2] Preferences are not an argument in profit maximi-

---

1. Emphasis in the original.
2. A more formal presentation of this model can be found in Singh, Squire, and Strauss (1986).

zation, while production optimization enters into the consumption budget allocation through the money value of the profits realized. Prices are in common in the two optimizations. Since production decisions are not based on consumer preferences, production is termed separable from consumption. On the other hand, the higher the level of profits, the higher the welfare of the household (as measured in economic terms). Therefore, consumption is conditional on production. This gives a recursive system linking the two sets of household decisions.

Separability simplifies the process of modeling the impacts of policy changes. The measurement of both producer and consumer responses can be achieved with linear techniques. Moreover, the one variable internal to the profit maximization that enters the consumption decision is measured in money, hence the output mix does not directly enter into the modeling of consumption.

A crucial assumption in this approach is that prices are exogenously given; neither the level of production nor the level of consumption influences the price a household has to pay or receive for a commodity. This condition would not prevail if producer prices differ from consumer prices, because the budget constraint differs depending on whether or not a household is a net consumer of the good produced. Changes in production, which lead to a shift of a household from being a net seller of a good to being a net purchaser (or at the corner), will affect consumption in a manner distinct from the change in profits. Changes in the level of sales or purchases that do not change the direction of flow, however, will not have such an impact.

When sales are between neighbors, as milk sales often are in rural India, then consumer and producer prices may not differ. Prices, however, also will be endogenous when purchased commodities are not perfect substitutes for home-produced goods. If a household would be unwilling to exchange a unit of a home-produced good for a unit of the "same" good on the market, then clearly the implicit or virtual prices are not the same. Indeed, the hypothetical exchange rate gives the ratio of virtual prices. The most important examples of imperfect substitutes in farm household models may be for inputs, such as family labor as opposed to hired labor (Singh, Squire, and Strauss 1986). The issue is similar, however, for outputs. Since few households would consider the quality of milk in the market to be identical to that produced at home, the virtual price of milk differs according to source.

In some situations, there is no local market for some outputs and inputs. For example, without refrigerated transport, there is often no market for milk, or at least not for the buttermilk component of fluid milk. In other situations, inputs such as firewood or water or bullock power are not traded. Such cases do not differ appreciably from the examples of near substitutes such as family labor. The endogeneity of prices is, however, more apparent. The level of production is simultaneously determined by the quantity consumed and the virtual (or shadow) price, which is household specific.

Moreover, even when households do not perceive a qualitative difference between produced and purchased commodities, these two values may be different in a dynamic context due to risk. This may provide an explanation for the crop mixes of smallholders.

The discussion of the consumption side of the farm household has so far dealt only with purchased commodities. There are, in addition, home-produced goods, such as nutrition and health, that are aspects of household consumption decisions. Such goods cannot be considered to be recursively determined with farm profits if there is no local market for female labor. When such local labor markets are absent, the opportunity cost of a woman's time in the production of health is simultaneously determined with agricultural production.

Economists usually assume tastes to be exogenous (Stigler and Becker 1977). Models of habit formation are somewhat of an exception to this generalization. Discussions of the nutritional impacts of commercialization often contain implicit assumptions about the endogeneity of preferences. Even when attempts are made to test such assumptions, for example, when marginal propensities to consume are allowed to vary according to sources of income, the mechanisms for such influences are seldom elucidated.

One plausible explanation for endogenous preferences is based on intrahousehold allocation of consumption. Household consumption is generally observed rather than individual preferences, which may vary across individuals in the household. A number of models attempt to explain the means by which differences in tastes are reconciled within a household, but for a number of technical reasons, these models of intrahousehold decision making and bargaining are difficult to distinguish and test empirically.[3] These models, nevertheless, provide one rationale for differences in consumption patterns according to sources of income.

Such differences in consumption patterns may also be generated by the relative instability of earnings. Economists generally assume that a greater percentage of a permanent source of income will be consumed than of a transitory source of income. There is no a priori reason to expect earnings from cash crops as a genre to be more or less permanent than profits from subsistence crops. It is logical, however, to expect that in any given period of observation, different sources of measured income

---

3. See Thomas (1989) and the references cited in that paper.

will contain different mixes of permanent and transitory elements, which can create spurious differences in spending out of income streams.

Note, also, that different production processes require different assets. As different types of assets are not equally acceptable as loan collateral (Binswanger and Rosenzweig 1986), different production processes indirectly influence credit constraints and, hence, consumption patterns.[4] Similarly, when credit constraints exist, changes in the seasonal flow of earnings may influence consumption.

Other models of changing preference ordering that may be relevant to the question of nutrition include models of demonstration and bandwagon effects as well as of the opposite snob effects. While some may be certain that city lights work their pernicious influence in this manner, there is little in economic theory that allows one to assess the welfare impact of a change in preferences. One can measure the nutritional impact, but it must be recognized that the direction of change in any measure of nutrition may not correspond to the direction of perceived change in welfare by some, or even all, members of the household.

As indicated above, relaxing the assumption that prices are exogenous to the household can have major implications for modeling the welfare impacts of changes in production. Similarly, any appraisal of an agricultural project must proceed differently if market prices are affected by the project than if they are not (Pinstrup-Andersen 1985). It is useful, however, to ask whether local food prices are changed when a new crop is promoted in the region or whether wages rise. The impact on the price of crops that are traded only in the local market depends not only on the magnitude of the shifts in supply and demand curves but also on their slopes.[5] The price of a crop that is traded in a larger market is, however, linked to that market and is generally not affected by shifts of supply in a relatively small region.

Nutritional Impact of Cooperative Dairy Development in Karnataka, India

The Karnataka Dairy Development Project is an example of an approach to agricultural development that emphasizes integrating rural households into a market economy by increasing their use of purchased inputs and increasing their marketed surplus.[6] While administratively

---

4. The relation of interlinked factor markets and nonseparability is also discussed in Singh, Squire, and Strauss (1986).

5. Pinstrup-Andersen, de Londoño, and Hoover (1976) present the relevant formula for calculating the nutritional impact of changes in food supply. The formula is analogous for changes in labor demand.

6. Details are presented in Alderman (1987).

distinct, the Karnataka Dairy Development Project is modeled closely on Operation Flood that aims to link 10 million isolated producers of milk into an urban marketing grid through a network of cooperatives, processors, and marketing unions. The debate on the merits of the Operation Flood approach has been contentious, in part because of the scale of Operation Flood and in part because of the broad range of equity as well as production goals of the program.[7]

The program has been criticized, for example, for stressing the marketing of milk to urban areas rather than increasing total milk production (Terhal and Doornbus 1983). While there is ample evidence that in certain regions the strategy has contributed significantly to increased production (Mergos and Slade 1987; Alderman 1987), it is useful to discuss the expected impact of a change in marketing alone. Marketing cooperatives can take advantage of economies of scale to reduce the costs of transportation and processing, which decreases the price wedge between the final consumer in the market and the producer, leading to higher farmgate prices. Milk prices, then, are also higher for rural consumers and would no longer be endogenous to either the household or to the community. The fact that producers gain while rural consumers lose (they may be the same physical entity in the farm household model) accounts for the fact that one critic of Operation Flood faults it for raising village milk prices (Crotty 1983), while another is equally dissatisfied with the program for failing to raise prices sufficiently (George 1983).

Tests of differences of producer prices between areas where cooperatives are present and where they are not were performed using data from Karnataka and Madhya Pradesh. The test controlled for animal type, season, and distance to the market (Madhya Pradesh only), and revealed significantly higher milk prices attributable to cooperatives (Alderman and Mergos 1987). The magnitude of the price change was small, however, ranging from 3.0 to 7.5 percent. To put it in perspective, this price change was less than the 10 percent premium paid for buffalo milk in both regions. In addition, retail prices in the villages studied were found not to differ from producer prices.

It is often argued that commercialization of dairying removes unmarketed buttermilk from the villages. A by-product of the manufacture of traded butter and ghee, buttermilk is a nutritious commodity with a sufficiently low opportunity cost as to be virtually free and reportedly given to laborers.[8] There is, however, no evidence of whether wages

---

7. An extensive review of the various debates is found in Alderman, Mergos, and Slade (1987).

8. George (1984) cites research from Haryana indicating such distribution of buttermilk but did not find buttermilk distribution to laborers in Kerala.

responded to the elimination of this in-kind component. While the Karnataka study indicated significantly higher production of buttermilk (and ghee) where cooperatives had not been formed, it did not find significant evidence that buttermilk was either traded or given away.

Once the assumption is relaxed that the only impact of the cooperatives was through higher producer prices applied to constant production, it is necessary to consider the impact on other prices due to changes in milk production, mediated either through prices or technological change. In Karnataka, as in much of India, there is little production of fodder, even where cooperatives are present. Moreover, the use of concentrates as animal feed is limited and based mainly on oilseed cakes and bran. Hence, there was no measured impact of commercialization of dairying on the price of either rice or coarse grains. Similarly, while an increase in the number of buffaloes and crossbred cows can be attributed to the presence of cooperatives in Karnataka, this increase was accompanied by only a small increase in labor—due to scale economies—and no changes in local wages.

Table 15.1 presents estimates of the impact of price and income changes attributable to cooperatives on nutrient intakes as well as the specific impact on the consumption of fluid milk.[9] The full matrix of price and income parameters used in this calculation are reported in Alderman (1987) and need not be reproduced here.[10] The calculations are based on a measured 8 percent increase in average expenditures, which was estimated to be the impact due to changes in production in villages with dairy cooperatives. Similarly, a 3 percent increase in milk prices is used, in keeping with the results referred to above.

Table 15.1 also contains information on producer and cooperative effects that is relevant to the theme of the possible importance of composition of profits and that requires further discussion. The cooperative effect is based on the coefficient of a dummy variable for control villages in a regression of calories and protein from purchased and home-produced foods. Other explanatory variables include the logarithm of total expenditures and its square, the logarithms of seven prices comprising the entire budget, variables for family size and the proportion of family

---

9. In this discussion, we focus on family intakes of calorie and proteins as a measure of nutrition. We recognize that nutrition is also affected by such factors as household distribution of food and the allocation of time and money to health care, but we have no data with which to study these aspects nor any reason to suspect that they work in a manner that would negate the general conclusions presented.

10. The price effects are from panel (fixed effect) estimates using four observations on each household. When combined cross-sectional and time-series approaches are used, the effect of an increase of milk prices on nutrient consumption is slightly positive. The effect in either model is inconsequentially small.

TABLE 15.1 Changes in milk, calorie, and protein consumption attributed to dairy cooperatives in Karnataka, India, January 1983–April 1984

Type of Consumption/Source of Change	Producers	Nonproducers
	(percent)	
Fluid milk consumption		
Due to price change	−3.0	−3.0
Due to increased income	3.5	—
Producer effect	7.0	—
Cooperative effect	−9.0	−9.0
Net change	−1.5	−12.1
Energy consumption		
Due to price change	0.0	0.0
Due to increased income	3.8	—
Producer effect	0.0	−2.1
Cooperative effect	−2.1	−2.1
Net change	1.7	−2.1
Protein consumption		
Due to price change	−0.1	−0.1
Due to increased income	3.8	—
Producer effect	0.3	—
Cooperative effect	−1.9	−1.9
Net change	2.1	−2.0

SOURCE: Alderman 1987.
NOTE: The producer effect is the impact of milk production on demand. The cooperative effect is the impact of the presence of a cooperative on demand. Income and price are held constant in both instances.

members under five years of age, and a dummy variable for vegetarians. T-statistics for the cooperative effect in the calorie and protein equations were 2.04 and 1.78, respectively. These are marginally significant for two-tailed tests and are indicative of a small impact of cooperatives on nutrient consumption, controlling for price and income effects. Similar regressions on the consumption of fluid milk as well as of ghee and buttermilk indicate a highly significant ($t > 5.0$) decrease in consumption of dairy products where cooperatives are present.

These effects do not appear to be due to membership in, as opposed to the presence of, a cooperative and were stronger for nonproducers in villages with cooperatives than for producers. No compelling evidence exists with which to distinguish whether this effect is due to preference shifts or to the effects of unmeasured transaction costs, including search costs (for nonproducers) and expected bonuses from the cooperative (for producers). Since nonproducers were found to reduce their dairy consumption, it is more likely that this impact is due to increases in unrecorded prices, in which case the direction of the welfare impact is apparent.[11] As discussed above, the welfare impact of the alternative interpretation—a preference shift—is indeterminant.

The producer effect represents the measured impact of a continuous variable for the proportion of total income derived from milk. This is measured for all producers, whether or not they were in a village with a cooperative. The estimation technique controls for potential biases attributable to omitted transaction costs. Each additional kilogram of milk produced leads to an additional 0.046 kilogram consumed by the household, controlling for income and prices. The t-statistic of this coefficient is greater than 10.0. Since cooperatives lead to a modest but significant increase in milk production, this production impact offsets much of the so-called cooperative effect on milk consumption for the average producer. While the magnitude of this effect is small, it is important in part because it challenges the separability paradigm. Also, while this effect has positive nutritional impacts when the crop being commercialized is also consumed by the household, it may have a different impact if it holds for other crops as well. For example, when a marketed crop partially or fully replaces a subsistence crop in a rotation, a producer effect would imply a decrease in consumption of the subsistence crop.

Finally, it should be noted that the income effect was found to be positive and sufficiently large to offset negative price and cooperative effects for producers. The implication for dairy policy is that the nutritional impacts will be enhanced to the degree that the project allows low-income households to increase their incomes and to the degree that it enables low-income households, including landless households, to participate in dairying activities.

Dairy Development and Incomes of the Poor

While we have argued that price effects and changes in purchasing patterns are real, albeit small, considerations when calculating the impact of the commercialization of dairying, we have also argued that the greater concern is whether the poor increase their incomes as a result of dairy commercialization. While it is unlikely that households will voluntarily adopt new practices that will reduce their absolute incomes, such decreases could come about if commercialization leads to a reduction in the demand for unskilled labor or for products produced by low-income households. Moreover, changes in the commercial environment may be accompanied by asset seizure and other nonmarket coercion. Similarly, as there are fewer potential markets for cash crops, there has been a historic tendency for commercial interests as well as governments to try

---

11. Utility is nonincreasing (generally decreasing) in the face of a price increase.

to establish positions of monopsony. While dairy marketing is less prone to such manipulation, the prototype cooperative for Operation Flood benefited from having monopoly rights over the sale of milk in Bombay during its formative years (Singh and Kelly 1981). In a related vein, milk prices in some of the major cities in India have been held low by municipal authorities—even to the point where rationing was necessary —in the name of nutrition and real wages.

The issue of nonmarket coercion or the relationship of cash cropping and market structure, however, goes beyond the scope of this chapter. We can, however, inquire whether dairy development influences either absolute or relative income and whether dairy assets are more or less evenly distributed than land.

A regression of total expenditures as a function of household assets and demography in Karnataka indicates that total expenditures were higher in all landholding categories, including the landless, if cooperatives were present (Table 15.2). In this community, dairying is more important, and the impacts of the cooperatives higher, the larger the landholding. In no case, however, is there evidence that any particular group lost absolutely following the introduction of the cooperative. There is also no evidence that nondairy earnings differ in villages with cooperatives compared to control villages.

A study of cooperative dairy development in Madhya Pradesh, which paralleled the Karnataka study, indicated the same basic results that income and producer effects on nutrition outweighed the price and nonsignificant cooperative effects in the sample (Mergos and Slade 1987). This study also found that project benefits increased with land ownership and education. Furthermore, low caste and landless households were less likely to engage in dairying, and hence were not receiving

TABLE 15.2 Difference in expenditures and earnings sources in cooperative and control villages, Karnataka, India, 1983–84

Landholding Class	Total Expenditures in Villages with Cooperative as a Percent of Expenditures in Control Villages	Ratio of Value of Milk to Total Expenditures	
		Cooperative Villages	Control Villages
Landless	107.4 (2.09)	0.164	0.107
0–1 hectare	103.2 (1.14)	0.160	0.120
1–2 hectares	117.1 (4.79)	0.218	0.125
More than 2 hectares	112.2 (3.28)	0.228	0.191

SOURCE: Alderman (1987), equation (12) and Table 19.
NOTE: Numbers in parentheses represent t-values.

direct income benefits due to the introduction of cooperative farming.[12] In the Karnataka villages without marketing cooperatives, the marketed share of milk is 25–40 percent. In contrast, the villages with cooperatives in Karnataka marketed over 60 percent of output as did both groups in the Madhya Pradesh study.[13]

Dairy products are generally luxury goods in rural South Asia, with budget shares increasing with income. This implies that the real income loss due to a price increase is roughly equal or slightly rising in percentage terms across income groups. A price ceiling or other attempts at keeping prices low for consumers would not only not be targeted to low-income consumers but would also, on the average, reduce dairy earnings for rural populations, holding production constant, by more than it would lower their cost of living. This holds for any landholding strata, except the highest, although it is clearly not true for nonproducers. Detailed studies from India suggest that there is no obvious justification on nutritional grounds to discourage commercialization of dairying. In general, the value of dairy development is neither appreciably augmented nor reduced by nutritional linkages. Dairying provides appreciable income streams to small and medium farmers; benefits to landless households are more modest, but not inconsequential. Nutritional impacts will be enhanced to the degree that commercialization and other dairy promotion increase income and its distribution.

APPENDIX

Measuring the Effects of Changes in Production on Consumption

It is often observed that the consumption of various agricultural commodities differs between producers and nonproducers (Horton 1985; Massell 1969). Quite often, and quite plausibly, an explanation for such an observation is that the producer effect picks up unmeasured differences between retail and farmgate prices. Even when a distinction is made between retail and farmgate prices, a dummy variable for producers may pick up the effects of other costs of marketing, including opportunity costs of time. These effects can be controlled with a fixed effects model (Hausman 1978).

---

12. Other studies, of course, have come to different conclusions (see Alderman, Mergos, and Slade 1987) reflecting, in part, the heterogeneity of India. The Madhya Pradesh study, however, is one of the more detailed studies of this topic and one of the few that separates out the impacts of landholding, asset transfers, and other causal factors.

13. The similarity of sales percentages in control and cooperative villages in Madhya Pradesh is in keeping with the absence of a cooperative effect on consumption patterns.

Suppose that there exists a proportional markup (or discount) in prices over that recorded in the data which can be attributed to marketing costs. Suppose, further, that these costs are household specific and fixed. Instead of the recorded P_h, the household faces $P_h\lambda_h$, where λ_h is the proportional price adjustment. By taking the logarithm of $P_h\lambda_h$, one has $\ln P_h + \ln\lambda_h$. In a cross-sectional model, a coefficient of a variable that correlates with $\ln\lambda_h$, say, a bivariate dummy variable for producers and nonproducers, will be biased. If $\ln\lambda_h$ is negatively correlated with production and consumption is negatively correlated with farmgate prices, a positive bias is expected. In the fixed effect model, the mean of all observations for a given household is subtracted from the observation in each period—that is, $\overline{Y}_{it} - Y_i = f(X_{it} - \overline{X}_i)$. This is mathematically equivalent to using a dummy variable for each household. When this is done, time-invariant factors such as the proportional markup drop out.

To investigate whether changes in production influence consumption patterns in a manner that differs from that predicted by changes in expenditure or prices, a fixed effect Engel curve equation is estimated for milk, which includes a variable MVV defined as the value of milk production divided by all expenditures. The coefficient of this term in a budget share equation is the unit change in the quantity consumed with a unit change in the quantity produced, holding total expenditure and prices constant. The full effect of a production increase, of course, must also include the income effects through changes in farm profits. The estimated results are

Milkshare = $-.0149 X' - .0022 H' + .0144 \ln P\text{milk}'$ (15.1)
       (.0027)   (.0007)   (.0068)

   $-.0258 \ln P\text{rice}' - .0007 \ln P\text{ragi}' + .0529 \ln P\text{pulse}'$
   (.0133)   (.0133)   (.0189)

   $+.0499 \ln P\text{meat}' - .0034 \ln P\text{otherfood}'$
   (.0255)   (.0118)

   $-.1637 \ln P\text{nonfood}' + .0416\ MVV'$
   (.0865)   (.0054) $r^2 = .101$,

where the price denotes deviations from the mean for each household over the four rounds; standard errors are in parentheses. X indicates the logarithm of total expenditures, H is household size, and the other terms are logarithms of prices. Ragi is a type of millet.

One may vary the assumption of a proportional price markup or discount and assume, instead, a fixed marketing cost. The opportunity cost would be $P_h + \lambda_h$ and a variant of equation (15.1) linear in prices would be more appropriate for the particular issue of the effects of changes in production, although it is not the preferred model for measuring price elasticities. Estimates from such a model changed neither the coefficient of MVV or its level of significance. The coefficient of MVV' in equation (15.1) implies an elasticity of milk consumption with respect to production of .22 at the mean. Holding expenditures constant, a 50 percent increase in household milk production would raise household consumption by 10 percent.[14]

The measurement of these effects is aided by the fact that fluid milk cannot be stored. In this study, farmgate prices were already included as regressors where appropriate. Furthermore, no significant difference between reported farmgate and reported retail prices were observed, probably because most milk sales to local consumers occur at the farmgate. Other sales are mainly to cooperatives, whose prices are based on urban demand and can be considered exogenous to the village as well as the household. The observed effect, then, is not likely due to differences in prices. While an analogous test with less perishable cereal grains is more difficult to perform, it is plausible that some—although not necessarily all—of the higher cereal consumption of producers compared to nonproducers commonly observed in cross-sectional regression can be attributed to this producer effect. For example, in cross section, this sample indicates significantly higher ragi and rice consumption by their respective producers.

One cannot dismiss such a result as reverse causality; families that prefer milk produce more. The model controls for any fixed differences in tastes. Furthermore, changes in dairy production over the short run are largely exogenous to the household. Any explanation in terms of habit formation merely begs the question, for any lagged relationship of milk production and consumption still requires an explanation of how production influences consumption.

14. Since budget shares must add to 1, MVV' must have a negative coefficient for at least one other commodity group. In this case, much of the adjustment due to changes in MVV comes in the budgets of rice and nonfoods.

16 Effects of Sugarcane Production in Southwestern Kenya on Income and Nutrition

EILEEN KENNEDY

Introduction

In 1984, at the request of the Government of Kenya, IFPRI initiated a study to evaluate the income and nutritional effects of a shift from maize to sugarcane production. The government was concerned that in areas undergoing this transition to commercial agriculture, particularly to sugarcane production, a deterioration in household-level food security and preschooler nutritional status was occurring.

The study was initiated in South Nyanza, a sugar-growing area, to evaluate the effects of cash crop production on agricultural production, income, and food consumption, and to assess the impact of cash cropping on the health and nutritional status of preschoolers and women.

The study site was served by the South Nyanza Sugar Company (SONY), which, typical of most sugar factories in Kenya, is primarily government owned. Smallholder producers who join the sugarcane outgrowers program are under contract to the company to produce sugarcane at a price determined by the government. Inputs such as seed, fertilizer, and hired labor are provided to the farmers, as needed, for a fee. The total costs of the package of inputs plus interest are deducted from the payment to the farmer for the sugarcane crop. The SONY outgrowers' program is typical of the way sugarcane production is organized throughout Kenya.

The first stage of data collection took place from June 1984 to March 1985. The sample included a representative sample of 504 households from the community; this research is one of the few studies on commercial agriculture to include a representative sample of agricultural as well as nonagricultural households from within the community.

A follow-up study was initiated in the same project area in December 1985 to build upon the information collected in the 1984–85 portion of the research.[1] The two studies together provided the opportunity to collect information on socioeconomic conditions, food consumption, and health and nutritional status for a cohort of households (called new entrants) prior to and after their entry into a smallholder sugarcane outgrowers' scheme, as well as up to and after payment for their first sugarcane crop. The 1984–85 and the 1985–87 studies, combined, provided a rare opportunity in which (1) baseline economic and health information is available concerning households prior to cash cropping and (2) farmers can be followed up to and after payment has been received. Sugar farmers, who received at least one payment for the cane crop, were also included, as were nonsugar farmers, merchants, wage earners, and landless persons (Kennedy and Cogill 1987; Kennedy 1989).

Of the 504 households in the 1984–85 study, 462 (92 percent) remained in the follow-up study. The group of households that was in both the 1984–85 and 1985–87 studies is called the cohort households and is the focus of the results presented in this chapter.

Income Effects

The results from the 1984–85 study indicate that incomes of sugarcane farmers who have received at least one payment for the crop are significantly higher than incomes of nonsugarcane producers (table 16.1). Sugarcane contributes to most of the difference in incomes between these two groups. Incomes per capita of new entrants (KSh 1,956 per capita)[2] were virtually identical to that of nonsugarcane producers (KSh 1,924 per capita) in the 1984–85 study. Interestingly, in the follow-up study, incomes per capita (both nominal and real) of the same new entrant group were significantly higher than those of the nonsugar group, and both new entrants and sugar farmers had incomes per capita significantly higher than of nonsugarcane producers (table 16.1).

Part of the difference in incomes between the cohort sample of new entrants and nonsugarcane producers arises from differences in marketed agricultural income (table 16.1). However, other sources of income also contribute to the difference in incomes between the two types of households. In the 1984–85 study, sugarcane production contributed 73 percent of the difference in income between sugar and nonsugar producers. The higher incomes of the new entrants in the 1985–87 study

---

1. For a detailed description of the survey design and research protocol for both studies, see Kennedy and Cogill (1987) and Kennedy (1989).
2. US$1 = 20 Kenyan shillings in 1989.

TABLE 16.1 Mean annual income per capita per year, by source and activity group, study 1 compared to study 2, cohort group, South Nyanza, Kenya, 1984–1985 and 1985–1987

	Agricultural Income				Nonagricultural Income				
	Used for Own Consumption		Marketed					Mean Nominal Income/Capita (KSh)	Mean Real Income/Capita[a] (1984 Level) (KSh)
Activity Group	Mean (KSh)	Share (percent)	Mean (KSh)	Share (percent)	Mean (KSh)	Share (percent)	N		
Study 1 (1984–85)									
New entrants	728	37	404	21	824	42	42	1,956	—
Sugar farmers	748	29	942[b]	36	901	35	139	2,591[c]	—
Nonsugar farmers	822	43	393	20	709	37	231	1,924	—
Merchants	51	2	17	1	2,141	97	29	2,209	—
Wage earners	171	8	45	2	1,821	90	18	2,037	—
Landless	163	13	48	4	1,079	83	43	1,290	—
Total sample mean	669	32	482	23	926	45	502	2,077	—
Study 2 (1985–87)									
New entrants	1,761[d,e]	46	791[d]	21	1,285	33	27	3,837[d]	3,070[d]
Sugar farmers	1,370[e]	40	625[e]	19	1,395[f]	41	146	3,390[f]	2,712[f]
Nonsugar farmers	1,302[d]	48	365[d,e]	14	1,041[f]	38	205	2,708[d,f]	2,166[d,f]
Merchants	571	11	49	<1	4,646[g]	88	15	5,265[g]	4,212[g]
Wage earners	972	30	233	7	2,017	63	14	3,222	2,578
Landless	841	36	162	7	1,336	57	33	2,338	1,870
Total sample mean	1,292	42	452	15	1,347	43	440	3,091	2,473

SOURCE: International Food Policy Research Institute (IFPRI), "Survey, 1984/85," South Nyanza, Kenya; and IFPRI, "Follow-Up Survey, 1985–87," South Nyanza, Kenya.

[a] 1985/86 incomes adjusted to 1984 levels using GDP deflator. World Bank 1986, 1987.
[b] Sugar farmers have significantly ($p < 0.05$) higher marketed agricultural income per capita than all other groups.
[c] Sugar farmers have significantly ($p < 0.05$) higher income than nonsugar and landless groups.
[d] New entrants versus nonsugar farmers ($p < 0.05$).
[e] New entrants versus sugar farmers ($p < 0.05$).
[f] Sugar farmers versus nonsugar farmers ($p < 0.05$).
[g] Merchants have significantly ($p < 0.05$) higher income than all other groups.

came from more varied sources. The new entrant group increased the proportion of income earned from commercial agricultural income while at the same time it increased subsistence income—that is, agricultural production used for own consumption. Table 16.2 provides an explanation of how this occurred. The new entrant group has the highest mean area per capita devoted to food crops. Of the KSh 1,129 difference in nominal income per capita between the new entrant and nonsugar cohort groups in the 1985–87 study, 41 percent is contributed by commercial agricultural income, 38 percent by subsistence income, and the remaining 21 percent by higher nonfarm incomes in the new entrant group (table 16.1).

The profitability of sugarcane compared to maize for small farmers is due in large part to the pricing policy pursued by the government of Kenya. The price paid to sugarcane producers is set by the government. Since 1980, the producer price of sugarcane has increased both in nominal and real terms; had the government used the world price of sugar to set the producer price for sugarcane, incomes of farmers would have been negative. The government's pricing policy has protected the incomes of sugarcane producers.

TABLE 16.2 Characteristics of farming patterns for agricultural households, cohort group, South Nyanza, Kenya, 1984 and 1986

Characteristic	1984[a]			1986[a]		
	New Entrants	Sugar Farmers	Nonsugar Farmers	New Entrants	Sugar Farmers	Nonsugar Farmers
Farm size (hectares)	5.0	5.6	3.7	5.0	5.1	3.4
Farm area devoted to all crops (percent)	51.7	66.9	56.6	55.6	55.6	44.4
Farm area devoted to food crops (percent)	36.4	36.0	52.1	31.4	31.4	40.3
Mean area per capita devoted to food crops (hectares)	0.19	0.18	0.19	0.17	0.14	0.14

SOURCE: International Food Policy Research Institute (IFPRI), "Survey, 1984/85," South Nyanza, Kenya; and IFPRI, "Follow-up Survey, 1985–87," South Nyanza, Kenya.
[a]Long rain seasons.

New entrants and sugar farmers put a larger proportion of their total land into production. The outgrowers' program, by providing factory labor to the sugarcane producers, has allowed farmers to cultivate more land. Subsistence food crop production has not been jeopardized by the entry into the sugarcane scheme.

Caloric Consumption of Households and Preschoolers

There were no significant differences in the caloric intake per adult-equivalent in the 1984–85 study across the different types of households. However, in the follow-up study, the daily per capita intake of 2,848 calories of new entrants was significantly higher than the 2,641 calories of nonsugar farmers and the 2,649 calories of sugar farmers (table 16.3).

More important than average energy consumption within the different activity groups is the distribution of caloric consumption within the groups. The households consuming 80 percent or less of their energy requirements are of particular concern, since it is unlikely that at this level of consumption these households have adapted successfully to low energy intake. The proportion of the new entrant sample that falls into this severely restricted group is only 24 percent compared to 30 percent of the nonsugar farmers and 27 percent of the sugar farmers in 1984–85.

TABLE 16.3 Comparison of household caloric intake for cohort and total sample, 1984–85 and 1985–86, South Nyanza, Kenya

	Kilocalories per Adult Equivalent per Day[a]		Households Consuming ≤ 80 Percent of Recommended Calories	
Activity Group	1984–85[b]	1985–86[b]	1984–85	1985–86
			(percent)	
New entrants	2,822	2,848[d,e]	24	30
Sugar farmers	2,689	2,649[d]	27	35
Nonsugar farmers	2,669	2,641[e]	30	33
Merchants	2,281	2,462	—	—
Wage earners	2,898	2,668	—	—
Landless	2,506	2,751	—	—
Mean	2,657	2,663	—	—

SOURCE: International Food Policy Research Institute (IFPRI), "Survey, 1984/85," South Nyanza, Kenya; and IFPRI, "Follow-Up Survey, 1985–87," South Nyanza, Kenya.
[a]For both studies, all-season average.
[b]No two groups significantly different at $p < 0.05$ level.
[c]All-round average.
[d]T-test for new entrant versus nonsugar group ($p < 0.05$).
[e]T-test for new entrant and sugar group ($p < 0.05$).

New entrant households not only had higher household energy consumption, on average, compared to the nonsugar farmer households, but a smaller proportion of new entrants fell below 80 percent of standard calorie consumption.

The household consumption function reported in table 16.4 indicates that total household income has a positive and significant effect on household caloric intake. Interestingly, different sources of income have different effects on household energy intake over and above the pure income effect. Nonfarm income has a negative effect on household caloric consumption. However, women's income has a beneficial effect on household caloric intake, holding overall income constant.

One major reason for the differential effects of various sources of income on energy consumption relates to control of income within the household. These data support one of the early hypotheses of the study, that it may not simply be total income but control of income and source of income that are important in influencing household-level food security.

Dietary patterns of preschoolers were also analyzed; the elasticity of household calories on a child's calories is 0.14, that is, for each 10 percent increase in household calories, there is a 1.4 percent increase in the child's calories.

TABLE 16.4 Household consumption function, 1985–87

Variable	β	t-Statistic	Sign
Women's income (percent)	18.6	2.69	0.007
Dummy Round 2[a]	−1,139	−2.89	0.037
Round 3	−1,975	−3.59	0.0003
Round 4	−1,824	−3.27	0.0011
Head of household schooling (in years)	−93	−1.7	0.08
Adult-equivalent units	2,278	46.1	0.0000
Income per capita	2.2	6.2	0.0000
Income squared	−1.43E-04	−4.2	0.0000
Nonfarm income (percent)	−31.4	−2.89	0.004
Relocated household[b] (1 = yes, 0 = no)	50	0.08	0.936
Sugarcane income (percent)	−5.974	−2.54	0.011
Constant	−665	−0.67	0.498
$R^2 = 0.62$			
Analysis of variance			
Regression 11 $F = 204$			
Residual 1,366 Sig $F = 0.0$			

SOURCE: Kennedy (1989).
NOTE: The dependent variable is total daily household caloric intake.
[a]Surveys were conducted in four rounds: round 2 = pre–long rains harvest; round 3 = harvesting; round 4 = post–long rains.
[b]Households that were relocated as a result of creation of the sugarcane factory.

258 Eileen Kennedy

Morbidity Patterns and Nutritional Status of Preschoolers and Women

The sugarcane scheme is one form of development assistance that was targeted to South Nyanza District with the expectation that the economic growth generated by the outgrowers' program would result in improved health and nutritional status for the population, and, in particular, for vulnerable groups such as preschoolers and pregnant and lactating women.

For the cohort sample of women and preschoolers, there is no significant difference in total time ill or time ill with diarrhea (preschoolers only) across any of the household groups. In addition, for both women and preschoolers, there are no significant differences across income quartiles in the total percentage of time ill in the follow-up study (table 16.5). This is similar to findings from the 1984–85 baseline study, which indicated that increasing income was not associated with a decreasing prevalence of illness, at least in the time period covered by the study. There are also no differences for preschoolers across income quartiles in the total time ill with diarrhea.

Findings on the nutritional status of preschoolers and women parallel the morbidity findings. Table 16.6 presents the average Z-scores[3] for height-for-age, weight-for-age, and weight-for-height for preschoolers for both studies and the mean body mass index[4] for women from both studies. There are no significant differences across the different catego-

TABLE 16.5 Total time ill and time ill with diarrhea of preschoolers and women, stratified by income quartiles, South Nyanza, Kenya, 1985–87

Activity Group	Income Per Capita Quartile			
	1	2	3	4
Preschoolers				
Total time ill[a] (percent)	27.6	30.6	31.9	31.0
Number[b]	(399)	(398)	(311)	(388)
Time ill with diarrhea[a] (percent)	3.7	4.8	3.9	4.3
Number[b]	(403)	(405)	(316)	(391)
Women				
Total time ill[a] (percent)	21.4	25.6	26.1	22.8
Number[b]	(170)	(162)	(140)	(159)

[a] No significant difference among groups.
[b] Number of women or children.

3. Z-score is defined as $\dfrac{\text{(Actual measurement} - \text{Standard)}}{\text{Standard deviation of the standard}}$.

Standards are based on National Center for Health Statistics 50 percentile standards.

4. Body mass index (BMI) = weight (in kilograms)/height^2 (in meters).

TABLE 16.6 Z-scores for children, in 1984–85 study and 1985–86 study, South Nyanza, Kenya

	1984–85			1985–86			Women's Body Mass Index (All-Round Average)
		Z-Score[a]			Z-Score[a]		
Activity Group	Length/Age	Weight/Age	Weight/Length	Length/Age	Weight/Age	Weight/Length	
New entrants	−1.46	−1.13	−0.27	−1.74	−1.06	0.005	22.1
	(90)	(90)	(90)	(61)	(61)	(61)	(58)
Sugar farmers	−1.34	−1.03	−0.22	−1.67	−1.14	−0.15	22.3
	(356)	(356)	(356)	(243)	(243)	(241)	(305)
Nonsugar farmers	−1.50	−1.17	−0.31	−1.76	−1.10	−0.04	22.2
	(556)	(556)	(556)	(349)	(353)	(349)	(390)
Merchants	−0.99	−0.86	−0.27	−1.05	−0.89	−0.26	22.6
	(62)	(62)	(62)	(29)	(29)	(29)	(24)
Wage earners	−1.65	−1.49	−0.59	−1.87	−1.49	−0.51	21.1
	(30)	(30)	(30)	(24)	(24)	(24)	(15)
Landless	−1.45	−1.06	−0.18	−1.99	−1.36	−0.16	22.3
	(77)	(77)	(77)	(40)	(40)	(39)	(44)
Sample mean	−1.42	−1.11	−0.28	−1.72	−1.13	−0.10	57.3
	(1,171)	(1,171)	(1,171)	(746)	(749)	(743)	(1,015)

NOTES: No two groups are significantly different. Numbers in parentheses are sample sizes.
[a] All-round average for each Z-score.

either women or preschoolers. There is also no difference in the prevalence of stunting (less than 90 percent of height-for-age), wasting (less than 90 percent of weight-for-height), or weight-for-age of less than 80 percent for the cohort sample of preschoolers.

Determinants of Preschooler Morbidity and Nutritional Status

Multivariate analyses for morbidity and nutritional status substantiate much of what has been presented in the descriptive analyses.

Morbidity models were specified for preschoolers and women.[5] Results for both women and preschoolers indicate that income is not a significant determinant of illness. In addition, health expenditures per capita are not significantly associated with the prevalence of illness in either women or preschoolers, at least in the short run. Both these findings may seem counterintuitive until one looks at them more closely. Income in the household tends to be spent on a mix of goods and services that, at least in the short term, do not impart a health benefit to women or children. Sugar-producing households spend a slightly higher proportion of their income on nonfood expenditures compared to non-sugar-producing households. Increased income in sugarcane scheme participant households is spent on items such as improved housing and education, which may produce health benefits in the long term, but, in the short run, are not associated with changes in morbidity patterns.

Policymakers are concerned about short-run morbidity because the acutely ill have a higher risk of mortality. In the study area, better-nourished children are, in fact, less likely to be sick (Kennedy 1989). Similar results are seen for women. Women with a higher body mass index are significantly less likely to be ill.

The lack of an income-morbidity relationship in either women or preschoolers should not be used to argue that income is not important, but rather that, at least in the short term and possibly in the medium term, planning for income-generating schemes should be coordinated with other health and sanitation initiatives. Complementarities between increased income and an improved health and sanitation environment should be stressed so that the potential effects of commercial agriculture on overall welfare can be enhanced.

Growth models for preschoolers were specified (table 16.7); one robust finding for all three Z-scores is the strong relationship between the baseline Z-score from the 1984–85 study and the anthropometric indicator from the 1985–87 study. Given that the time span between the

---

5. See Kennedy (1989) for model specification and empirical results.

TABLE 16.7 Regressions of preschooler Z-scores for height-for-age, weight-for-age, and weight-for-height, 1985–87 study

Independent Variables	Height-for-Age		Weight-for-Age		Weight-for-Height	
	β	t-Statistic	β	t-Statistic	β	t-Statistic
Head of household away	−0.134	−0.74	−0.137	−0.99	−0.03	−0.20
Household size	−0.012	−1.51	−4.28E-03	−0.70	1.44-03	0.216
Sex (1 = boy)	0.090	1.09	0.03	0.40	−0.03	−0.47
Female head of household	−0.211	−1.45	−0.14	−1.29	−0.07	−0.54
Average Z-score (Study 1)	0.605	20.67	0.55	20.09	0.39	10.64
Child's calories	2.42E-04	1.96	2.02E-04	2.13	1.10E-04	1.07
Percent time with diarrhea	−0.02	−3.36	−0.014	−3.03	−7.09E-03	−1.46
Mother's height	0.026	3.69	7.80E-03	1.46	−6.3E-03	1.09
Area (hectares)	2.98E-03	0.29	0.01	1.36	0.01	1.29
Age	0.025	8.10	0.01	6.21	1.27E-03	0.51
Constant	−5.60	−4.56	−2.03	−2.16	2.02	2.04
R^2	0.51		0.49		0.20	
Degree of freedom	512		514		512	
F-value	53.4		49.3		13.0	

SOURCE: Kennedy (1989).

baseline period and the end of the 1985–87 study is two and one-half years, an effect this strong might not be expected. However, an increase in the height-for-age Z-score of 1.0 in the baseline period is associated with a 0.61 increase in the 1985–87 height-for-age Z-score. The corresponding values for weight-for-age and weight-for-height are 0.55 and 0.39, respectively. This would indicate that children who were doing well in the 1984–85 baseline study had a high probability of doing well in the follow-up period. Conversely, children who were not doing well earlier had a high probability of continuing to have less than optimal nutritional status. These data also give credence to those who advocate growth faltering as a major criterion for identifying children who are at risk of long-term growth problems.

Preschooler caloric intake is a significant positive determinant of children's height and weight but not weight-for-height. Similarly, the percentage of time ill with diarrhea has a negative effect on height and weight but not weight-for-height. Finally, in general, as children get older, their Z-scores for height-for-age and weight-for-age improve.

Conclusions

The results from the two studies in Southwestern Kenya suggest some very positive impacts of commercial agriculture on household income. The sugarcane outgrowers' program, as it is implemented, is associated with a significant increment in income for both new entrants and sugar farmers. This income, in turn, produces some positive effects on household energy consumption of sugar farmers. However, this benefit at the household level does not appear to have had a dramatic influence on preschooler morbidity patterns or growth. There is a growing awareness that family-level factors may be poor predictors of a child's nutritional status.

Many governments and international agencies are putting increased emphasis on income-generating schemes as a way of achieving health and nutrition objectives. While income may be a necessary condition, increases in income, by themselves, may not be sufficient to alleviate malnutrition, at least in the short term.

Data from the 1984–85 study suggest that the health and sanitation environment had the greatest impact on preschoolers' growth. In the 1985–87 study, one of the major determinants of child growth is the growth pattern of this same child in the baseline study; this earlier growth of the child was influenced significantly and negatively by the preschooler's morbidity patterns and the health and sanitation environment. Children who were not doing well earlier continued to have inadequate growth, which suggests that without improvements in factors that influence children's health, their growth will not be substantially improved. Greater emphasis needs to be placed on the health implications of agricultural policies and projects, with particular attention on ways to improve the health infrastructure in a given community.

Finally, there is an issue that was not touched on directly in the study but warrants discussion: nutrition education. The community in which the outgrowers' scheme has been implemented is one where malnutrition is endemic. There may not be an awareness on the part of households that malnutrition is, in fact, a problem, since their children look like all other children in the community. The outgrowers' program involves approximately 30 percent of the households in the community. This would be an excellent and visible way to reach a significant portion of the community regarding the nutritional needs of the maternal and child population. Nutrition education integrated into a primary health care delivery system could have a significant effect on the health and sanitation environment of children.

To date, most of the farmers who joined the outgrowers' scheme have remained. It is naive to think, given the way the program now operates, that there will be a mass exodus back to food crop production. Farmers are making a profit and will probably remain in the scheme. Some fine tuning of the program will help maximize the potential impact of the increased income on household and preschooler nutritional status.

17 Commercialization of Rice and Nutrition: A Case from West Kenya

RUDO NIEMEIJER

JAN HOORWEG

Introduction

The Kenyan government has actively encouraged cultivation of cash crops as part of its past and present development policies. Cash crop production of industrial and food crops has grown steadily, although the share of agricultural production that is marketed is still less than 50 percent (Kenya, Central Bureau of Statistics 1988). This chapter studies the cultivation of commercial rice at large irrigation schemes in Nyanza Province, western Kenya.

In Kenya, the pressure on arable land is already high (Kliest 1985). Irrigated agriculture offers an important means of agricultural intensification. Kenya's potential for irrigated agriculture is estimated at 500,000 hectares or more, of which, in 1985, some 40,000 hectares were actually under irrigation (Ruigu 1988). About 60 percent of the irrigated area belongs to large commercial enterprises and is used for cultivation of cash crops, such as pineapples, and horticultural crops. Another 10 percent of the irrigated area is accounted for by smallholder schemes, which have grown substantially in recent years (Ruigu 1988). The National Irrigation Board (NIB) manages another 10,000 hectares, with seven large schemes in different parts of the country, where mainly cotton, rice, horticultural crops, and maize are grown. At the NIB schemes, individual farmers are allocated plots to cultivate scheduled crops, and nothing but these crops. The schemes are centrally managed and the produce is centrally purchased. Clearing and preparation of the plots are taken care of by the management. Tenants have limited freedom in organizing their production and are largely dependent on deci-

---

The research reported here was made possible under a program of cooperation between the Ministry of Planning and National Development, Nairobi, and the African Studies Centre, Leiden, and was funded by the Netherlands Ministry of Development Cooperation. The authors are grateful for the assistance of Frank Noy.

sions and practices ordained by management. Tenants at these schemes are faced with a twofold transition—to commercialization and to collective production under management of a parastatal body. This chapter focuses on this combined change and its impacts on household food security and child nutrition.

The NIB has, over the years, suffered various problems and setbacks and generally has not managed to achieve its initial objectives. By 1985, only one scheme had managed to pay its way, whereas the others still received government subsidies (Ruigu 1988). Production at the schemes has generally stayed below projected levels because of technical problems, while pests and diseases have further lowered yields. Considerable inequality exists among households within the schemes. For instance, analysis of production records at Ahero Irrigation Scheme revealed that because of differences in water delivery to certain parts of the scheme, some blocks consistently had lower productivity than others (Noy and Niemeijer 1988). In addition, factors such as the time when the plot is prepared, which is beyond the control of farmers, together with differences in individual farming skills, cause considerable income disparities among tenant farmers. Thus, a disaggregated view of irrigation and farm characteristics is needed to trace technological change and commercialization effects on household consumption and nutrition.

Commercialization and Nutrition

Commercialization, as understood here, means increased production for the market, notably through the production of cash crops. Commercialization means that households increasingly use a significant proportion of their resources, such as land, labor, and capital, for cash crop production. Commercialization is not, especially in the early stages, an irreversible process; production for the market is usually the outcome of choices made by the producers, and households may decide on other crops from one year to the other or, at times, may even revert to subsistence production. Cash crop production also frequently means modernization and intensification of cultivation, notably through the use of fertilizers, insecticides, hired labor, and fixed capital investments, including, for instance, irrigation. Once the modernization process is under way, it is no longer easily reversible and, in the present case of tenants at large irrigation schemes, there is little element of choice left.

Commercialization can influence household nutrition in a positive, negative, or neutral way. Commercialized agricultural production may entail higher output of food crops or higher incomes to secure nutritional needs, or both. However, there is evidence of situations where productivity increases have been realized at the expense of the nutritional status of the farming population or by increased maldistribution of wealth at regional, community, and household levels, or both (Fleuret and Fleuret 1980; Tosh 1980; Pinstrup-Andersen 1985; Pacey and Payne 1985; Maxwell and Fernando 1989).

Findings from different studies in Kenya are equivocal, with different results reported for different crops in different parts of the country. Korte (1969) found nutritional deficiencies among rice growers in Mwea Irrigation Scheme, despite economic progress. Elsewhere in Central Province, a positive relation was found between involvement of households in coffee cultivation and the nutritional status of their children (Hoorweg, Niemeijer, and Steenbergen 1983; Hoorweg and Niemeijer 1989). An analysis of national survey data in two areas studied pointed at a negative correlation between involvement in sugar cultivation and nutritional status in one area and no such correlation in the other (Kenya, Central Bureau of Statistics 1979). Cotton and pyrethrum production, however, were found to be neutral to preschooler nutrition, while coffee and tea production showed no consistent relation. Haaga et al. (1986), analyzing the same survey data, concluded that the cultivation of export crops (coffee and tea) is not generally associated with a higher percentage of stunting. Instead, they concluded that a general association exists between increased landholdings, increased income, and better nutrition and suggest that the effects of cash cropping may possibly be damaging among farmers with very small holdings. However, in households with malnourished children in Central Province, the percentage of the holding devoted to cash crops was not higher than among the general population; this was also the case for households with very small plots (Hoorweg and Niemeijer 1982, 1989). Another study among sugar growers found that food crop production was not negatively affected by sugar cultivation nor was there any negative influence on the nutritional state of young children (chapter 16).

Irrigated cultivation at large schemes in Kenya implies not only changes in agricultural techniques and farm management practices for the farmers concerned but also changes in patterns of income generation and food supply. Scheme production implies a significant transition for the smallholders concerned, with many farm families becoming dependent on agricultural sales to secure their livings. This transition has not gone without problems. In the schemes in Nyanza Province, the first paddy harvests yielded some 6,000 kilograms per hectare, but yields soon became much lower and went down to 2,000 kilograms per hectare. Since then, the situation has improved and yield levels have doubled. Moreover, the planned double-cropping system was not realized; instead, the cropping rate in recent years has been nearer to 1.5 (Noy and

Niemeijer 1988). This pattern of decline after initial success was also observed among rice irrigation schemes in The Gambia (chapter 22).

Nutrition conditions at some of the large schemes in Kenya are reasons for concern and have repeatedly received publicity in the national press. Low income levels of the tenants, diseases associated with stagnant water, unbalanced diets because of a possible change to rice as a staple food, and unbalanced spending of money incomes are speculated to be among the causes of the nutritional problems at these schemes. This chapter assesses the nutritional conditions prevailing among the farming households at two rice schemes in West Kenya: the Ahero Irrigation Scheme and the West Kano Irrigation Scheme, situated on the Kano Plain.

The Study Area

The Kano Plain covers an area of about 650 square kilometers, located near the town of Kisumu in West Kenya. The landscape consists of a wide alluvial plain through which a number of rivers run west toward Lake Victoria. The plain is bordered by hills to the east and steep escarpments to the north and south. The climate is relatively dry with high average temperatures during the day. The soils, of the black cotton type, are fertile but difficult to drain, and seasonal floodings and waterlogging limit agricultural potential.

The population density in the area was 177 persons per square kilometer in 1979 (Kenya, Central Bureau of Statistics 1981), with households mostly living in scattered compounds or in homesteads on the slightly higher grounds. Households go through a distinct cycle, beginning with the establishment of a compound. With an increase in the number of children and with children growing older, there is a period of expansion. There is further expansion when the husband marries additional wives who also move into the compound and when sons marry and bring their wives to live at the compound. In a later phase, the parental household may decrease in size when some sons start building their own compounds and most of the girls are married off. When the head of the parental compound dies, new cycles are started on the new compounds of the sons.

Smallholder farmers make up more than 80 percent of the active population of the Kano Plain. Crop production is mainly geared toward subsistence of the farm family. Maize and sorghum are the main food crops, supplemented, in some cases, by sweet potatoes and cassava. Cotton, rice, and sugarcane are the main cash crops. Agricultural techniques are mainly traditional in nature and, on the whole, are less advanced than in other parts of Kenya. Consequently, yields are low. Crop production is usually accompanied by livestock rearing. Agricultural activity is supplemented with off-farm wage labor, for example, in the adjacent sugar estates north of the Kano Plain, in Ahero trade center, or in Kisumu. The NIB irrigation schemes also provide opportunities for casual labor to the surrounding population.[1] Even so, the Kano Plain with its high natural population increase offers only limited economic possibilities, resulting in considerable outmigration since the 1950s (Kliest 1984).

Irrigation Schemes

The NIB established two large irrigation projects in the Kano Plain area: the Ahero Irrigation Scheme in 1969 and the West Kano Irrigation Scheme in 1976. Land was expropriated and the scheme plots were subsequently distributed among the previous owners and smallholder farmers from neighboring areas. In both schemes, 1.6 hectares of irrigated land are allocated per farming household. Together, the schemes in the mid-1980s covered a total area of 4,800 hectares, out of which 840 hectares were cultivated by 519 tenants in Ahero, while 553 tenants farmed 880 hectares in West Kano. In Ahero, nearly all irrigated farmland is used for paddy cultivation, whereas in West Kano, irrigated farmland is divided between paddy and sugarcane production.

Besides cultivating irrigated crops, tenants also grow rainfed crops, usually on a small area around the house but also on plots outside the schemes. Substantial differences exist with respect to access to rain-fed farmland; some tenants possess relatively large plots, while others have none. When the schemes were first set up, all tenants were obliged to live in designated villages within the schemes. Over time, these villages have become very crowded, compounds are close to each other, and there is little space for gardens. In these circumstances, it is difficult to build extra houses and there is little room for the traditional expansion of the homestead. In fact, sons, on reaching adulthood, are no longer allowed to reside in the scheme according to scheme regulations. Tenants are not allowed to keep cattle at the schemes. The lack of space is an important obstruction and a major reason for people's desire to move outside the scheme. Later, it was tolerated for tenants to take up residence outside the scheme and still retain their scheme plot.

Nowadays, tenants can be distinguished as either "resident tenants," who live within the schemes and have no or relatively little land outside

---

1. It has been estimated that about 40 percent of the agricultural labor in the schemes is provided by hired laborers (Houtman 1981).

Commercialization of Rice and Nutrition: A Case from West Kenya 269

the schemes, or "nonresident tenants," who live outside the schemes and have more sizable tracts of nonscheme land. Depending on the scheme, an estimated 30-50 percent of tenants belong to the latter category (Sterkenburg, Brandt, and von Beinum 1982; Noy and Niemeijer 1988). There are also smallholder schemes in the area that were started by the farmers themselves and are controlled by farmers' committees. Participating farmers individually cultivate a plot, privately owned or rented. Plots are much smaller in size than at the NIB schemes, and farmers generally start to cultivate only when and if they have labor available, thus facing considerably lower labor costs.

Study Design and Findings

An initial survey was conducted during March-April 1984, which is the season of the long rains prior to the harvest of the main staple foods, maize and sorghum, and is a time when foodstocks are usually at their lowest level and nutrition problems are most manifest.

The study compares four groups of farmers, differing in their degree of participation in, and dependence on, irrigated rice production:

- nonrice growers
- resident tenants at the large-scale irrigation schemes
- nonresident tenants at the large-scale irrigation schemes who also farm sizable plots of land elsewhere
- individual rice growers who have a combination of resources similar to that of the third group but who usually cultivate only small rice plots.

Data were collected during a single-visit survey. They covered household composition and conditions, household resources (agricultural production, livestock, and off-farm employment), food consumption (household and individual consumption), and nutritional status (of young children and their mothers). Food consumption data were collected by the 24-hour recall method, and supplemented with observational data from a smaller survey.[2]

A summary of the main economic characteristics of the four groups is presented in table 17.1. Not surprisingly, in view of the residency regulations, the household size of the resident tenants is lowest. The household size of the nonresident tenants, in turn, is highest, with the two other groups in between. Nonresident tenants have the largest land resource base, combining four acres under cash crops with considerable

---

2. A detailed description of this survey is presented in Kliest (1984) and Niemeijer et al. (1985 and 1988).

TABLE 17.1 Resource base of sample rice farmers in West Kenya, 1984

Characteristic	Non-Rice Growers	Individual Rice Growers	Nonresident Tenants	Resident Tenants
Number of households	134	54	64	83
Household size[a]	7.9	8.8	10.3	7.3
Migrant worker[b]	0.8	0.8	0.6	0.1
Cash crop area (percent)[c]	21	17	68	78
Cropped area (acres)	2.8	3.5	5.9	5.1
Cattle (percent)[d]	38	41	31	8

SOURCE: Niemeijer et al. (1985).
[a]Average number of people per compound.
[b]Average number of migrant workers per compound.
[c]Combined average area share for rice, sugarcane, and cotton in total land use.
[d]Percent of households with cattle present.

participation in other economic activities. Individual rice growers and non-rice growers derive considerably less income from cash crops, but have more food crops under cultivation and more off-farm income, and are more involved in keeping cattle. Comparing these four groups, it is clear that resident tenants are an atypical group. They have a large area under cash crops, but in other respects—food crop area, livestock, and off-farm income—they are far behind the other groups and are mostly dependent on their rice cultivation, although some of the resident tenants have a small income from horticultural crops grown on the bunds between the rice fields. Rice production of the two tenant groups is virtually identical, but that of the individual rice growers is smaller.[3]

Income Diversification

Table 17.2 gives estimates of the annual incomes of the four groups, based on the survey findings and where necessary complemented with information from other studies.[4] Included in the total income is the

---

3. The paddy harvests of the two groups of tenants, as reported for the period March 1983–March 1984, were 62.5 and 64.0 bags, respectively. For the purposes of this study, the cultivation of sugarcane by farmers in the West Kano scheme is relatively unimportant—it averaged only 0.5 acre per farmer over the two groups. Furthermore, at the time, the results for sugarcane cultivation were poor. Marketing problems caused serious delays in harvesting, and the returns to sugarcane in the early 1980s were even lower than those to rice cultivation (Houtman 1981).

4. The starting point for all estimates was the survey data (Niemeijer et al. 1985), together with field notes on household budgets collected for a small number of households (Veenstra 1987). Further data concerning farming budgets were available from Jaetzold and Schmidt (1982) for non-rice growers, and Houtman (1981) for tenants at the NIB schemes, while Sterkenburg, Brandt, and von Beinum (1982) gave results of incomes obtained in a single-visit survey for all groups. Incomes from rice farming were further verified from the yearly reports of the NIB schemes and Noy and Niemeijer (1988).

TABLE 17.2 Estimates of annual income of four groups of farmers in West Kenya, 1984

Income Characteristics	Non-Rice Growers	Individual Rice Growers	Nonresident Tenants	Resident Tenants
Share of total income (percent)				
Cereals/legumes	24	20	14	11
Livestock/horticulture	22	22	10	8
Off-farm income	50	38	27	14
Cash crops	4	21	49	67
Income/consumption unit[a] (KSh)	1,163	1,315	1,275	1,317
Income diversity	Medium	Diverse	Medium	Specialized
Income diversity index[b]	19	8	17	28

SOURCE: Survey by the authors.
[a]In adult-male equivalent.
[b]Standard deviation of proportions of above-mentioned income characteristics.

income from cash crops (rice, sugar, and cotton), rainfed crops (cereals and legumes), livestock produce (milk and ploughing), horticultural crops (bund cultivation), and off-farm sources (casual labor, regular wage employment, and remittances). Total annual income estimates strongly reflect the description of the resource base given above. There are important differences among the four groups. Non-rice growers and resident tenants are groups with comparatively low total household incomes; individual rice growers and nonresident tenants have incomes that are 25-40 percent higher. However, these income differences largely disappear when income is corrected for household size. Income per consumption unit is more or less stable over the four groups. However, the four groups do differ by the composition of their incomes. Income is quite diversified in the case of individual rice growers but is largely restricted to one income source in the case of resident tenants. Non-rice growers and nonresident tenants take an intermediate position in this respect.

Consumption and Nutritional Differences

In table 17.3, the four groups have been rearranged according to the degree of income diversification, ranging from specialized to diversified. With regard to food consumption, there is no change in the staple diet, and maize remains as important as elsewhere. This lack of change is understandable as calories from rice are more expensive than those from maize and sorghum by about a factor of two (Noy and Niemeijer 1988).

TABLE 17.3 Nutritional conditions among farmers in West Kenya, by degree of income diversification, 1984

Nutritional Indicators	Resident Tenants	Non–Rice Growers	Nonresident Tenants	Individual Rice Growers
Income diversity	Specialized	Medium	Medium	Diverse
Average household food consumption per consumption unit (kilocalories)	2,494	2,592	2,681	2,767
Percent of households with less than 60 percent of recommended intake[a]	26	21	23	13
Average food consumption of preschool children (kilocalories)	552	658	695	684
Percent of children with less than 60 percent of recommended intake[b]	44	39	34	32
Percent of preschool children with less than 90 percent of height-for-age standard	41	18	19	14
Percent of school-aged children with less than 90 percent of height-for-age standard	30	22	14	18

SOURCE: Niemeijer et al. (1985).
[a]Recommended intake = 2,600 kilocalories per consumption unit.
[b]Recommended intake on the basis of per kilogram body weight (FAO/WHO 1974).

There is a general improvement in nutritional indicators moving from incomes that are derived from one source to incomes that are more diversified. Other indicators also point in the same direction, although less consistently. Besides income diversification, no alternative explanations offer themselves for the identified differences between the nutritional characteristics of the four groups. There are no differences in health conditions: incidence of child health complaints and percentage of school-aged children with low weight-for-height were similar across the different groups, a finding similar to the other study from Kenya reported in chapter 16. The hypothesis that farming families at the schemes eat only rice and, therefore, have an unbalanced diet, was not confirmed; only small quantities of rice are consumed in addition to the main staple food, maize.

One group stands out, though in a negative way—namely, the resident tenants, who score lowest on all indicators, with 40 percent of the preschoolers and 30 percent of the school-aged children showing signs of stunting. The difference in nutritional indicators between this group and the groups with more diversified incomes is high. However, the difference in nutritional indicators between households with moderately diversified and highly diversified incomes is small. Apparently, negative nutritional effects of highly specialized incomes at low-income levels come into effect at a certain threshold, below which these effects become quite pronounced.

One explanation is that varied resources facilitate households to spread risks across years, and ensure that income is evenly distributed and basic household needs receive balanced attention throughout the year. Where this is not the case, misfortunes can start a downward spiral toward nutritional deterioration.

To gain a better understanding of the mechanisms operating in such households, a second in-depth study was conducted during October–December 1986 in Ahero Irrigation Scheme among a subsample from the group of resident tenants—the most specialized group—studied in the initial survey described above. The detailed design of this study is presented in Noy and Niemeijer (1988). While average production of this group of resident farmers over the past 15 years equaled that of the scheme as a whole, individual plot production was determined to a large extent by the block in which the plot was situated and the season in which planting occurred, factors which individual households can do little about in a scheme with agriculture under supervision. Production factors under more household control, notably available household labor and income from off-farm employment, were positively related to paddy production, but livestock ownership was negatively related in regression analysis (Noy and Niemeijer 1988).

Analysis of food expenditures by income source within this rather specialized group revealed a striking difference between the use of male and female incomes for food. Per capita (per consumption unit) food expenditures were increased with income from casual labor by women and from local (illegal) sales of paddy and bund crops, incomes also in the female domain. Even within the altered household economy of the resident tenants, a traditional division of economic responsibilities prevails, in which women are mostly responsible for day-to-day matters, in particular, the provision of food, even when most food has to be purchased. Yet, in this scheme with specialized agriculture under central supervision, women's freedom to raise the resources for these day-to-day and food provision functions is severely constrained. Again, these findings are much in line with those from the other study on sugarcane in Kenya reported in chapter 16.

Conclusions: Income Diversification Matters

Essential changes in farming practices for agricultural modernization include introduction of new crops and improved crop varieties, modern farming techniques and production methods, and alternative land tenure arrangements. These changes imply a large degree of commercialization of agriculture coupled with different forms of production, such as the corporate and private estates that prevailed in colonial times, the agro-industrial complexes and state farms of today, as well as smallholder farmers, whether or not organized in large-scale schemes or other forms of collective production (Hinderink and Sterkenburg 1987). In this chapter, we are dealing with large-scale schemes in which autonomy of individual farmers is quite limited. Management of such schemes has generally assumed that maximization of tenant incomes can be achieved by maximizing rice production and that this, in turn, will lead to greater well-being of the tenants. From the described experiences in Kenya, it has become clear that this process cannot be taken for granted. Moreover, tenant households are not as homogeneous as they are often perceived. Interests of the household heads, the official tenants, may differ from those of their wives and sons or other relatives who cultivate part of the rice plot or who provide part of the labor. Even when food has to be purchased, women are still responsible for provision of food and men are responsible for the lump-sum expenses, such as school fees, clothing, and economic investments.

From the findings of this study, an interesting mix of results emerged: a group with quite unfavorable nutritional conditions and a group with relatively favorable nutritional conditions are both tenants at the schemes, while individual rice growers outside the scheme also do relatively well. It is the group that is mainly dependent on the specialized income from the plots in the schemes—the resident tenants—that shows symptoms of severe nutritional stress. Resident tenants have less land for rainfed agriculture, fewer employed migrants remitting funds to the household, and if they own livestock, they do not have usufruct, a major source of income for non-rice growers. Nonresident tenants, on the other hand, avail themselves of several resources in addition to the rice plot. This group of nonresident tenants has fewer nutritional problems, and this appears to be so because of diversification of income. In fact, in the income range observed here, composition of income appears to be more important than level of income for household nutrition. This finding runs counter to one of the basic assumptions underlying the organization of the rice irrigation schemes—namely, that tenants will be able to earn their living from commercial cropping as their only economic activity. The poor nutritional conditions among the resident tenants puts this assumption under question. On the other hand, in cases where commercial rice cultivation is only one of the available income sources, such negative nutritional differences do not stand out.

18 The Triple Role of Potatoes as a Source of Cash, Food, and Employment: Effects on Nutritional Improvement in Rwanda

JÜRGEN BLANKEN
JOACHIM VON BRAUN
HARTWIG DE HAEN

Introduction

Potatoes, of all food crops, had the fastest average annual growth rate of production in Rwanda between 1970 and 1985: 6.8 percent. This increase in potato production was mainly brought about by a rapid expansion of the cultivated and harvested acreage, which grew by 6.7 percent per year during the same period, whereas average yields only slightly increased by 0.1 percent per year. In 1983, potato production covered an estimated 3.3 percent of the total harvested food crop acreage, and contributed to 3.1 percent of total calorie production (Delepierre 1985).

Given the extreme scarcity of arable land and the rapidly increasing demand for food due to a high rate of population growth, the intensification and diversification of domestic food production are the main objectives of the national agricultural development strategy (Ministère du Plan de Rwanda 1983). Potatoes, in particular, are slated to play a major role in this respect in the future. According to the Ministry of Agriculture's projections, potato production will expand by 5.2 percent per year between 1980 and 2000. Higher yields (4.1 percent per year), rather than area expansion, are assumed to be the major source of future growth.

Within Rwanda, both production and consumption of potatoes are concentrated in only a few growing areas. Following the Rwandan agro-ecological classification system, two major regions can be distinguished —the Volcanic Highlands and the Central Zaire-Nile Divide (Jones and Egli 1984).

In the Volcanic Highlands, prevailing environmental conditions are very favorable for potato cultivation: deep and fertile soils, abundant rainfall following a continuous distribution, and mild temperatures permit cultivation almost throughout the year. It is in this zone that potato production has expanded most rapidly in recent years and that potatoes have attained the highest share in total food crop acreage. Besides the favorable environmental conditions, both input and output market infrastructure are relatively well developed and have led to a more intensive use of modern production technology such as improved seeds, fungicides, and spraying equipment.[1]

The percentage of marketable surplus is also highest in the Volcanic Highlands. and has been estimated at 45 percent of total potato production. This region also provides the bulk of the potatoes that are distributed and marketed in the non-potato-growing regions of Rwanda, particularly the urban centers.

Compared to the Volcanic Highlands, both environmental and market conditions are much less favorable in the Central Zaire-Nile Divide. In general, poorer soils and prevalence of very steep slopes render crop production more difficult, causing severe erosion problems. This zone can be regarded as quite remote, since infrastructure and the transport system, in particular, are only poorly developed. Consequently, availability of modern potato production technology is more limited, and interregional marketing of production surplus is more expensive due to high transportation costs.

Within the Central Zaire-Nile Divide, a special and recent case of land use is the expansion of potato production into the former natural forest of Gishwati. In the northern part of the Gishwati area, a reforestation and pasture improvement project financed by the World Bank has been under way since 1980. Originally, farmers and project employees were allowed to cultivate potatoes in the cleared and reforested parts of the project area in order to keep weeds down. But in the last few years, potato cultivation has expanded rapidly and in an uncontrolled manner, even into areas outside the project.

The access to additional farmland, although probably temporary, has had a tremendous impact on the rural economy of the area. For participating households, such access to land provides an extra source of off-farm income and raises their potential food supply for subsistence. For the region, it generates additional employment and stimulates growth of food supplies on local markets. On the other hand, the Gishwati case may also have some rather problematic implications, whatever the future prospects of the reforestation project. If the authorities stick to the official plan and enforce regrowth of trees, the impact on the regional economy will come to an end and will later appear only as a short-term

1. There can be no doubt that the rapid expansion of potato cultivation in this area was, to a large extent, encouraged by the foundation of the National Potato Program in 1979. The headquarters are located in Ruhengeri. This program was supported by the International Potato Center (CIP). The main activities include selection, multiplication, and distribution of improved potato varieties, including research and extension (Scott 1986; Haugerùd 1985).

effect. If, as appears more likely, farmers continue to cultivate the area and grow annual crops on steep slopes and marginal land, environmental damage from soil degradation and changes of the microclimate cannot be ruled out.

Whatever the final outcome, some interesting questions are raised: (1) How does availability of new technology and additional land (for potatoes) affect production patterns and commercialization? (2) What are the employment effects? (3) How is the incremental income used? and (4) What are the effects on food consumption and nutrition?

Potatoes in the Agricultural Production Systems

The study was conducted in Giciye commune in the Central Zaire-Nile Divide in northwestern Rwanda. A detailed survey of production, marketing, income, consumption, nutrition, and health was undertaken during 1985–86, covering 200 households.

The farming systems—that is, land use patterns and technology—prevailing in the study area are exclusively based on smallholder agriculture. Land is extremely scarce at this location and, in 1985–86, average farm size was only 0.75 hectare. With an average family size of approximately five persons, the resulting average person-land ratio of four adult person-equivalents per hectare is extremely high by international standards. Due to high rates of population growth and rapidly increasing food requirements, former wood and pasture lands have been transformed into permanent arable land. This land transformation process also includes the cultivation of very marginal areas on steep slopes, and a reduction in fallow periods. As a consequence, the scope for livestock production is limited and the agricultural production systems are predominantly based on crop production.

Table 18.1 reveals that land use is dominated by the main cereals—maize and sorghum, which together accounted for half the total farm acreage. They are cultivated in both pure stands and mixed crops. Sweet potatoes and peas each averaged approximately 15 percent of land use. Forty-three percent of sample households had land in the Gishwati area to grow potatoes. While average farm size of households with and without access to potato fields in the Gishwati was identical, the average size of the Gishwati plots was 0.27 hectare, thus increasing the average size of holdings by approximately 36 percent from 0.74 to 1.01 hectares for households with Gishwati land.

Observed differences in land use shares of different crops on the core farms between households with and without Gishwati land cannot be related to differences in access to Gishwati land, but to the effect of different agro-ecological conditions. Farm households with access to

TABLE 18.1 Land use of households with and without additional Gishwati land, Rwanda, 1985–86

Basic Resources and Demographic Indicators	Households With Gishwati Land	Households Without Gishwati Land	Average of All Sample Households
Number of households	18	24	42
Average farm size (hectares)			
Excluding Gishwati land	0.74	0.75	0.75
Including Gishwati land	1.01	0.75	0.86
Number of adult person equivalents	2.80	2.70	2.70
Number of consumer-equivalents	3.60	3.30	3.50
Altitude (meters)	2,428	2,244	2,322
Land use (as percent of total core farm)[a]			
Maize (including mixed)	37.0	24.8	30.0
Sorghum (including mixed)	13.2	19.4	16.8
Beans	2.0	9.4	6.3
Peas	21.9	10.5	15.4
Sweet potatoes	9.4	20.0	15.5
Potatoes (non-Gishwati)	0.0	2.0	0.2

SOURCE: Based on a survey conducted by the International Food Policy Research Institute.
NOTE: Based on a subsample of 42 households where detailed data on cropping systems have been collected.
[a]Core farm refers to the farmland, excluding Gishwati land.

Gishwati land were located on higher altitudes and grew relatively more maize and peas, in particular, and less sorghum, sweet potatoes, and beans.[2]

Farmers with access to Gishwati land did not grow any potatoes in pure stands on their own fields, although they did so on the Gishwati fields. These farmers continued to grow a considerable amount of potatoes in mixture with maize. Thus, it can be concluded that the introduction of improved potato seeds and the increasing availability of other modern production technology for potatoes neither motivated a change in technology nor led to higher shares of potatoes in the average core farm outside the Gishwati land. On the contrary, in the sample households, modern potato cultivation was almost exclusively restricted to the

---

2. Multiple regression showed, for example, that the land use share of maize increases, *ceteris paribus*, by 6 percentage points per 100 meters of altitude. The analysis further revealed that, within the same altitude group, the person-land ratio is the most important determinant with respect to different crop shares in total farm size.

Gishwati area. This leads to the more fundamental question of whether the farms outside the Gishwati area are more resistant against innovations and changes in the external economic environment. As will be shown, there are some obvious adjustments of the labor economy of the farm households, but not of the subsistence-oriented production system per se.

Factors Explaining the Stability of Production Patterns on Core Farms

An obvious explanation for why farmers are eager to grow potatoes on the new land but not to expand the potato area on their own farms could be that potatoes are the only crop permitted on the new land but are not competitive to other crops. Yet this is not so evident; at least in terms of returns to scarce land, potatoes are rather competitive.

Table 18.2 summarizes the average land and labor productivity for the most important crops and cropping systems. Cropping systems at the study location are extremely labor intensive compared to other African countries. Potatoes have the highest labor input, an average of 622 person-days per hectare. Potatoes, with 5,526 million calories per hectare and season, yield the highest gross land productivity of all crops. The gross margin per hectare is also highest for potatoes. However, due to extremely high labor requirements, the labor productivity of potatoes is relatively low and ranks only fifth after maize (both in pure stands and mixed crops), peas, and sweet potatoes. In terms of gross margin per person-day, potatoes tend to be more profitable than cereals but less profitable than beans, sweet potatoes, and, in particular, peas.[3]

Given that potatoes show by far the highest land productivity, why do farmers at this extremely land-scarce location not shift to potato production on a larger scale on their own fields? Different environmental conditions cannot be regarded as a factor inhibiting a more widespread expansion of potato production outside the Gishwati area, at least for the higher-altitude zones of the study area. A closer look at technical and financial characteristics of potatoes as well as explanations by farm families suggests some other reasons why farmers have so far not specialized in this apparently profitable commodity: high cash requirements,

---

3. It has to be added that the data shown in table 18.2 relate to land and labor productivities per crop and season. Since the length of the vegetative cycles of the crops under consideration varies substantially from eight to nine months in the case of maize, sorghum, and sweet potatoes to five months for leguminous crops and potatoes, the possibility of double cropping exists, at least theoretically, for the latter. Especially for potatoes, two crops per year on the same plot have often been observed. Thus, the comparative advantage of potato cultivation with respect to land productivity in both physical and monetary terms might be even greater than the data in table 18.2 suggest.

TABLE 18.2 Average land and labor productivity of different cropping systems, Rwanda, 1985–86

Item	Maize (Pure Stands)	Maize (Mixed Stands)[a]	Sorghum (Mixed Stands)	Beans (Pure Stands)	Peas (Pure Stands)	Sweet Potatoes (Pure Stands)	Potatoes (Pure Stands)	Tea (Smallholder)[b]
Total labor input (person-days per hectare)[c]	261.6 (0.42)	260.5 (0.38)	346.9 (0.64)	315.1 (0.24)	101.5 (0.56)	398.0 (0.63)	621.8 (0.78)	470.0 (0.40)
Yield (kilogram per hectare)[c]	1,037 (0.45)	— —	— —	875 (0.50)	476 (0.55)	5,055 (0.42)	9,615 (0.87)	2,692 (0.49)
Yield (1,000 calories per hectare)	3,383	3,789	3,649	2,655	1,488	5,472	5,526	—
Gross revenues (US$ per hectare)[d]	161	281	226	292	159	281	642	329
Variable costs (seeds; fungicides for potatoes only) (US$ per hectare)[d]	7	127	75	17	17	0	222	98
Gross margin per hectare (US$ per hectare)[d]	155	154	151	275	142	281	419	232
Labor productivity (kilocalories per person-day)	12.9	14.6	10.5	8.4	14.7	13.7	8.9	—
Gross margin per person-day (US$)[d]	0.62	0.59	0.44	0.87	1.40	0.71	0.67	0.65
Number of observations	21	24	19	18	35	6	18	12

SOURCE: Based on a survey conducted by the International Food Policy Research Institute.
NOTE: Based on a subsample of 42 households only where detailed data on cropping systems have been collected.
[a] Most frequently maize, beans, and potatoes.
[b] Average for 1985–86 (annual).
[c] Figures in parentheses are coefficients of variation.
[d] US$1 = FRw 89.9 (1986 exchange rate; average of January–August).

higher risk of crop failure, seasonal labor bottlenecks, tastes and preferences, and insufficient information.

As indicated in table 18.2, potatoes have the highest per hectare cash requirement for fertilizer, seeds, fungicides, and other inputs. Furthermore, farmers often have to overcome seasonal labor shortages by hiring nonfamily labor, which causes additional liquidity problems. Risk of crop failure, due to diseases in particular, is also high for potatoes. Traditional dietary preferences tend to constrain change, since many households would have to reduce production of other subsistence crops in order to expand potato production, which would imply either an increase in the consumption share of potatoes or a higher dependence on food purchases in exchange for potato sales. While both types of adjustment do occur, elasticities of response are rather low. Although production patterns on the farms have turned out to be quite inflexible, the labor economy has responded much more noticeably to the new situation.

Impacts on Employment and Division of Labor

Adoption of potato production in the Gishwati led to some noteworthy changes in the division of labor to agricultural production at both the interhousehold and intrahousehold levels. While the contribution of nonfamily members to the total annual labor input to potato production (23.0 percent) did not differ greatly from that to all crops (26.3 percent), the composition of the nonfamily labor input was quite different. Employment of daily wage laborers was very important in potato production. Wage labor contributed, on average, 44.5 percent of total nonfamily labor input to potato production. In general, the main source of nonfamily labor was reciprocal exchange with relatives or neighbors or both, which provided about 78.2 percent of the total nonfamily labor input, while the share of paid daily wage labor was only 17.8 percent.

Land preparation, harvest, and postharvest activities are the main operations in potato production for which wage labor is hired. In absolute terms, wage labor was hired for approximately 64 person-days per hectare and season. Hence, potato production in the Gishwati area significantly contributed to rural employment creation in the agricultural sector. It will be shown later that this aspect of employment creation is of paramount importance, since the vast majority of farm households at the study location are too resource poor to earn a living from agriculture alone and, therefore, depend heavily on off-farm income.

At the intrahousehold level, potato production significantly affects the gender division of labor in agricultural production. Women's contribution to total family labor input by crop and activity was always much higher than that of men, and it is only in potato cultivation that adult males provided a substantial part, 41.7 percent, of the total family labor input.[4] While certain operations, like weeding, were gender specific and almost exclusively women's domain for most crops, men participated to a larger extent in all cropping activities for potato production. In households with potato fields in the Gishwati, men provided an average of 33.7 percent of total family labor input, compared to only 22.0 percent in households without access to Gishwati land.

The access to additional land outside the farms had two further interesting effects: it mobilized extra labor inputs by male family members, and it motivated a substitution of labor by capital on fields of the farms. Labor input of adult males for nonagricultural on-farm activities and off-farm work did not vary substantially between households with and without access to Gishwati land. Hence, the higher male adult labor input in agricultural production in potato-growing households led to an absolute increase in total annual labor input for all activities, from 162 to 175 days per male adult (an increase of 8 percent). Only marginal differences can be observed with respect to absolute total labor input per female adult and distribution of female labor input into agricultural and nonagricultural activities between the two groups of households (table 18.3). Thus, it can be concluded that adoption of potato production increased the overall level of employment per male adult, without reducing the labor input for other activities and off-farm work, in particular.

Another important impact of the incremental potato production is reduction in labor input intensity on the farm. Labor input intensity averaged 211 person-days per hectare and year in those households with potato fields, significantly lower than the labor intensity of 283 person-days for other households. Surprisingly, shifting labor from the farm to potato production was not detrimental to overall calorie production and average land productivity on the farm, because of increases in the capital intensity of production technology. Apparently, income from Gishwati production was partially invested in better quality tools for field work.

Food Availability and Commercialization

The majority of the sample households cannot meet their subsistence requirements from own production on the farms and consequently depend to a large extent on additional market purchases. Although average farm size without the Gishwati land was, as pointed out earlier,

---

4. Adult women above 15 years of age contributed 74.1 percent of total annual family labor input, whereas the share of adult men averaged 25.3 percent, and the remaining share of 0.6 percent was provided by children of both sexes below 15 years of age.

TABLE 18.3 Labor allocation in households with and without additional Gishwati land, by gender, Rwanda, 1985–86

	Labor Allocation			
	Households Without Gishwati Land		Households with Gishwati Land	
Labor Input per Year	Days	Share of Total (percent)	Days	Share of Total (percent)
Adult male				
Agricultural production	46	28.4	62	35.4
On-farm, nonagricultural activities	40	24.7	40	22.9
Off-farm work	76	46.9	73	41.7
Total	162	100.0	175	100.0
Adult female				
Agricultural production	106	60.6	110	60.8
On-farm, nonagricultural activities	69	39.4	71	39.2
Total	175	100.0	181	100.0

SOURCE: Based on a survey conducted by the International Food Policy Research Institute.
NOTE: Based on a subsample of 42 households only where detailed data on cropping systems have been collected.

almost identical for households with and without access to the Gishwati land, calorie production per household and per consumer-equivalent on the farm was higher, by 15 percent, for households with additional land in the Gishwati.[5] On average, potato production in the Gishwati provided an additional 40 percent calorie production, which raised daily calorie supply per adult consumer-equivalent of these households to 3,587 versus 2,184 calories in the non-Gishwati farm households. Assuming average annual requirements of approximately 2,800 calories per consumer-equivalent per day, potato production in the Gishwati thus enabled these households to become surplus-calorie producers.

Table 18.4 shows that production systems in the study area were very much subsistence oriented. In 1986, the average subsistence degree in agricultural production, defined as the value of subsistence production over the value of total agricultural production (including sales of animal products and home-produced sorghum beer), amounted to 73.7 percent.

5. The reason for the higher calorie production on farms with Gishwati income is partially due to higher altitude and related differences in cropping patterns. It is also the result of capital-labor substitution, as indicated before.

TABLE 18.4 Overall degree of subsistence orientation in agricultural production and share of marketed surplus by crop, Rwanda, 1985–86

Crop	Households Without Gishwati Land ($n = 105$)	Households with Gishwati Land ($n = 81$) (percent)	Average of All Sample Households ($n = 186$)
Share of marketed surplus			
Maize	3.2	1.4	2.5
Sorghum	6.5	1.8	4.5
Beans	0.8	0.0	0.4
Peas	0.6	2.4	1.4
Sweet potatoes	6.3	2.6	4.7
Potatoes			
Non-Gishwati	0.5	0.8	0.6
Gishwati	—	28.7	—
Subsistence degree[a] (1986)	76.8	69.3	73.7

SOURCE: Based on a survey conducted by the International Food Policy Research Institute.
[a] Defined as the value of subsistence production over the value of total agricultural production.

The subsistence degree was somewhat lower in households with access to Gishwati production.

The degree of market integration of all crops grown on the farms is rather small. Differences of minor importance can be observed for the marketed shares of sorghum and sweet potatoes, on the one hand, and for peas, on the other hand, but these might be attributable to differences in land use. Potatoes on the farm appear to be grown predominantly for subsistence consumption, as indicated by the low commercialization degree of 0.8 percent only. With a commercialization degree of 28.7 percent, Gishwati potatoes are the only crop marketed to a larger extent.

Access to additional land resources also affected interfarm equity within the group of households with access to Gishwati land. The size of the additional potato plots varied considerably from 0.02 hectare to 2.8 hectares. Grouping these households by the size of the core farm shows that households in the lower quartile were able to increase total farm size from 0.18 to 0.25 hectare, or by 38.8 percent, whereas, in the upper quartile, average farm size increased from 1.3 to 1.73 hectares, or by 33.1 percent. The inequalities with respect to access to potato fields in turn had an important impact on the use of the additional calorie production from potatoes (table 18.5). When households with access to Gishwati land are grouped by the level of calorie production per consumer-equiva-

TABLE 18.5 Increase in farm size and calorie production in households with additional Gishwati land, by quartile of calorie production per consumer-equivalent, Rwanda, 1985–86

	Calorie Production per Consumer-Equivalent Quartile			
Item	First	Second	Third	Fourth
Farm size (hectares)				
Without Gishwati land	0.50	0.59	0.60	0.86
With Gishwati land	0.61	0.71	0.75	1.32
Increase (hectares)	0.11	0.12	0.15	0.46
Increase (percent)	22.0	20.3	25.0	53.5
Daily calorie production per consumer-equivalent				
Without Gishwati land	942	1,510	2,367	5,079
With Gishwati land	1,189	2,016	3,025	7,707
Increase (percent)	26.2	33.5	27.9	51.7
Total potato production sold (percent)	7.6	31.5	29.2	45.4

SOURCE: Based on a survey conducted by the International Food Policy Research Institute.

lent on the core farms, only minor differences existed among the first three quartiles concerning additional land in the Gishwati: the increase in total farm size ranged between 20 and 25 percent, whereas, in the upper quartile, total farm size increased drastically from an already high level of 0.86 hectare to 1.32 hectares (by 53.5 percent).

Although the additional calories from potato production substantially increased total calorie production of the first three quartiles, it did not allow these households to become surplus producers. The additional calories from potato production were primarily used for subsistence consumption, as indicated by the percentage of total potato production that was marketed. The marketed surplus in the first quartile, in particular, was extremely low and averaged only 7.6 percent of total Gishwati production. In the upper quartile, a considerable marketable surplus exists, and the average for the whole sample was 28.7 percent (table 18.4). Thus, it can be concluded that among the group of Gishwati farmers, it was the relatively richer households that participated in potato production to the greatest extent and thereby were able to expand their production and income potentials even further.

Another important aspect of potato production in the Gishwati is its impact on income control within the household. It is frequently found that in many Sub-Saharan countries women are mostly responsible not only for cultivating food crops but also for their marketing. Women in this setting in Rwanda provide the bulk of the labor input in agriculture, but men participate to a larger extent in the new potato production. It was found that the higher male labor input in potato production also led to more male control over the use of that production. While women were responsible for most of the marketing of food crops in 75 percent of the households, they exerted dominant control over potato sales in only 25 percent of the households.

Finally, with respect to long-term implications, a warning must be expressed. Although the Rwanda case can be used to examine the short-term impact of extra agricultural off-farm income, it cannot be generalized as a case study of a stable long-run external source of extra employment and income. While the expansion of cultivated land provided a rather steady flow of additional income in earlier periods of low population pressure and extended forest areas, today the situation is drastically different, as Rwanda's agriculture has nearly reached a cropping frequency that approaches permanent cultivation year round. This has already caused yields of major crops to decline and forest areas to shrink rapidly. Given the engagement of farmers in potato production on newly cleared forest land, it is not unlikely that they will tend to keep the land cultivated and prevent the regrowth of trees, against the original intention of the project. This would certainly not only cause yields of potatoes to decline but would also establish severe ecological risks, a change toward a less humid microclimate, in particular.

While the production story of area expansion and new technology in potato production is certainly central to consumption potentials at household levels, behavioral factors come into play that permit us to assume only to a limited extent a direct translation of the production effects into food energy consumption effects, even in the very subsistence-oriented farm households. In the following two sections, we therefore make an attempt to explain calorie consumption and preschooler nutrition at the household level and, in this context, evaluate the role of commercialization versus subsistence orientation of production.

Calorie Consumption: Deviating Effects of Commercialization

The calorie consumption model shows a strong relationship between income and calorie consumption: a 10 percent increase in income raises calorie consumption at the sample mean by 4.7 percent (table 18.6). Substantial differences exist between the top and bottom income quartiles' use of incremental income for food energy: an additional FRw 100 (US$1.11) of per capita monthly income would raise household-specific calorie consumption by 6.5 percent in the bottom quartile house-

TABLE 18.6 Estimation results for relationship between calorie consumption and subsistence orientation

Explanatory Variable[a]	Parameter	t-Value	Mean of Variable	Standard Deviation
TOEXCA	1,286.151	23.01	6.70	0.55
POTPRICE	−23.605	−2.54	8.56	3.55
POTSWEET	−114.406	−1.21	0.84	0.37
SUBFOOD	13.667	10.41	49.23	21.50
CAPITA	−373.914	−5.63	1.61	0.46
CHSHARE	1,400.815	9.64	0.29	0.20
FEMHEAD	376.931	4.13	0.11	0.32
ROUND1	384.618	5.25	0.33	0.47
ROUND2	390.543	5.24	0.33	0.47
Constant	−6,480.784	−14.681	—	—
(CALADEQ)	—	—	2,611.84	1,102.30
$\bar{R}^2$		.657		
F-value		120.3		
Degrees of freedom		551		

NOTE: The dependent variable is CALADEQ = Calories per day per adult-equivalent person.
[a]Definitions of variables:
TOEXCA = income proxy; logarithm of total expenditure per capita per month in respective survey round (in FRw)
POTPRICE = price of potatoes in FRw per kilogram
POTSWEET = ratio of potato price over sweet potato price
SUBFOOD = consumed own-produced food (value in percent of total expenditures)
CAPITA = household size (number of persons)
CHSHARE = percent of children under five per capita in households
FEMHEAD = female-headed households = 1, else = 0
ROUND1, ROUND2 = dummy variable for survey rounds 1 and 2 = 1, else = 0

holds and by 5.5 percent in the top quartile households (these estimation results are not listed in table 18.6). The propensity to consume more food with rising income is high, even among the "rich" in the top quartile (total expenditure per capita is US$224 per year in this "rich" group).

Increased subsistence orientation (lower commercialization) raises calorie consumption over and above the price and income effects. This increase is statistically significant. The influence of that effect is also quite substantial: a reduction in subsistence orientation by 10 percentage points from, say, 50 percent to 40 percent of value share of consumption from own production in total expenditures leads to a 5.2 percent drop in calorie consumption, holding other variables constant.

The model analysis confirms that female-headed households — ceteris paribus — consume higher levels of staple food on per capita terms than do other households. In bottom quartile households, female-headedness of households implies a 17 percent higher per capita calorie consumption level than found in other households. This effect is confined to female-headed households among the poor. In the top quartile expenditure class, the effect of female-headedness of households on calorie consumption is not significant. This result could be interpreted that in the poorest households where women are in control of resource allocation, relatively more is spent and allocated in terms of household production resources toward nutritional improvement, but this effect is no longer apparent when basic food requirements are satisfied as in the top expenditure quartile.

Nutrition Effects of Commercialization: Health and Sanitation Effects Dominate

Nutritional effects of commercialization are not only driven by income-consumption linkages but also by the health and sanitation environment of the household. The nutritional status of children aged between six months and seven years is evaluated by using anthropometric measurements of weight and height. WHO-NCHS standards were used as reference population statistics to identify prevalence of malnutrition and nutritional status of individual children. It must be remembered that child growth is not an indicator that enables us to distinguish food deprivation from infection as an initiating event (Payne 1987).

The child population in the survey households was weighed and measured before the initial survey and during each survey round. In the beginning of 1986, 21.5 percent of children were growth retarded, that is, they were below 90 percent of the reference height-for-age. About 10 percent of the children were underweight, that is, they were below 80 percent of the threshold level of weight-for-age, and 5 percent were showing symptoms of wasting, that is, their weight-for-height was below 90 percent of reference standards.

A straightforward comparison of nutritional status in households that are and are not involved in cultivating new potato fields in the Gishwati shows a lower prevalence of malnutrition in households with commercial potato fields.[6] This may largely reflect a positive relationship between income and nutritional improvement. Multivariate analysis to explain differences and short-term changes in children's nutritional status sheds further light on this.

The results are presented in table 18.7. There is a statistically significant nutritional improvement effect of the incremental food consump-

---

6. Height-for-age Z-scores: − 1.22 with Gishwati, − 1.68 without Gishwati. Weight-for-age Z-scores: − 0.46 with Gishwati, − 0.71 without Gishwati. Weight-for-height Z-scores: 0.31 with Gishwati, 0.28 without Gishwati.

TABLE 18.7 Multivariate analysis of determinants of nutritional status of children, 6–72 months

| | MODEL 1 | | | | MODEL 2 | | | | MODEL 3 | | | |
| | Height-for-Age (HAZ) | | | | Weight-for-Age (WAZ) | | | | Weight-for-Height (WHZ) | | | |
Explanatory Variable[a]	Estimated Parameter	t-Value	Mean of Variable	Standard Deviation of Mean	Estimated Parameter	t-Value	Mean of Variable	Standard Deviation of Mean	Estimated Parameter	t-Value	Mean of Variable	Standard Deviation of Mean
CALORIES	1.69E-04	3.50	2,561	1,035	9.36E-05	2.52	2,561	1,035	3.08E-04	2.58	2,578	1,044
CALORIES SQUARED	—[b]				—[b]				-4.88E-08	-2.41	7732319	6112009
SICK	-4.80E-04	-0.07	3.51	6.87	-0.0182	-3.44	3.51	6.87	-0.0208	-4.64	3.50	6.82
WORMS	-0.2521	-2.68	0.50	0.50	-0.0671	-0.93	0.50	0.50	0.0791	2.30	0.51	0.50
CLEAN TOILET	0.5295	5.55	0.60	0.49	0.2151	2.94	0.60	0.49	—[b]			
ln CAPITA	1.1396	3.81	1.82	1.78	0.9359	4.08	1.82	1.78	0.3046	1.57	1.82	1.75
BORDER	-0.1356	-2.65	3.34	0.31	-0.1613	-4.12	3.34	0.31	-0.0725	-2.19	3.36	0.31
SEX	0.3004	3.18	1.53	0.50	0.2573	3.56	1.53	0.50	0.0722	1.19	1.54	0.50
AGE	-5.35E-03	-1.99	44.39	20.86	-3.39E-03	-1.65	44.39	20.86	1.19E-03	0.69	43.96	20.95
HEIGHT OF MOTHER	0.0296	3.65	158.13	5.83	0.0169	2.72	158.13	5.83	5.85E-03	1.30	158.16	5.79
HEIGHT OF FATHER	0.0358	3.41	166.65	4.58	0.0199	2.48	166.65	4.58	1.96E-03	0.29	166.60	4.62

	Regression 1			Regression 2			Regression 3		
Dependent variable	HAZ			WAZ			WHZ		
Constant	−14.61	−6.34	—	−8.3254	−4.72	—	−1.8473	−1.49	—
HAZ	—	—	−1.50	—	—	−0.65	—	—	—
WAZ	—	—	1.23	—	—	—	—	0.92	—
WHZ	—	—	—	—	—	—	—	0.26	0.77
$\bar{R}^2$	0.13575			0.09401			.05643		
F-value	10.78547			7.46429			4.79772		
Degrees of freedom	613			613			625		

[a] Definitions of variables:

- CALORIES = calories per adult-equivalent per day
- CALORIES SQUARED = (calories per adult-equivalent per day)^2
- SICK = number of days sick last month
- WORMS = dummy for medium or heavy load of worms in stool examination, 1 = positive results, 0 = no or low infestation
- CLEAN TOILET = dummy for clean toilet, 1 = clean, else = 0
- ln CAPITA = logarithm of household size in number of persons
- BORDER = birth order of child, 1 = first born, 2 = second, etc.
- SEX = sex of child, 1 = male, 2 = female
- AGE = age in months at time of taking anthropometric measures
- HEIGHT OF MOTHER = height of mother in centimeters
- HEIGHT OF FATHER = height of father in centimeters

[b] Not in equation.

tion and, in the case of the short-term nutrition indicator (weight-for-height), the effect is also decreasing at the margin.[7] Although the calorie consumption effect is highly statistically significant, the net effect in improving nutritional status is small. For example, a 10 percent increase in calories in a household that consumes 2,000 calories per adult-equivalent person would increase the Z-score weight-for-height value of children in this household, *ceteris paribus*, by 0.021 (8 percent of the mean value of the sample's weight-for-height), the Z-score height-for-age value by 0.035 (2.3 percent), and the Z-score weight-for-age value by 0.019 (2.9 percent). Nevertheless, other than in the case from Kenya (chapter 16), a clear relationship between the (largely potato-driven) income-expenditure–food consumption linkages to the nutritional status of children is established with this study.

Similarly, as in the case from Kenya (chapter 16) and several other cases (chapters 12, 14, 22), health and sanitation-related variables overshadow the consumption impact on nutrition. As expected, current underweightedness (weight-for-height and weight-for-age below standard) is substantially a result of current or recent morbidity or episodes of morbidity. This is not the case for the height-for-age model, in which there is a strong adverse effect of worm infestation on the long-term growth performance of children. Children severely affected by worm infestation (which is actually the case for half of the child population in the survey) have a Z-score for height-for-age that is 17 percent lower than the mean.

Improved household sanitation significantly improves children's nutritional status. The CLEAN TOILET variable might be interpreted as a proxy for more generally improved household sanitation conditions. The CLEAN TOILET parameter estimate suggests that improved sanitation conditions, in comparison to poor conditions found in 40 percent of the households, lead to an approximately 33 percent improvement in both height-for-age and weight-for-age indicators.

The results from this multivariate analysis underscore that malnutrition in this environment is to a very large extent a health problem that requires tackling the health and sanitation side of rural services in tandem with the employment and income problems. Clearly, for this population as well as for other similarly poor ones (see chapter 20 for Malawi), income matters for nutritional improvement, as the earlier aggregate analysis showed (table 3.3).

---

7. The latter result was not found in models 1 and 2 for the height-for-age and weight-for-age model, respectively, where the squared calories variable was, therefore, dropped from the model.

Conclusions

The major findings of this study are summarized:[8]

1. Due to extreme land scarcity and market risks, farmers tend to be rather reluctant to adjust their production systems to a changing economic environment. Subsistence production has a high priority in this setting, which will be the case as long as infrastructure deficiencies imply high price risks, and off-farm employment (which already provides 55 percent of average household income) is risky and unstable.
2. While crop production patterns are quite stable, technologies are adjusted more elastically, as exemplified by the case of potatoes. Off-farm incomes are partially used to invest in farm equipment and improved tools. As a result, yields improve and some family labor is released for other activities.
3. New potato production had a double impact on employment of family labor: mobilization of extra male labor capacity, which was thus far underemployed, and capital-labor substitution. Also, the rural wage labor market expanded.
4. Land scarcity and low productivity in the study region make off-farm incomes indispensable for meeting farm families' minimum food requirements. New employment in specialized potato production had the extra property of combining increased purchasing power with additional supply of staple food.
5. Access to additional land resources did not reduce existing inter-household differences in income and food supply; in fact, inequities were further widened.
6. In evaluating the overall role of expanded potato production, direct benefits have to be balanced against external costs over the short and long run. Benefits result from the triple role of potatoes—extra employment, income growth, and increased food supply. However, these benefits might sooner or later decline as a result of environmental degradation. Compared to the original forest, soils under potatoes are subject to erosion and nutrient run-off. Moreover, the loss of forest area may have negative implications for the regional climate and biodiversity.
7. Calorie consumption is found to increase rapidly with rising income in this very poor area. However, increased commercialization (reduced subsistence orientation in production) results in lower calorie consumption per adult-equivalent person, holding income constant.

---

8. These conclusions also include some generalizations documented in greater detail in von Braun, de Haen, and Blanken (1991).

This latter effect seems to relate to male control over incremental cash income from the potato production and lump-sum cash income from the crop. However, the positive income effect of potato production was larger than the deviating effects of the control and form of income for calorie consumption at the margin. Thus, food consumption could increase.

8. Nutritional improvement of children at the study location is largely determined by health and sanitation conditions. Yet, incremental calorie consumption in food-deficient households is also found to improve child nutritional status. To the extent that incremental potato production increased household calorie consumption and household ability to invest in human welfare and health directly and indirectly, it added to nutritional improvement. Growth in employment and income of the poor needs to move in tandem with provision of rural public health and sanitation services in order to maximize the benefits of both—private income growth and public services expansion—for nutritional improvement.

19 Maize in Zambia: Effects of Technological Change on Food Consumption and Nutrition

SHUBH K. KUMAR

CATHERINE SIANDWAZI

Introduction

This chapter examines household food consumption patterns and nutrition status of the population in Eastern Province, Zambia, as well as the effects of adoption of new maize technologies on their characteristics. This analysis, mainly based on surveys carried out by IFPRI in the mid-1980s, has three primary aims:

- Identification of differences in these conditions between areas and households with and without a high level of improved agricultural technology use;
- documentation of characteristics of some determining factors of nutritional status, such as household income, food availability, prevalence of diseases, and intrahousehold decisionmaking;
- identification of the significance of alternative determinants of the food and nutrition situation and assessment of the significance of technological change and commercialization of agriculture.

Zambia today is faced with the challenge of resuscitating its economy through developing the agricultural sector. Until the late 1970s, Zambia's economy was heavily dependent on the mining sector, and the economy was strong enough to subsidize both the small-farm sector and consumers as well as to support food imports for the economically and politically important urban populace.

Due to recent economic changes, Zambia has had to reassess its agricultural policies. Current structural adjustment programs are attempting to provide greater support for agricultural growth. This policy change has the potential for improving rural incomes, food consumption, and the nutritional status of the population. This chapter examines the process of agricultural growth as it has occurred so far, with particular attention to technological change in the main staple, maize, to assess

Place of Maize in the Diet

Maize is the main staple for most Zambians. Besides maize, which is grown and consumed nationwide, sorghum, bulrush millet, finger millets, and cassava are other major staples. Maize is predominantly consumed in urban areas and in rural areas along the rail line, where the bulk of the country's population is located. In addition, maize is the main staple in the Eastern Province, which still has some sorghum- and finger millet-producing areas. In the Western and Northwestern Provinces, sorghum, bulrush millet, and cassava, together with maize, form the staple foods, whereas in the high-rainfall Luapula and Northern Provinces, tapioca and finger millet practically replace maize as the staple food. Since the mid-1970s, the outlying provinces that had large shares of nonmaize staples in their diet have shown rapid increases in maize production, which is likely to have increased the significance of maize in their diets.

Technological Change in Maize

Until recently, the main improved-maize varieties available to Zambian farmers were versions of SR52, which was first developed in Southern Rhodesia (now Zimbabwe) in the 1950s. This variety was imported on a limited scale primarily for large-scale commercial farmers until 1965-66, when it was first produced in Zambia. It was only in the early 1970s that acreage under this maize hybrid increased substantially with its adoption outside the commercial or estate farming sector, as it began to reach the village farming sector (Celis, Milimo, and Wanmali 1991.) By the late 1980s, about 60 percent of maize area was estimated to be under improved varieties. Acreage under hybrid maize in the Eastern Province is reported to have increased from between 10,000-20,000 hectares in 1975-80 to over 40,000 hectares in 1982-85.[1] Reliable estimates of trends in adoption since then have been difficult to obtain.

However, over 75 percent of hybrid maize area was and still is cultivated by the larger farmers. Household surveys, including those carried out by IFPRI with collaborators in 1981-82 and 1986, confirm this pattern. The two main reasons for this dominance of larger farmers are the high cost of hybrid seeds that have to be purchased each year and

---

1. Based on assessments by J. Waterworth, EPADP Project, Msekera Research Station.

the grain's poor storage characteristics under local conditions. Until recently, nearly all farmers in the smallholder sector[2] that grew maize planted local maize for home consumption and hybrid maize primarily for sale. Consequently, only those farmers who were in a position to expand cultivated area beyond that required for subsistence needs adopted hybrid maize. Access to, or ownership of, oxen for ploughing is closely associated with farmers' ability to expand cultivation in Zambia and also with hybrid maize adoption. Even though total area under maize appears to have increased rapidly during the 1980s, a recent IFPRI survey showed that even those farmers with the largest farms (above five hectares) grew hybrid maize on only half of their total maize plantings (table 19.1), although nearly all of them planted the hybrid maize. At the median farm size of about two hectares, somewhat less than half the farmers grew the hybrid maize, and this crop constituted less than one fifth of their area under maize. These surveys also indicate that almost all of the maize sales are from hybrid maize production. For instance, in 1981–82, 23 percent of all maize produced was sold, but local varieties made up less than 1 percent of the sales. However, not all hybrid maize produced is sold; in 1981–82, approximately 40 percent of hybrid maize was kept for home consumption. Further investigations show that when households keep hybrid maize for home consumption, it is only meant to last until the start of the rains, as after that it is likely to deteriorate rapidly.

Research initiated by the International Maize and Wheat Improvement Center (CIMMYT) on high-yielding open-pollinated, hard-kernel, and short-maturing maize varieties has been under way in Zambia since the late 1970s. The spread of these new varieties is limited by local seed production capacity and by distribution channels that make it difficult for farmers in the smallholder sector to get access to these seeds (Keller and Mbewe 1988).

Pattern of Adoption of Hybrid Maize Production and Tracing Its Nutritional Outcome

As observed earlier, hybrid maize adoption has accompanied farmers' ability to expand area cultivated beyond that required for subsistence needs. Since use of ox plough cultivation enables this expansion, there is a high degree of overlap between hybrid maize cultivation and use of ox cultivation. As producer support prices for maize improved relative to other major crops during most of the 1980s, hybrid maize

---

2. These farms are generally smaller than ten hectares but are more than 90 percent of all farms.

TABLE 19.1 Characteristics of farms, by size, Eastern Province, Zambia, 1986

Characteristic	<1.00	1.00–1.99	Size of Farm (Hectares) 2.00–2.99	3.00–4.99	>5.00
Distribution of household (percent)	23.6	31.7	15.4	18.7	10.6
Distribution of total area (percent)	5.7	18.6	14.7	28.1	32.9
Average size (hectares)	0.6	1.5	2.4	3.8	8.0
Fertilizer use					
Farmers using fertilizer (percent)	51.7	67.9	65.8	67.3	100.0
Crop area fertilized (percent)	45.5	49.8	52.3	43.5	72.3
Average rate of application (kilograms/hectare)	141.1	124.4	105.3	139.4	120.9
Hybrid maize cultivation					
Farmers growing hybrid maize (percent)	6.9	30.8	42.1	52.2	96.1
Maize area under hybrid varieties (percent)	3.1	15.1	17.6	24.2	49.2
Oxen cultivation					
Farmers using oxen (percent)	25.9	56.4	65.8	87.0	96.2
Farmers owning oxen (percent)	17.2	17.9	36.8	58.7	76.9
Area cultivated by oxen (percent)	18.4	46.0	58.4	70.3	93.2

SOURCE: Jha, Hojjati, and Vosti (1991).

production increased in areas with good market access around the country, even though agro-ecological conditions may have been more suitable to other crops.

To the extent that larger farmers are substantial adopters of hybrid maize, the direct benefits are likely to be obtained by them alone. At the area level, however, there could be some benefits for other households via employment and price (or food availability) effects, often referred to as growth linkage effects. In Eastern Province, these patterns of adoption and distribution of economic benefits appear to hold. Households that grow hybrid maize have farms twice the size of those of nongrowers. Per capita income (measured by total consumption expenditure) is about 25 percent higher in adopting households. More striking are the income differences between high-adopting areas and low-adopting areas. Nonadopters in areas that have a high degree of adoption of hybrid maize have incomes nearly 45 percent higher than nonadopters in areas with a low level of adoption, even though adopters' farm size is smaller. Adopters in high-adoption areas have farms that are larger and per capita incomes that are higher (by about 33 percent) than adopters in low-adoption areas.

Since larger farmers in Zambia are more likely to be adopters, it is difficult to compare them with nonadopters to assess the benefits of adoption, since one may expect the larger farms to be doing better with respect to food consumption and nutrition simply by virtue of being better off to begin with. This comparison may be further complicated if there are any linkage effects for other households in the high-adopting areas. This complication is taken into consideration by setting up the tabular analysis in such a way that comparisons can be made both between adopting households and nonadopting households and between similar areas that have different degrees of adoption.

This study was carried out in the Eastern Province, which has two main agro-ecological zones, plateau and valley. Hybrid maize adoption is limited to the plateau areas, since the presence of tsetse flies in valley areas prevents cattle ownership and, hence, ox plough cultivation. Of ten study locations, two were in the valley areas and eight on the plateau. The plateau sites can be classified into high- and low-adopting areas, based on the extent of hybrid maize adoption observed during the study year. Overall, the plateau sites have good access to physical infrastructure such as all-weather roads and major markets, whereas the valley areas generally have poor access to this infrastructure. Variations in hybrid use in the plateau sites seem to have been largely determined by the capacity of the local cooperative unions for providing credit and making inputs such as fertilizer and seeds available in a timely manner to the farmers. In the past, the government channeled subsidized inputs for the small-farm sector through provincial cooperatives linked to farmers through primary cooperatives at the local level.

Food Consumption Levels and Adequacy

A comparison of food consumption levels by area- and household-level technology adoption shows that there are substantial differences between low adopters and high adopters (classified in table 19.2 by farm size below and above the median for each area) as well as between high- and low-adopting areas. Per capita caloric consumption is higher in the high-adoption areas, for both small and large farm households, in comparison with the low-adoption areas (table 19.2). Larger farms in high-adoption areas have about 32 percent more calories per capita than do larger farms in low-adoption areas. Smaller farmers, even though they have very low adoption levels, have about 27 percent more calories per capita in high-adopting areas than in low-adopting areas. However, the consumption difference between small and large farms is slightly higher in the high-adoption areas: it increases from about 12 percent to 22 percent higher consumption for larger farmers in high adopting areas, as a result of greater dietary gains made by larger farmers.

The area-level difference in intake level suggests either that better-off areas (in terms of food consumption) are better able to adopt improved technologies than worse-off areas or that widespread adoption of improved agricultural technologies results in a broad distribution of

TABLE 19.2 Daily per capita consumption of nutrients, by high- and low-adoption areas in plateau region, Eastern Province, Zambia, annual sample average, 1986

Nutrient	High-Adopting Area		Low-Adopting Area	
	Small Farm	Large Farm	Small Farm	Large Farm
Energy (calories)	2,299.71	2,772.78	1,770.78	2,035.67
Protein (grams)	65.23	78.34	47.80	57.62
Fat (grams)	36.41	39.16	20.53	27.18
Carbohydrates (grams)	403.50	494.11	329.58	381.03
Calcium (milligrams)	383.42	444.55	278.41	434.04
Iron (milligrams)	18.39	21.13	13.48	18.02
Thiamine (milligrams)	1.15	1.26	0.72	1.14
Riboflavin (milligrams)	1.09	1.32	0.93	1.06
Niacin (milligrams)	15.62	16.68	9.34	12.41

SOURCE: IFPRI surveys, 1986.

NOTE: Median farm size in each area is used as a cut-off point for small/large farm classification.

benefits in the area, resulting in an overall improvement in diets even though inequity appears to increase somewhat. Area-level differences in dietary intake are not fully consistent with area-level differences in nutritional status of preschool children, as seen later. High-adoption plateau sites had a lower short-term preschool child malnutrition but a higher longer-term (height-for-age) indicator compared with low-adopting areas. For older children, aged between five and ten years, high-adopting areas and households were worse. Also inconsistent are results from the valley sites where consumption levels, on aggregate, are similar to the low-adoption plateau sites, but levels of preschool child nutrition, much better during the lean season, are worse during the dry season, when the situation is improving on the plateau. Part of the explanation for the anomaly may lie with differences in employment levels and, therefore, effective caloric requirements for households in the valley and plateau and in high- and low-adoption areas. It has been shown that adoption of improved agricultural technologies raises labor input and employment substantially. The differences in caloric intakes may, therefore, be more apparent than real.

Cereal composition of the diet shows that households that do not grow hybrid maize rely much more on purchased maize meal than do hybrid maize growers. Although there does not appear to be much of a difference between small and large farmers in the amounts of purchased maize meal per capita, in percentage terms small farmers rely more on purchases for their consumption. In both absolute and percentage terms, it is the small farmers who do not grow hybrid maize who are most dependent on purchases. This distinction is relevant in considering the impacts of the structural adjustment program in Zambia, under which net buyers of maize meal would be hurt most by the elimination of the consumer subsidy for maize.

The seasonal pattern of food consumption differs between high- and low-technology adoption areas. Again, the pattern for valley and low-adoption plateau areas is relatively similar, but different from that for the high-adoption plateau areas. In the low-adoption areas, intake levels are lowest in the March–June period, then rise until around January, when they start to fall again. In the high-adoption areas, consumption is lowest in the January–February period and highest in the immediate postharvest period of July–August, after which it declines slightly until December (figure 19.1).

This pattern of consumption fluctuation suggests that low-adoption areas have a more precarious food supply situation even after the harvest, and therefore conserve their resources for when they would need to increase consumption to meet increased work loads later in the year. In the high-adoption areas on the other hand, where the food situation is

FIGURE 19.1 Daily per capita consumption of calories, by high- and low-adoption areas in Plateau Region, Eastern Province, bimonthly sample average, 1986

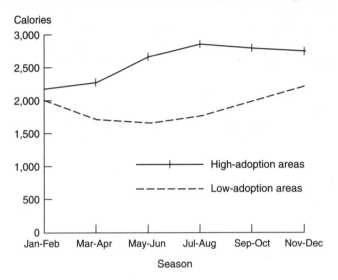

more comfortable, the increase in consumption following the harvest is quite pronounced. However, since most of the maize surplus is marketed in the postharvest period (primarily due to the maize-marketing policy in effect at that time), these areas may also experience food shortages during the lean season. In both low- and high-technology adoption areas, however, the extent of caloric fluctuation between low and high points is similar on a percentage basis—consumption at the high point increases about 30 percent from the low point. In terms of daily per capita caloric increase, this represents about 500 calories in low-adoption areas and 640 calories in high-adoption areas.

Child Nutritional Status

The nutritional status of all household members was measured four times during the study, in February, May, September, and December 1986. In terms of the agricultural cycle, these times represent the late planting/weeding period, the early harvest period, the postharvest period, and the land preparation/early planting period. This chapter examines the nutrition status of children under five years of age and those between five and ten years of age, and variations in the nutrition status by season and degree of adoption of improved maize varieties.

The overall nutrition situation is worst during February for both the under-fives and the five-to-ten-year-olds (table 19.3). For the under-fives,

TABLE 19.3 Child malnutrition, Eastern Province, Zambia, February 1986

Anthropometric Indicator	Total Sample		Valley Sites		Plateau Sites	
	<5 Years	5–10 Years	<5 Years	5–10 Years	<5 Years	5–10 Years
			(percent below −2 Z-score)			
Round 1 (February 1986)						
Weight-for-age	24.2	27.9	5.6	13.3	26.2	29.9
Height-for-age	53.0	59.8	44.6	40.0	53.9	62.6
Weight-for-height	5.0	5.1	0.0	7.1	5.6	4.9
Round 2 (May 1986)						
Weight-for-age	12.4	14.9	23.7	8.0	10.5	16.3
Height-for-age	53.1	38.5	45.9	24.0	54.3	41.5
Weight-for-height	2.4	2.8	2.7	54.3	2.3	2.5
Round 3 (September 1986)						
Weight-for-age	15.4	15.1	19.5	18.2	14.5	14.4
Height-for-age	48.5	30.2	41.5	30.3	50.0	30.2
Weight-for-height	0.0	3.6	0.0	6.7	0.0	2.9
Round 4 (November 1986)						
Weight-for-age	20.6	7.5	16.7	2.8	21.6	9.3
Height-for-age	51.3	19.5	41.7	11.1	53.8	22.7
Weight-for-height	1.3	5.6	4.3	12.9	0.5	3.2

SOURCE: IFPRI surveys 1986.

there is a marked improvement in the two weight-related indicators (weight-for-age and weight-for-height) between February and May, but not in the height-for-age indicator. Between May and September, the height indicator also shows an improvement. By November, however, all indicators begin to show deterioration. The pattern is slightly different for the five-to-ten-year-olds, whose improvement in weight between February and May is accompanied by an improvement in the height indicator, suggesting a possibly faster height response to dietary improvement for this age group. The height improvement also appears to be sustained for a longer time during the year for this age group as compared with the under-fives, in both valley and plateau sites.

Valley areas do better at all times of the year with respect to height-related nutrition indicators for both age groups, suggesting better long-term nutrition. In the shorter-term weight-related indicators the pattern varies by age group and location. For the under-fives, the worst time in the valley is November, as compared to February in the plateau. This difference between the two areas may be explained by differences in ecology and, hence, in cropping patterns.

Comparison of differences by level of hybrid maize adoption is limited to the plateau sites, since adoption in the valley was practically nonexistent. Table 19.4 compares the prevalence of malnutrition in under-fives and five-to-ten-year-olds by area- and household-level hybrid maize adoption in February, the time when the situation is at its worst on the plateau sites. At the area level, high-adoption areas do better with respect to the shorter-term weight-related measures for the under-fives, but this is not as evident in the older children. The latter finding could mean that either older children face greater stress in the high-adoption areas, or the improvements are recent and more pronounced in the faster-growing younger children. At the household level, adopting households (and larger farmers) have slightly lower levels of preschool child malnutrition. However, this pattern is evident only in the short-term nutrition indicator of weight-for-height in the older children. Longer-term height-for-age and weight-for-age indicators are poorer in adopting households for these five-to-ten-year-old children. These observations suggest that households adopting improved agricultural technologies may be undergoing changes, possibly in child-care, that are stressful for children, despite the apparent improvement in diets. This effect is more evident in the five-to-ten-year-olds, whereas the preschool-aged children have marginal improvements in their nutrition.

Prevalence of Diseases

Morbidity was reported every month for each household member. The two main diseases were diarrhea and malaria, with respiratory and

TABLE 19.4 Child malnutrition, by adoption of hybrid maize, Eastern Province, Zambia, February 1986

	<5 Years			5–10 Years		
	Weight-for-Age	Height-for-Age	Weight-for Height	Weight-for-Age	Height-for-Age	Weight-for-Height
			(percent below −2 Z-score)			
Area-level technology adoption						
High[a]	22.8	58.8	0.0	38.0	64.0	10.4
Low[b]	29.4	49.4	11.0	22.8	61.4	0.0
Household-level technology adoption						
High[c]	25.0	50.0	4.5	34.1	70.5	2.4
Low[d]	26.6	56.8	6.5	28.3	55.0	6.9

SOURCE: IFPRI surveys.

[a]Top half of the villages with the highest frequency of households adopting hybrid maize and fertilizers.
[b]Bottom half of the villages with the lowest frequency of households adopting hybrid maize and fertilizers.
[c]Adopters of both hybrid maize and fertilizer.
[d]Non-adopters of hybrid maize, but may adopt fertilizer.

miscellaneous fevers comprising the third significant group of illnesses. For both diarrhea and malaria, children under three years of age had the highest number of episodes reported. For the group of respiratory and other fevers, rates were highest again for children less than three years, as well as adults.

Diarrheal diseases were found to be most prevalent in the valley sites, possibly because of the higher water table and greater use of ponds and streams, which can be contaminated more easily. In the plateau sites, children from higher-adopting areas were surprisingly more prone to diarrhea than those from low-adopting areas. At the household level, however, adopting households had a lower diarrheal rate than non-adopting households, which would suggest that at the area level, non-adopters fare worse in areas that are predominantly adopting improved technologies.

Malaria cases were lowest in the high-adopting plateau areas. Prevalence of malaria in the low-adopting plateau areas was about the same as in the valley areas. It is possible that improvements in economic status in the high-adopting areas, accompanied by possible improvements in housing, may be a factor in the lower case rates for malaria in the high-adopting plateau areas. This pattern is also reflected at the household level with lower malaria rates for all age groups except the five-to-fourteen-year-olds in those households adopting hybrid maize.

The pattern observed for the group of respiratory illnesses and fevers was similar to that for malaria, with the high-adopting areas doing best. However, valley areas seemed to be doing somewhat worse than low-adopting plateau areas. These differences were not evident with technology-adoption differences at the household level.

Examination of household water sources shows that there is an overall better availability of improved water supply (protected wells or boreholes) and sanitation in areas with a high-adoption rate, although it is not evident whether there is any difference between adopters and nonadopters within each type of area. This suggests that there are some linkage effects at the area level that make the adopting areas have a better water supply and sanitation situation than the nonadopting areas.

Women's Role in Agricultural Decision Making

It is postulated that women's role in household decision making is associated with their access to resources and a reduced work burden, both of which can have favorable effects on the nutrition of young children. Since mothers are the primary caretakers of children, the time and material resources at their disposal are expected to benefit children via quality of child care, foods available and prepared for child feeding, and ability to utilize health and nutrition education and services.

For all households, women provided 54 percent of household farm labor input. This share was highest in the valley, 67 percent, whereas it was 51 percent in the plateau, with about the same share in high- and low-adopting areas. Households that adopted hybrid maize increased household labor input substantially, by 14 percent in low-adopting areas and 46 percent in high-adopting areas. Males provided a larger share of the increased labor input, increasing their input by 24 percent in low-adopting areas and by 56 percent in high-adopting areas. However, women still provided about half of the labor input even though their relative contribution declined with hybrid maize adoption. Women's absolute amount of labor input also increased with hybrid maize adoption, by 5 percent in low-adopting areas and 38 percent in high-adopting areas.

Another facet of women's role in agriculture deals with their direct responsibility in farming. It is often argued that though women may provide much of the agricultural labor input, they do not have any real decision-making role. In order to check this out, we examined two different facets of decision making. In the first, we requested households to identify individuals who were primarily responsible for cultivation decisions of different plots of land.³ Based on this information it was ascertained that women independently farmed varying amounts of land, with additional land farmed under joint management with male members of the household. Even though the head of household was involved in decisions for most of the plots, he or she was by no means the sole independent decision maker. Women independently farmed only 14 percent of the land in the valley and 24 percent on the plateau. However, they also jointly managed another 46 percent of the land in the valley and 35 percent on the plateau, together accounting for 60 percent of land in both areas.

Looking further into variations in women's involvement with hybrid maize adoption, we find that both the share of their independent and jointly farmed land declines with adoption. Low-adopting areas have 31 percent of land farmed independently and 43 percent jointly by women, in comparison to only 17 percent of land farmed independently and 28 percent jointly in high-adopting areas. Similar declines are evident at the household level in both areas. These differences could mean either that households adopting hybrid maize are changing their decision-making patterns or that households with women as major decision makers are for some reason less likely to be among the adopters. It is likely that the way in which the new technology is disseminated and supporting inputs are provided are contributing factors in both cases.

Similar results were obtained for another facet of intrahousehold decision making. In this part of the study, we examined women's involvement in transactions associated with agricultural operations such as hiring of labor, purchasing of inputs, and selling of produce. Since there are two predominant ethnic groups in Eastern Province, results were analyzed not only by agricultural technology adoption but also by ethnic group. Results indicate that the highest level of women's involvement in agricultural decision making exists in the valley and low-adoption plateau sites. In these areas, women were closely involved in a larger share of all labor-hiring decisions, including payment. Decisions involving purchase of inputs were made over 50 percent of the times by women in the valley households, and about 20 percent of the times in the low-adoption sites. Overall, labor-hiring decisions were made about 25 percent of the time by women, and input purchase decisions about 45 percent when such transactions occurred. There was a marked reduction in women's involvement in agricultural decisions in the high-adoption sites, especially in the patrilineal ethnic groups (Ngoni and Tumbuka).

3. This notion as well as obtaining women's labor input separately was strongly resisted by expatriate and Zambian male economists on the team who asserted that this information was (1) not relevant for economic analysis of production, and (2) the head of household was responsible for all production decisions.

Women's labor-hiring decisions as well as decisions involving input purchases were cut by more than half. In the patrilineal ethnic households, the extent of reduction was even more significant.

Conclusions

The analysis conducted in Eastern Province shows that areas and households that are substantial adopters of improved agricultural technology, as represented by hybrid maize production, are better off in terms of household consumption expenditure and food available for consumption. Some improvement is also observed in water supply and sanitation and in several morbidity indicators in hybrid-maize–adopting households and areas. High-adopting areas and households also have better nutrition indicators for children under five years of age. For older children between five and ten years of age, however, their nutrition is somewhat worse in the high-adopting areas and households. The absence of an unequivocal improvement in child nutrition, despite substantial gains in income, food availability, and several health-related indicators, could possibly be explained by other changes taking place, such as a higher work load for household members, including women, and a decrease in the agricultural decision-making role of women, which could be detrimental to child nutrition.

20 Tobacco Cultivation, Food Production, and Nutrition among Smallholders in Malawi

PAULINE E. PETERS

M. GUILLERMO HERRERA

Introduction

Malawi is one of the poorest countries in Sub-Saharan Africa, with a per capita income of about US$170 (World Bank 1990). Although Malawi has an enviable record in achieving national food self-sufficiency, it has achieved that at a very low level of the supply-demand balance; the aggregate picture masks widespread food insecurity at the household level. National statistics indicate that over half of the children under five years of age suffer from malnutrition and almost one in five dies before reaching the age of five. The causes of this grim situation are low incomes, chronic and seasonal food shortages, inadequate diet (especially for weaning infants and children between one and two years of age), and high morbidity rates. To date, policy reforms have included restructuring the Agricultural Development and Marketing Corporation (ADMARC), which is the government marketing agency, and agricultural pricing policies. Smallholder response to higher producer prices has increased sales to ADMARC, but this seems to indicate switches among crops rather than an aggregate increase in output (Lele 1989b). This study was conducted in 1986–87 to assess the impact of the commercialization of maize and tobacco in an area in southern Malawi.

Research Site and Methods

The research site, set in the southern part of Zomba District and about 15 miles from the town of Zomba, is just within the area that had been covered by one of the largest white-owned estates in the Shire Highlands, the Bruce Estates (White 1987; Vaughan 1987). Here, cotton and, later, tobacco had been grown mainly by tenants and, to some extent, by laborers. Many of the older respondents in the sample had worked on the estates, and some of them had learned their skills in tobacco cultivation there. In addition, farmers have long produced vegetables and grain for sale in the local markets, which are held on different days of the week in different locations, or in the bigger market of Zomba. Today, smallholders who grow tobacco are supposed to be registered with the state's agricultural extension staff, but, in this study, about half of the tobacco growers sold their tobacco through other growers who were registered or through estates. The area is one where sales of crops are a well-established practice, with numerous local markets within a day's travel of two major urban centers, and one where most labor exchanges are remunerated in cash or in kind.

Tobacco- and non-tobacco-grower households with varying sizes of landholdings were purposely selected for this study; the study sample overrepresents larger landholders. Fourteen percent of sample households have landholdings of less than 0.7 hectare, compared with 35 percent of households nationally; 52 percent of sample households have between 0.7 and 1.5 hectares, compared with 40 percent nationally; and 34 percent have over 1.5 hectares, compared with 25 percent nationally (national figures are taken from Hiwa 1988). The mean holding for the sample is 1.5 hectares. The average holding in the Liwonde Zomba Project area just to the north of the research site is 0.8 hectare, slightly under the average for the whole district.[1]

Cash Crops, Production, and Income by Gender

Smallholders have highly diversified incomes, which entails juggling multiple sources of income in trying to achieve family food security and welfare. Malawian smallholders are poor by international or even African standards; mean total income per capita for the sample households over a ten-month period was K 78 or US$35 at 1986–87 exchange rates (table 20.1).[2] The household economy is highly commercialized in that, on average, just over 30 percent of the total income comes from marketed agricultural production, and a further 39 percent comes from off-farm sources. The latter include transfers, mostly from relatives working elsewhere, which make up 15 percent of total income, on average. Some of this transfer income provides the cash for buying fertilizer or paying for hired agricultural laborers, while for poorer house-

---

1. For more details on sample selection and survey design, see Peters and Herrera (1989).

2. Income, other than the value of retained own-crop production, is based on income data collected during the monthly income and expenditure survey. Similar studies often prefer to use expenditure data for estimating income levels, since income is often underreported. In this survey, however, reported income actually exceeded reported expenditures by an average of 9 percent, so income rather than expenditure is used for the analysis.

TABLE 20.1 Per capita income for different farm households, by type of income, Malawi, 1986–87

Household	Number of Households	Value of Home Consumption[a]	Market Income[b]	Off-Farm Income	Total Income
			(kwacha/capita)		
Total sample	210	23.26	23.95	30.58	77.78
Nontobacco households	148	23.11	17.37	32.30	72.78
Tobacco households	62	23.62	39.65	26.46	89.73
Small tobacco households	25	19.30	27.05	20.72	67.07
Tobacco specialists	37	26.54	48.17	30.34	105.05

SOURCE: Surveys conducted by the authors.
[a]Value of maize retained for home use.
[b]Income from sale of crops and agricultural products.

holds, transfer income supplements the low income realizable from small landholdings.

On average, sample households depend for some 30 percent of their total income on own-produced maize. Increasing total income is not correlated with an increasing proportion of income from market income. In this sample, the proportion of market income drops for the two middle quartiles, indicating a greater propensity to retain the maize harvest as income increases. It is only in the top income quartile that market income provides a larger proportion of income than does the value of retained production.

As a percentage of market income and wage income, sales of all food crops are more important for the poorest quartile, as they comprise 20 percent of total income compared to 12 percent for the top quartile. For many of the poorest, casual agricultural wage work (called *ganyu*) is a major income strategy, especially during the rainy season, when food is scarce but local work is more available. Households with more land, more food stores, or more cash to pay wages hire such workers. In this way, the income strategies of the poorer households are joined through the local labor market to the production strategies of the richer households. Some remuneration is in kind, usually in the form of food crops (maize, cassava, and sweet potatoes), but relatives and neighbors are as likely to pay in cash as in kind.

Female-headed households[3] have lower income per capita than male-headed households—K 65.84 compared with K 84.15. However, the highest mean incomes are earned by the female-headed TEBA households, where the husband is a labor migrant in South Africa. These households rely on remittances sent by the husbands. Some of the remittances are used to hire agricultural laborers and to purchase fertilizer. Such households earn a smaller proportion of their income from selling maize than other households. They sell few other crops, so their total crop income comprises only 5 percent of their total income, compared with 18 to 19 percent for other female-headed households and 25 percent

---

3. On the basis of a very detailed analysis of the composition of households and the pattern of absence or presence by husbands, we deciphered three types of female-headed households: *de jure female-headed households* (40 percent of the female-headed households), where an adult woman is currently without a husband, that is, she is widowed or divorced; *male-absentee households* (41 percent), where the husband was not present for at least half the time during the research period and was engaged in employment activities within Malawi; and *TEBA households* (19 percent), where the husband is a labor migrant in South Africa. TEBA is the acronym for The Employment Bureau of Africa, the recruiting agency for South Africa. Migrants are said in the sample villages to "have gone to TEBA," hence the label used here.

for male-headed households. The other female-headed households fit the usual stereotype of being poorer than male-headed households, with the male-absentee households being the poorest, with a mean per capita income of K 51. Land shortage makes it difficult to earn a sufficient income from farm production alone; thus men tend to be absent from home a great deal, looking for other income opportunities (such as selling crops, catching and/or selling fish, agricultural and nonagricultural manual labor, construction, rough carpentry, and selling firewood or grass).

In the sample as a whole, female income is equal to male income. In male-headed households, however, the share of household income in women's hands is less (37 percent of mean household income), and it declines with increasing total income and with the incidence and scale of tobacco production. It is important to recognize that female-controlled income is that reported by individual women to be in their hands; it is not necessarily the income earned by women. In the sample area, the conventional behavior expected of married men is to hand over all or some of their income to their wives. This area is one where matrilineal succession and inheritance and uxorilocal residence are the norm both statistically and normatively.[4] In addition to women being perceived in a very real way as "owners of the land" and as bearing authority in their own village, along with their brothers, there appears to be a strongly held convention that income in a conjugal home should be as much the wife's affair as the husband's, irrespective of how that income was earned. This does not mean, of course, that there are no differences of opinion over the disposition of income; there are. Nor does it mean that the earner of income does not have a potentially stronger say in the disposition of that income. There is no doubt that women have far fewer options for earning income outside own production or self-employment than do men, and that their wages are often lower. On the other hand, if noncausal wage or salaried work is excluded, women's earnings may equal or exceed those of men who do not have skills to sell in the labor market. Several women who brew beer or *kachasu*, a local gin, were able to make more money than did their husbands working in casual farm labor or other temporary odd jobs. Women selling tomatoes or grains could earn as much or more cash as their husbands doing similar work.

---

4. All sons, except the one designated successor to his mother's brother and the guardian of his sisters (*mwini mbumba*), have to leave on marriage. Hence, men use land in their wives' villages. Since marriages do take place within a village, some married men live in their own natal villages.

Food Security and Cash Cropping

"Maize is our life," (*chimanga ndi moyo*) say the people of Zomba. Maize is used to make the staple meal of stiff porridge (*nsima*) that is equated with "food" and eaten with a relish, usually vegetables. Similar to reports from Guatemala, the Philippines, and Zambia (see chapters 12, 13, and 19), all households try to provide as much of their own maize supply as they can from their land. Maize occupies most of the land and labor of the area. Eighty-four percent of the cultivated area is planted with maize, virtually all of it intercropped with legumes (cowpeas, pigeon peas, and beans), groundnuts, pumpkins, and other crops, which are used for own consumption and for sale. The mean household area cultivated (1.5 hectares) is a close measure of the land available, since most land is in almost permanent cultivation. The mean household maize harvest of 880 kilograms, or 145 kilograms per capita, falls below the government of Malawis' (GOM) estimate of annual maize needs of 157 kilograms per capita. As with landholding and income data, however, the mean figure masks considerable variation across the sample.

There are two measures of relative subsistence achieved by households. One is the share of own total production income directed to the household's own use, called the income subsistence ratio, which is 34 percent for this sample. The other measure, the agricultural subsistence ratio, is the share of own agricultural production directed to the household's own use, which is 56 percent for this sample (table 20.2). As might be expected, the agricultural subsistence ratio measure is higher, on average, for poorer than richer households, for smaller than bigger landholdings, and for nontobacco growers than tobacco growers, indicating the greater use of agricultural production for own consumption than for marketing. The drop in the agricultural subsistence ratio for the middle landholding group (0.7 to 1.5 hectares) indicates that households feel more able to use some of their agricultural product for earning a cash income, while a rise in the ratio for the larger holdings (over 1.5 hectares) reflects their ability to increase the proportion of maize retained for own use as well as to have a market income. The degree of agricultural subsistence is highest not for the poorest income quartile but for the middle 50 percent, before dropping for the richest quartile.

Overall, the degree of commercialization of maize production is low—only 11 percent of the maize harvest, on average, is sold. The only significant variation across the sample is that the poorest income quartile has the lowest percentage of retained maize (78 percent), for reasons that will be discussed below. Retained maize forms the bulk (70 percent) of the total maize available to the households and half of their total food expenditures over the ten-month survey period. The reasons behind

TABLE 20.2 Subsistence ratios for main sample groups, Malawi, 1986–87

Group	Agricultural Subsistence[a]	Income Subsistence[b]	Consumption Subsistence[c]	Food Consumption Subsistence[d]
			(percent)	
Total sample	56.0	33.6	32.6	51.0
Head of household				
Male	53.1	33.8	33.0	52.2
Female (TEBA)[e]	75.1	31.3	30.4	54.4
Female (Malawi)[f]	57.7	30.7	29.0	42.9
Female (de jure)[g]	58.5	36.9	35.4	52.2
Nontobacco households	61.8	34.4	32.3	49.5
Tobacco households	44.6	31.8	33.3	—
Small tobacco growers[h]	46.2	30.9	31.8	57.6
Specialist tobacco growers[i]	43.6	32.5	34.4	56.8
Landholding class				
<0.7 hectare	59.3	25.4	24.7	37.7
0.7–1.5 hectares	54.1	32.4	31.2	48.2
>1.5 hectares	57.6	39.0	37.9	60.5
Income quartiles				
First (<K44)	55.2	35.6	32.2	46.8
Second (K44–62)	59.1	33.3	33.7	50.7
Third (K62–85)	56.6	33.9	33.3	53.2
Fourth (>K85)	53.0	31.8	31.2	53.3

SOURCE: Surveys conducted by the authors.

[a] Value of home consumption (that is, retained maize)/value of agricultural production (market income plus home consumption).
[b] Value of home consumption/total income minus value of agricultural inputs, that is, hired labor and chemicals.
[c] Value of home consumption/total expenditures (including value of home consumption).
[d] Value of home consumption/total food expenditures (including value of home consumption).
[e] Husband was labor migrant in South Africa.
[f] Male-absentee household where husband was not present for at least half the survey period and was working within Malawi.
[g] Adult widowed or divorced women.
[h] Received less than one third of their crop income from tobacco sales.
[i] Received more than one third of their crop income from tobacco sales.

people's wish to produce as much of their own maize as possible are related to their assessment of their ability to pay the higher prices in the deficit period (December–January) as measured against their needs for cash immediately after the maize harvest; the insecure set of income opportunities available to them, which are more constricted for the poor; the fluctuation in supply at local ADMARC selling centers; and the costs in time as well as in cash involved in purchasing maize. There are also taste preferences—the locally produced maize is pounded to a fine white flour, *uf a woyera,* and cooked to a white, glassily smooth paste, which is considered far superior to the *ngaiwa,* or whole maize flour, made from hybrid maize purchased from ADMARC. Finally, respondents considered that a store of maize was a more secure source of food than dependence on cash for purchased maize. Factors contributing to this opinion included the difficulties of finding cash when it was needed, the tendency for cash to be used for purposes other than food purchases, the stronger claims by other relatives and neighbors on cash as opposed to household maize stores, and various cultural mechanisms for regulating the amount of maize taken out of the store, including strict control by women.

For these reasons, the more staple food supplies that people are able to produce for themselves, the less the risk they face of increased prices or of short supplies. These reasons, combined with the insufficiency of land, even among this better-off sample, leads to the fact that while only 56 percent of the sample sells maize, 99 percent purchases maize. The larger a household's harvest, the more likely some of it is sold. Sellers as a group have significantly higher maize harvests (188 kilograms, compared with 130 kilograms per capita for nonsellers). Forty percent of households with over 200 kilograms per capita of maize harvests are net sellers, compared with 8 percent of those with less than 200 kilograms per capita harvests. On the other hand, although households with smaller maize harvests are less likely to sell, there is no simple relation between the size of harvest and grain sales. As harvests rise, there is first an increase in the percentage of households selling, but then the percentage drops off considerably before rising again at the highest harvests. Households with low maize harvests but few other income opportunities are forced to sell some grain, while an increasing total income enables greater withdrawal from the grain market for sale.

How well, then, do the sample households do in obtaining a supply of maize (that is, harvest minus sales plus purchases)? The most significant correlations in both size of harvest and total maize supply are with income and landholding size. This holds for both total household and per capita maize supply (table 20.3). The top income quartile has over two times the total maize supply achieved by the lowest quartile. The

TABLE 20.3 Total maize supply available to sample households, Malawi, 1986–87

Group	Number of Households	Household Maize Supply (kilograms)	Per Capita Maize Supply
Income quartile			
First (<K44)	52	702	103
Second (K44–62)	53	848	156
Third (K62–85)	53	1,196	189
Fourth (>K85)	52	1,546	318
Landholding class			
<0.7 hectare	4	579	147
0.7–1.5 hectares	28	911	180
>1.5 hectares	29	1,524	228
Nontobacco households	138	1,014	194
Small tobacco growers	25	1,242	158
Specialist tobacco growers	37	1,346	231
Male-headed households	137	1,152	195
Female-headed households			
TEBA[a]	14	1,441	242
Malawi[b]	30	823	157
De jure[c]	29	777	186
Total sample	210	1,072	191

SOURCE: Surveys conducted by the authors.
[a]Husband was labor migrant in South Africa.
[b]Male-absentee household where husband was not present for at least half the survey period and was working within Malawi.
[c]Adult widowed or divorced women.

largest landholders have total maize supplies that are over three times those of the smallest landholders, but because household size tends to increase with land size, in per capita terms the largest landholders have one-and-a-half times the maize supply of the smallest landholders.

Households in the poorest income quartile have the lowest mean hectarage cultivated and reap lower mean harvests (548 kilograms, compared with the sample mean of 880 kilograms). Despite this, they are more dependent on grain sales than the richer households, earning 5 percent of their total income from this source. These poorest households retain a smaller share (78 percent) of their maize harvest, thus selling twice the proportion that other households do. The poor are in a classic double bind: they need cash (for immediate needs or to repay debts incurred in the difficult preharvest period) and so have to sell some of their harvest, knowing full well that they will later have to find other means to purchase grain.

Some families are unable either to generate enough income from their own production to retain their labor on their farms or to raise even the small capital needed for entering small-scale retailing. They are dependent on the poorest paid self-employed activities and on providing casual labor for others to tide them over until the harvest. Households in the bottom income quartile are able to obtain only 100 kilograms of maize (for ten months) per person. This is under a third of the amount obtained by households in the top income quartile and well below conventional assessments of adequacy.

How do these households cope? The government policy in place over the past 15 years and more of controlling maize prices through the ADMARC system can be argued to have been one means of keeping the deprivation in some check. On the other hand, general economic conditions, agricultural policies that favor larger-scale producers, and a wage freeze have essentially meant declining standards of living for rural households (Kydd and Christiansen 1982). Other coping mechanisms include following a diversified household income strategy. A most important way by which the poor cope is by relying on the social network in which they are embedded, which acts as a kind of insurance system for those in cycles of phases of want. Some households tide themselves over the deficit period by sharing with relatives who have bigger reserves. There is a clear pattern, especially in December and January, of poorer families eating together with other relatives. Even within households, the number of fires (and, hence, cooks) is frequently reduced during the deficit period. Another strategy is to move people rather than food. While it is much more common for children to be moved between households for their meals at different points in their families' cycles and in the seasons, some adults, especially the aged, are also moved.

Tobacco Growing, Food Crops, and Food Security

A central question in this study is the effect of tobacco production on income, food security, and nutritional status. Tobacco is the highest value crop in the area and thus has a potentially positive effect on income. On average, tobacco growers have a higher household and per capita income than other households (K 90 compared with K 73 per capita; see table 20.1). There is little difference between tobacco growers and other households in the proportion of income derived from food crop sales. Apart from tobacco sales, the major difference between tobacco growers and other households in sources of income is that the latter derive a higher proportion of their income from wages. However, there is a great degree of variability in the income earned from tobacco. Over half of the tobacco growers earn over one-third of their crop income from tobacco sales. These households (specialist tobacco growers) have a per capita income of K 105, earn 24 percent of their total income from tobacco, and allocate 24 percent of their land to tobacco cultivation. The remainder, or "small tobacco growers," concentrate far more on food crop production and sale and add tobacco growing to their activities. They do not have less land on average but have larger households. They allocate 17 percent of their land to tobacco, but earn only about 4 percent of their mean per capita income of K 67 from tobacco sales.

A simple measure of returns to tobacco production (gross sales minus costs of chemical inputs and hired labor) suggests a considerable risk for small-scale producers of not achieving a positive return on growing tobacco; 40 percent of the small tobacco growers reported no cash income from tobacco sales, and half of these (that is, 20 percent of the total tobacco growers) had a loss. In contrast, the specialist tobacco growers did better: only three (8 percent) incurred a net loss.

Do tobacco growers reallocate their resources away from maize production to tobacco, or do they add tobacco to the former? What are the causes and consequences of these actions? Tobacco growers have more land, on average, than households that do not grow tobacco. However, their larger household size results in their per capita landholdings and per capita maize harvests, on average, not being larger than those of nontobacco households. On average, tobacco farmers plant less of their land with maize than do nontobacco households (72 percent, compared with 89 percent). Similar per capita maize harvests, however, indicate that, on average, tobacco production does not displace maize below a certain level.

Tobacco growers as a group are more engaged in the market as far as grain sales are concerned; 72 percent of small tobacco households and 62 percent of specialist tobacco households sell grain compared with 47 percent of nontobacco households. However, there is no difference in consumption behavior—a majority of both tobacco-growing and non-tobacco-growing households are net purchasers of grain. Moreover, tobacco households do not have a lower consumption subsistence ratio (the value of retained maize to total expenditures, see table 20.2) than nontobacco households; nor do tobacco households have a significantly lower proportion of their total maize supply derived from own retained maize (73 percent, compared with 69 percent).

In summary, tobacco growers, who on average derive a much higher proportion and absolute value of income from crop sales, use resources not only to grow crops for sale but also to provide a higher proportion of their own staple food. One might almost say that far from consumption subsistence (for staples) being an index of less commercialization in total agricultural production, for these smallholders it is one of more commercialization. The paradox is only apparent: households with sufficient land and other resources to do so opt for both increased crop sales and high levels of own staple food provision.

The risk in tobacco production demonstrated in the variable earnings derives from the following: first, a failed crop means not only a lack of cash income but also a negative return because of the high cost of inputs required for tobacco production. Second, the use of available land, labor, and cash for inputs necessarily diverts these from other crops, especially maize. And third, the production and processing of dark-fired tobacco is intrinsically more demanding and risky than that of other crops in the area. For these reasons, tobacco has the potential not only for profit but also for loss.

The proportion of household land allocated to tobacco is slightly higher for those with holdings under 0.7 hectare (25 percent) than for those with larger holdings (19 percent for holdings between 0.7 and 1.5 hectares, and 22 percent for holdings over 1.5 hectares). It is obvious that statistical analysis is severely limited with such small totals, but an inquiry may give some clues for future consideration of the relationship between food security and cash cropping for land-scarce households. The small group of six households growing tobacco with under one hectare of land received more than one-third of their crop income from tobacco and were all male-headed households. Using male and female labor, they combined tobacco growing and vegetable growing, the highest value crops in the area, with off-farm employment. All but one of these households derived a higher proportion of income from off-farm sources than the sample mean. Crop sales replace the dependence on grain sales found among other land-scarce households. Only two of the households managed to reap maize harvests above the sample mean. Of the remaining four households, three were able to purchase maize to increase their total supply to almost double their harvest, whereas the fourth increased the harvest stored by only one-third.

In summary, this very small group of six land-scarce households invested its labor in the intensive production of high-value crops of tobacco and vegetables, combined with off-farm income. The contribution of tobacco to land-scarce households with enough labor can be positive, given its relatively high value. But, for this group, the outcome in terms of purchasing power of combining tobacco production in this way is variable, since 50 percent still attained less than 157 kilograms per capita of maize (over ten months).

Expenditure Patterns and the Role of Own-Produced Food

The acquisition of food depends not only on own production, of course, but also on the use of cash income from marketing crops and from other sources. Overall, this sample of smallholders has budgets that are dominated by women's expenditures. More of women's expenditures go to food purchases than men's, although men do buy food, including maize—13 percent of male expenditures are on maize, compared with 25 percent of women's expenditures. Men's expenditures are directed more to nonfood items, including education (though only tiny amounts are spent on this), health costs, and agricultural inputs, as well as on alcoholic drinks (mostly beer and *kachasu*).

Increasing income is associated with a decrease in the budget share allocated to purchased foods, although while the share to purchased maize goes down, that to meat, fruit, and drink rises. The pattern of expenditures among tobacco growers is similar to that among other households. To the extent that tobacco growing results in higher income, the effects are similar to higher income among other households, that is, more is spent on the more expensive foods such as meat, fruit, and fish and on more drinks and agricultural inputs and less is spent on grain, vegetables, and roots. Moreover, the higher the tobacco income, the lower the share of expenditures in women's hands.

In these findings, we see the commonly expected phenomenon of a declining share of expenditures to food as income rises (Engel's law). However, once the value of retained maize is included in total expenditures, the decline in the budget share of food with an increase in income is far less marked. Thus, the share of home-produced food does not decline with increasing income: the differences are slight, but if anything, the share rises before dropping slightly in the top quartile (see table 20.4). As discussed earlier, retained maize makes up the bulk of household maize supply. The strategy followed by all households to achieve as high a proportion of total staple food supply with own production results in high subsistence ratios on a consumption basis, even for the most commercialized households in production terms. The lack of a decline in the share of home-produced food in total budget as income increases again suggests a population with a chronic food shortage that is reluctant to reduce its grain supply proportionately with a rise in total income.

Nutritional Outcomes

Overall, women and preschool children in the sample were found to be stunted but not acutely malnourished. Weight-for-height of both preschoolers and adult women were within normal limits, while children showed moderate to severe deficits in height-for-age and weight-for-age. The low prevalence of severe malnutrition contrasts with findings from other studies in Malawi (Malawi, Centre for Social Research 1988) and may be due to the fact that the study sample did not include completely

TABLE 20.4 Absolute value of per capita nonfood purchases, food purchases, and home consumption, and their share of total expenditure (including home consumption), Malawi, 1986–87

Group	Number of Households	Nonfood Purchases		Food Purchases		Home Consumption[a]	
		Value (kwacha)	Share (percent)	Value (kwacha)	Share (percent)	Value (kwacha)	Share (percent)
Total sample	210	30.65	36.74	21.54	30.67	25.03	32.59
Head of household							
Male	137	33.19	38.00	21.21	28.98	25.79	33.02
Female	73	25.89	34.39	22.16	33.83	23.61	31.77
TEBA[b]	14	50.65	42.33	28.02	27.31	35.42	30.35
Malawi[c]	30	21.38	33.14	22.87	37.91	15.11	28.95
De jure[d]	29	18.99	31.85	19.73	32.77	24.79	35.38
Nontobacco households	148	29.28	35.43	22.48	32.30	25.26	32.27
Small tobacco growers	25	27.46	38.74	18.31	29.43	20.18	31.83
Specialist tobacco growers	37	38.30	40.63	19.96	24.99	27.40	34.38
Landholding class							
<0.7 hectare	29	23.97	34.90	23.96	40.39	15.69	24.71
0.7–1.5 hectares	110	28.38	36.34	22.53	32.42	22.39	31.24
>1.5 hectares	71	36.90	38.12	19.01	23.98	32.95	37.90
Income quartiles							
First (<K44)	52	13.00	32.60	13.47	35.22	12.03	32.18
Second (K44–62)	53	21.04	34.80	18.01	31.55	19.18	33.65
Third (K62–85)	53	28.55	38.02	20.78	28.73	24.28	33.25
Fourth (>K85)	52	60.24	41.58	33.97	27.19	44.78	31.24

SOURCE: Surveys conducted by the authors.

[a] The value of retained maize harvest.
[b] Husband was labor migrant in South Africa.
[c] Male-absentee household where husband was not present for at least half the survey period and was working within Malawi.
[d] Adult widowed or divorced women.
[e] Received less than one-third of their crop income from tobacco sales.
[f] Received more than one-third of their crop income from tobacco sales.

landless households or those people employed as full-time laborers on estates. The early deficit in height found among young children in the sample, together with the short stature and normal weight-for-height of their mothers, suggests a pattern of intergenerational stunting.

Income was positively associated with nutritional outcomes (tables 20.5 and 20.6). Per capita income (including value of retained maize harvest) and per capita expenditures were the most powerful correlates of child stature. Similarly, the results of the dietary intake surveys show that, within a context of generally low caloric levels, households with more resources are able to provide more adequate levels of energy intake during the deficit period (December and January). However, child energy intake showed fewer correlations with resource measures during this period, probably indicating that morbidity, which is at its highest in this period, constrained child food consumption independently. After the harvest, when morbidity was reduced, child energy intake correlated with the retained maize harvest. Regression analysis also indicates that per capita income explains a significant proportion of variance in household, maternal, and child energy intake in the deficit period but not in the postharvest survey. These findings and the much lower energy intake elasticity to postharvest income suggest that households that have small maize supplies and low cash resources to satisfy energy requirements during the season of scarcity are highly vulnerable to food scarcity, whereas immediately after the harvest, differences across households

TABLE 20.5 Correlations of child size (in terms of Z-scores) with household variables

Variable	Height-for-Age	Weight-for-Age	Weight-for-Height
Total household income	0.14520*	0.20178*	0.15543*
Total expenditures per capita	0.20646**	0.22630**	0.11777
Production used for own consumption	0.13790*	0.20927**	0.15953*
Income from tobacco as a percent of crop income	−0.04289	0.00049	0.06642
Hectares cultivated by households	0.07037	0.13702**	0.14433*
Households' harvest (kilograms)	0.11581	0.21628**	0.20346**
$n = 215-217$			

SOURCE: Surveys conducted by the authors.
* = Significant at 5 percent level.
** = Significant at 1 percent level.

TABLE 20.6 Correlations of child size with household resource variables

Variable	Height-for-Age Z-Score	Weight-for-Age Z-Score	Weight-for-Height Z-Score
Total food expenditures	0.20851**	0.25048**	0.14850*
Female share of expenditures	−0.03044	−0.01125	0.00675
Off-farm income as percent of total income	0.05109	−0.02591	−0.07766
Number of preschoolers	−0.04176	0.00066	0.06200

SOURCE: Surveys conducted by the authors.
* = Significant at 5 percent level.
** = Significant at 1 percent level.

diminish. Once the relative "plenty" of the postharvest season passes, the poorer households are once more dependent on their lower cash incomes to secure food.

There is a strong seasonal pattern in nutritional status. Moderate to severe malnutrition was more prevalent among preschool children during December and January, when food is most scarce, work is hard, and morbidity is at its highest. Household, maternal, and child energy intakes were all low in comparison with recommended intakes, particularly in this deficit period. This seasonal variation is exacerbated by low income. Thus, only the top income quartile had a significantly higher energy intake in this period to compensate for the extra demands. In contrast, children in the bottom quartile consumed significantly fewer calories on average during the deficit period than during the postharvest period. This seasonal change in prevalence of malnutrition implies that the poor are vulnerable to fluctuations in food supplies and are unable to meet seasonal energy needs. The findings are typical of a population with chronic malnutrition due to insufficient dietary intake, excessive morbidity, or both, during early childhood.

While the anthropometric measures did not predict the subsequent incidence of disease, the incidence of diarrhea among preschool children did predict subsequent nutritional outcomes. Thus, the Z-scores for height, weight, and weight-for-height during the first anthropometric survey in October 1986 did not correlate with the incidence of diarrhea or other symptoms during the first period of observation (November 1986 to February 1987), whereas the incidence of diarrhea correlated with height-for-age and weight-for-age in the second anthropometric survey in February 1987.

Study results clearly show the effect of infectious disease, in the context of chronic food insecurity, on child growth. Each bout of diarrhea or infectious disease that is not followed by compensatory feeding during convalescence results in additional stunting. Normally nourished children had disease symptoms 30 percent of the time during the preceding three months, while those with mild malnutrition had them 35 percent, those with moderate malnutrition had them 37 percent, and those with severe malnutrition had them 54 percent.

The positive effects of income on nutritional status have already been noted; in contrast, morbidity was found to be largely independent of measures of income and wealth. For example, there was a significant difference in the mean Z-scores of weight between those children in the top income quartile who had experienced diarrhea throughout the survey period and those who had not. Hence, even within the highest income group, freedom from diarrhea improved the capacity for growth. In spite of the higher income and larger landholdings of tobacco growers, their children were not significantly different in nutritional status from those of nontobacco households (table 20.5; see correlation of Z-scores with tobacco as percent of crop income).

In summary, higher income provides for higher calories for households, especially those in the top 30 or 35 percent of the income distribution. The poor households are particularly vulnerable to food shortage in the deficit period of preharvest and have lower levels of caloric intake. On the other hand, the high incidence of morbidity in an area of endemic malaria, respiratory illness, and diarrhea acts as "a grand leveller," affecting nutritional status independently of income levels and having a strong influence on anthropometric outcomes.

Consumption expenditures and the share of tobacco in land use and income are probably endogenous, thus a function of many of the same characteristics that simultaneously determine calorie intake and child growth. Using the data set described above, Sahn and Shively (1991) used a linear combination of exogenous variables in order to instrument per capita expenditures and tobacco share to resolve the endogeneity problem. With a calorie consumption and a height-for-age function, they then estimated income-nutrition relationships, using consistent two-stage least-square estimates. Calorie elasticities with respect to expenditures derived from these estimates range from 0.59 for the upper expenditure tercile to unity for the poorest households. The tobacco share variable and its interaction with expenditures were significant. When the total derivative was taken, the tobacco share elasticity was very near zero when evaluated at the mean (about $-.03$). Thus, while of negligible magnitude, the impact of tobacco intensity on calorie consumption shifts signs with income (Sahn and Shively 1991, 22). This significant but small adverse effect did not carry over into nutritional status differences: while per capita expenditures show the expected positive effects of additional income on child nutrition, no significant effect of cash cropping was found by Sahn and Shively (1991) in the height-for-age function. Consistent with findings from Kenya (chapter 16), decreased levels of stunting in female-headed households compared to male-headed households were found in separate model analyses for the two household categories (Kennedy and Peters 1992).

Conclusion: Policy Issues Relevant to Cash Cropping and Food Security

In a country as poor as Malawi, anything that promises an increase in income to people is welcomed. Hence, high-value crops such as tobacco are attractive. Income plays an important role in influencing household calorie consumption and preschooler nutrition in this poor area. Furthermore, if that income is earned through growing tobacco rather than through producing maize, adverse impacts are not found for nutrition, as pointed out on the basis of the Sahn and Shively (1991) analysis with the data set collected for this study.

This study revealed wide variability in income realized from dark-fired tobacco production. Farmers' strategies indicate their awareness of risk. Thus, while tobacco producers have more land available to them, on average, they, like nontobacco farmers, aim to produce a high share of their own maize needs, and they do not have a lower consumption subsistence ratio (the value of home-produced retained maize to total expenditures). To that extent, cash crop income is part of the strategy, not the whole strategy, for production and income.

The implications for food security are particularly critical in the promotion of cash cropping in a land-scarce area or for land-scarce households. Since the risk is higher for low-resource farmers, the attendant risks to cash crop production have to be taken seriously by policymakers and donors and not ascribed to "backwardness" or "tradition." The conditions governing the supply of necessary inputs (seeds and fertilizers), credit, and information, and those conditions controlling the collection and marketing of the crops are essential to lessening the risks for the small farmer.[5] Equally important is the supply of food available

---

5. A different issue, not considered here, is the protection of such high-value crops by governments regulating production and marketing. In Malawi, smallholders are barred from growing burley and flue-cured tobacco and have to register to grow the dark-fired, air, or sun-cured tobacco allowed them. Currently, production of burley tobacco is being opened up to smallholder farmers.

relatively secure, the canny farmer will not change the strategy of producing a high share of the family's food supply.[6] Where policy changes and "policy reform" are in full swing, as in Malawi and other countries in Africa, then it is doubly important to consider these conditions.

---

6. The liberalization of the maize market in 1987, which closed down many rural supply centers and opened the marketing of grain to an underdeveloped private trade, resulted in a four- to fivefold increase in the price of purchased maize in the following season. Although the conditions in 1988 and 1989 have prevented such a price increase, the farmers are doubtless reinforced in their reluctance to rely on purchased supplies of their basic food.

21 Smallholder Tree Crops in Sierra Leone: Impacts on Food Consumption and Nutrition

FRIEDERIKE BELLIN

Introduction

Tree crops such as coffee, cocoa, and oil palm have been promoted for many years in Sierra Leone. Tree crop production was believed to be associated with improved nutritional status; significantly lower rates of malnutrition have been reported in Eastern Province (26.0 percent),[1] which has been the leader in terms of tree crop production, than in Southern Province (33.4 percent), which has only recently started promoting tree crops (Sierra Leone, Government of 1978). Whether it is actually tree crop production or other factors that are mainly responsible for the improved nutritional status in Eastern Province has not yet been investigated, and, therefore, simple conclusions should not be drawn. The purpose of this study is to evaluate the impact of tree crop promotion on other agricultural production, income and expenditure patterns, and nutrition.

The Role of Agriculture in the National Economy of Sierra Leone

Agriculture remains the most important sector in Sierra Leone, providing employment for nearly 80 percent of the rural population. Moreover, agriculture is charged with the responsibility of meeting the food needs of the increasing population as well as of producing export crops to earn foreign exchange.

Due to a steady decline in export earnings from the mining industry after 1975, agriculture has increasingly contributed to foreign exchange earnings in relative terms. In 1977, agriculture contributed 44.1 percent of total exports. Thereafter, due to decreasing world market prices for

1. Based on weight-for-age less than 80 percent.

This study was sponsored by the Bo-Pujehun Rural Development Project in Sierra Leone.

coffee, cocoa, and oil palm products (Sierra Leone's major export crops), the agricultural contribution declined considerably and was only 25 percent in 1987, with coffee and cocoa each making up about half of the agricultural export earnings. As the share of oil palm products in Sierra Leone's agricultural exports declined and then disappeared completely in 1987, the share of coffee and cocoa became more prominent. Over the same period, Sierra Leone was importing increasing quantities of rice, the cost of which in some years even exceeded export earnings from the agricultural sector.[2] A frequently asked question is whether promotion of tree crops can in the long run increase export earnings to the extent that the balance between agricultural exports and food imports can remain positive, at a minimum.

The Study Area and Research Design

The study was conducted in the Bo and Pujehun Districts of the Southern Province in Sierra Leone. The provincial headquarters are situated in Bo, about 150 miles from Freetown, the country's capital. Transportation facilities and district road networks are rather poor.

The climate is defined by two distinct seasons—a rainy season from May to October, followed by a dry season from November to April. The common farming system is rotational bush fallow, which involves heavy work at the beginning of the farming year when the bush has to be brushed, burned, and cleared. This work as well as ploughing is traditionally done by men. Women's responsibility lies in weeding the farm two to three times a year. Women are also responsible for vegetables and roots that are intercropped with upland rice. Tree crop plantations require regular pruning and involve a lot of labor during harvesttime, which often leads to labor bottlenecks due to high labor requirements for upland rice. Swamp rice production has been promoted to farmers during the last decade, but adoption of swamp rice has not been as high as expected.

Bo-Pujehun Rural Development Project

In 1981, the Bo-Pujehun Rural Development Project (RDP) was launched with the overall objective of improving the socioeconomic conditions of the rural population in the region. The first and biggest sector program launched was the agricultural program, with the objectives of strengthening existing ministry services through training and mobility and of providing improved planting materials and extension

---

2. According to the Bank of Sierra Leone, this took place, for example, in 1984 and 1986.

services. Other sector programs, such as health and nutrition, fisheries, community development, rural roads, water supply, and a women's project component, were included later.

Coffee, cocoa, and oil palm technology packages have been supplied to farmers since 1983. As of 1987, the program had provided packages for 2,350 hectares of oil palm, 315 hectares of coffee, and 350 hectares of cocoa. While coffee and cocoa are pure export crops, the oil palm package caters to both domestic and foreign demand; oil palm products can be used for home consumption and for sale on the local market. There are considerable palm oil shortages throughout the year, which means that demand is by no means satisfied. Tree crop packages were targeted to small farmers.

Study Design

Tree cropping entails long-term investment (in perennial crops), which results in fixed capital and may impact adversely on short-term household liquidity in an initial stage. The related dynamics for income, consumption, and nutrition cannot be captured fully in comparative analyses. However, comparisons of rural households with different degrees and duration of involvement with tree crops allow some insights into these issues.

In order to assess the impact of tree crop farming on nutrition and income, this study was conducted on three survey subgroups:

1. "Old" tree crop farmers, who are farmers already involved in tree crop production prior to the Bo-Pujehun RDP. These farmers did not receive any project inputs and their plantations were already yielding output.
2. Subsistence farmers, who are farmers not yet significantly involved in tree crop production.
3. "Young" plantation or new adopter farmers, who are farmers who received tree crop seedlings from the Bo-Pujehun RDP about two to four years ago. At the time of this survey, they had just invested in the plantations but were not receiving any benefits, since the tree crops were not yet producing. These farmers did not have any old plantations.

Thirty villages situated in seven chiefdoms where the Bo-Pujehun RDP distributed coffee, cocoa, and oil palm seedlings to some farmers from 1983 to 1985 were identified. A preliminary survey was conducted in these villages to verify project information on participation and to prepare a list of farmers to be sampled. Farmers in the three subgroups were selected from the same villages in order to avoid locational influences. The final sample surveyed in 1988–89 included 52 young farmers, 47 old farmers, and 40 subsistence farmers. Details of the sampling design are described in Bellin (1991).

Adoption and Adoption Constraints

A PROBIT model, comparing subsistence farmers with new adopter farmers, found no significant differences between them in household characteristics such as total area under production, age of household head, labor availability, or household composition. An interesting difference was, however, found in the educational level of the household heads, which was significantly better in the case of new adopters: 23 percent of young farmers had been to secondary school, compared to only 4 percent of the old farmers and none of the subsistence farmers. We shall return later to the role of education for tree crop adoption.

Tree crop production is expanding in the survey area. Fifty percent, or 20, of the subsistence farmers surveyed stated that they had started or were about to start a plantation. Eleven of these 20 farmers received their seedlings from local sources, while the other nine farmers were supplied by the Bo-Pujehun RDP. Five of the nine farmers supplied by the project belonged to the low-expenditure tercile, and two farmers each belonged to the medium- and high-expenditure groups. These observations indicate that, first, apart from project-based promotion of tree crops, there is already an independent movement toward adoption of tree crops, and that, second (and more interesting in this context), the low-expenditure group has benefitted from the project, which may be due to the subsidized seedlings.

Adoption of tree crops absorbs labor and land from food crops, which can be expected to lead to a decrease in food crop production. The investment of time and money in tree crops that do not have cash returns for the first five to seven years causes a cash flow problem. This cash flow problem is partly compensated for by off-farm activities that play a much more important role among young farmers. Cash is needed, especially in the absence of a functioning credit market, for farmers to invest in new plantations. All farmers received their loans from the informal credit market. The poorer farmers in each group had higher amounts of loans derived from informal sources.

Food Crop–Tree Crop Competition for Land and Labor

The most frequent criticism expressed in connection with promotion of tree crops is that food crop production and thus food availability will decrease at the household level. The study results indicate that farmers newly adopting tree crops have a much smaller area under food crops; although their total area under production is 20 percent higher than that of subsistence farmers, the new adopters' area under food crops is only half of the area that subsistence farmers cultivate (table 21.1). Established tree crop farmers have 60 percent more total area under cultivation and cultivate 20 percent less area with food crops than subsistence farmers. Thus there clearly appears to be a partial replacement of food crop area by tree crops among the old plantation growers that is accentuated among the new adopters. However, higher intensification is notable on the new adopters' food crop fields: their productivity per unit of land in food crops is 51 percent higher than that of subsistence farmers and 42 percent higher than that of old adopters. Thus the area substitution effects induced by tree crop adoption may be at least partly compensated by higher productivity in subsistence crop cultivation. Unlike the case of the food crop yield effects of cash crop adoption in Guatemala (chapter 12), the productivity differences in this case study cannot be traced over time and may have existed before adoption.

Competition between food and tree crops arises not only in terms of land but also in terms of labor allocation. In fact, since land is still in excess supply in most of the project area, labor constraints are the main concern of farmers. We thus ask, to what extent does tree crop production in old plantation and new adopter farm households lead to labor reallocation between food and cash crops, to changes in division of labor between men and women, and to changes in overall labor time? Regarding the last, an almost equivalent total allocation of average work time is found among the three groups (table 21.2). As women's time allocation to household chores was not accounted for in the survey, it is not surprising that their total work time (4.6–4.8 workday-equivalents per

TABLE 21.1 Household resources of old plantation farmers, subsistence farmers, and new adopters, Sierra Leone, 1988–89

Resources	Old Plantation Farmers	Subsistence Farmers	New Adopters
Sample size	47	40	52
Mean household size	8.1	8.7	8.5
Labor use for farm work (percent)	78.3	80.6	73.8
Total area under cultivation per capita (acres)	0.8	0.5	0.6
Food crop area (percent)	49	100	42
Tree crop area (percent)	51	0	58
Income per capita (Leone/year)	4,419	4,448	3,894

SOURCE: Survey made by the author, Sierra Leone, 1988–89.

TABLE 21.2 Labor allocation to food crop cultivation, tree crops, and other work, by farm household and gender, Sierra Leone, 1988–89

Work	Old Plantation Farmers		Subsistence Farmers		New Adopters	
	Men	Women	Men	Women	Men	Women
			(workday-equivalents/week)			
Food crops	2.7	4.0	3.5	3.7	2.7	3.5
Tree crops	1.5	—	0.2	—	1.1	—
Other	0.8	0.8	1.4	1.0	1.4	1.1
Total[a]	5.0	4.8	5.1	4.7	5.2	4.6

SOURCE: Survey made by the author, Sierra Leone, 1988–89.
[a]Excluding household work.

week) appears somewhat less than men's (5.0–5.2 workday-equivalents per week).

More noteworthy is the labor reallocation between food and cash crops that is observed. Comparison between the subsistence and old plantation groups gives indications of longer-run effects: among old plantation farmers, men's work in food crops and other work is substantially less than among subsistence farmers in order to permit the allocation of 30 percent of their time to tree crops. Women's work in food crops among old plantation farmers is slightly higher (+ 8 percent), while time for other work is slightly lower (− 20 percent) than among subsistence farmers, in order to fill the gaps. Short-run effects are highlighted by comparing the subsistence and new adopters groups: again, new adopters' male members of households spend 23 percent less time on food crops to free up time for tree crops. New adopters' female household members work in food crops is not increased but slightly decreased and other work is about the same. The reason for this distinctly different short-term pattern versus long-term pattern is constraints in food crop land availability. Men's time constraints prevent incremental land clearance for food crops. Thus there are short-term effects in area allocation that transmit into time allocation. Women may not have much opportunity to spend more time on food crops, even if they desired—or felt pressed—to do so. This seems to change, however, in the long run, as comparison between the old plantation farmers and new adopters indicates: women have taken over some of men's work in food crop production.

Figure 21.1 shows the seasonal patterns of work of farmers. Note that men's workload appears more stable throughout the year than women's, and tree crop–farming households in general have a more stable work pattern than subsistence farm households.

FIGURE 21.1 Seasonal work time profiles by type of work and gender.

<u>Men</u>

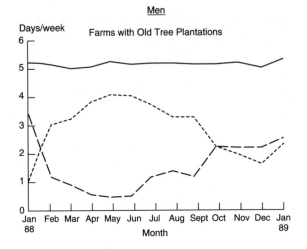

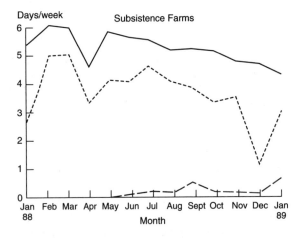

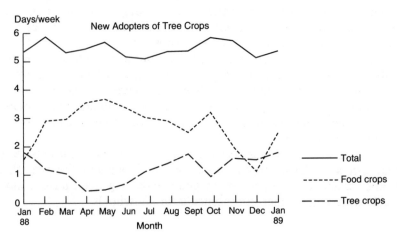

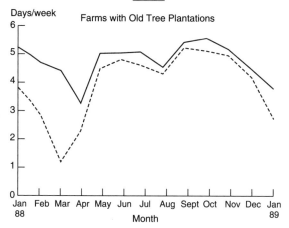
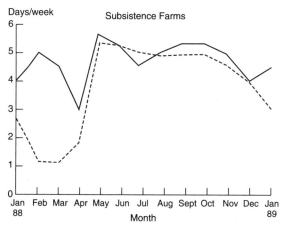
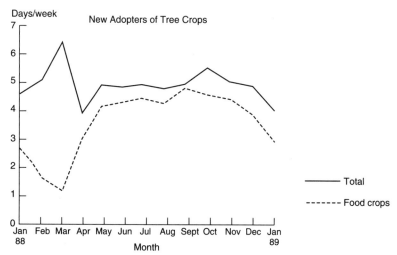

335

Income and Income-Source Patterns

Mean annual income per capita was about the same in the old plantation group and the subsistence group but about 12 percent lower in the new adopters group (table 21.1). Twenty-five percent of the old plantation group is in the bottom income tercile of the joint income distribution of the three groups as are 45 percent of the subsistence group and 30 percent of the new adopters group. Thus income distribution among the subsistence group is most skewed. To net out income distribution differences when looking at income-source patterns, we look at income-source patterns of the middle tercile of the joint distribution of the three groups. In these households, tree crop income accounts for 13 percent of income in the old plantation farmers group and for 2 and 3 percent in the subsistence and new adopters groups, respectively. The old plantation farmers and new adopters are more commercialized in a general sense because of higher nonagricultural income than the subsistence group (14 percent in old plantation farmers group and 20 percent in new adopters group versus 6 percent in subsistence group).

Women and Tree Crops

Even though women play an essential role in the rural economy, most development activities have bypassed them. Since tree crops are controlled mainly by men, women's responsibilities in food crop production are increasing but, as indicated above, at a slower rate. Apart from food crop production, women are highly involved in the processing of tree crops, which in fact is mainly their domain. Marketing of tree crops, however, is left to men, who in return will give some compensation to women in the form of clothes or a small amount of money. Only two women in the old plantation farmers group complained that they were not getting anything for processing the tree crops. Women also compensate themselves through hidden channels, such as putting part of the processed oil or coffee aside and selling it themselves. This, of course, is well known, even to their husbands, but difficult to assess quantitatively.[3] It is, therefore, not surprising that women in most cases support the idea of tree crop plantations.

Tree Crops, Consumption, and Nutrition

Data assessed in this survey indicate a very high percentage of total expenditures on food. On average, 83 percent of all expenditures were on food. A relative decline in food expenditures (including value of home-

---

3. It is evident that this leads to problems in terms of assessing the income of households.

produced food) with increasing expenditures was observed only in the young plantation farmers group. The comparatively higher share of food expenditures might be partly due to the problem of assessing sensitive expenditures such as expenses on secret societies, social commitments, and court cases, all of which play a very important part in Mende culture and involve considerable expenditures as well.[4] The fact that the food expenditure share remains high with increasing income is due to more expensive calories or, in other words, to better-quality food (table 21.3). Comparing the middle terciles shows that the old plantation farm households purchased substantially higher shares of their food (27 percent versus 18 percent for the two other groups). Table 21.4 summarizes dietary as well as anthropometric indicators for children in the survey. There is an obvious increase in calorie and protein intake—obtained from 24-hour recalls—with increasing expenditures (income) and shrinking household size. Note that in the same expenditure classes, the subsistence group consumes notably less calories than the other two groups.

Anthropometric measurements, however, show no such clear-cut patterns. The old plantation farmers and new adopters farmers, who show the higher calorie and protein intake patterns, also show a higher prevalence of undernutrition than subsistence farmers for some of the anthropometric indicators. Nutritional status is obviously not influenced just by calorie and protein intake but also by other variables (table 21.4). We explore this further in multivariate analysis.

Two different models are specified to evaluate the possible impacts of tree crops on consumption and nutrition. The first model tests whether calorie consumption is linked to household size, income, and income-source patterns (off-farm income and tree crop income shares). It is further tested whether the group of new adopters was different from the rest of the survey sample due to their recent adoption of the perennial crops, which may have put their household economy under stress (Table 21.5). Hence, the model is estimated for the total sample with special reference to the new adopters.

There is no significant difference in calorie consumption between the subgroup of new adopters and the total sample. The results show that a 1.0 percent increase in income would increase calorie consumption by 1.4 percent, whereas a 10.0 percent increase in off-farm income, on the other hand, would decrease calorie intake by 1.7 percent (weakly significant). The impact of household size on calorie consumption is strong:

---

4. These observations have been confirmed by a British anthropologist who, at the same time, did a one-year case study in a Mende village in the Eastern Province (Leach 1990).

TABLE 21.3 Annual expenditure and expenditure share of farm households, by expenditure tercile, Sierra Leone, 1988–89

Item	Old Plantation Farmers			Subsistence Farmers			New Adopters		
	Low	Medium	High	Low	Medium	High	Low	Medium	High
Number of households	12	21	14	18	7	15	16	19	17
Total expenditures (Leone/year)	3,582.4	5,600.1	9,553.0	3,612.9	5,537.3	8,745.4	3,486.5	5,483.8	8,184.6
Expenditure shares (percent)									
Food	83.3	80.1	84.6	84.4	85.9	87.7	83.5	79.1	79.4
Purchased food	27.6	26.9	22.1	15.2	18.2	17.3	24.0	17.5	20.3
Subsistence food	55.7	53.2	62.4	69.3	67.7	70.4	59.5	51.6	59.0
Nonfood	16.7	19.9	15.4	15.6	14.1	12.3	16.5	20.9	20.6

SOURCE: Survey made by the author, Sierra Leone, 1988–89.
NOTE: Low, medium, and high refer to expenditure terciles.

TABLE 21.4 Per capita calorie and protein intake and anthropometric indicators of children aged less than three years, by farm household type and expenditure tercile, Sierra Leone, 1988–89

Item	Old Plantation Farmers			Subsistence Farmers			New Adopters		
	Low	Medium	High	Low	Medium	High	Low	Medium	High
Average daily per capita calorie consumption	1,723.1	1,991.0	2,197.9	1,696.2	1,767.9	1,946.9	1,713.5	1,824.0	2,061.8
Average daily per capita protein consumption (grams)	41.0	44.5	52.7	37.7	42.2	45.2	39.3	43.7	50.9
Anthropometric indicators									
Weight-for-height[a]	3.4	2.9	0.0	0.0	0.0	2.7	3.7	1.9	2.9
Height-for-age[b]	48.3	29.4	29.4	33.3	38.9	29.7	48.1	28.3	35.3
Weight-for-age[a]	58.6	41.2	44.1	31.1	22.2	43.2	37.0	26.4	47.1

SOURCE: Survey made by the author, Sierra Leone, 1988–89.

NOTE: Low, medium, and high refer to expenditure terciles.

[a]Percent of children below 80 percent of standard.
[b]Percent of children below 90 percent of standard.

TABLE 21.5 Determinants of calorie consumption

Explanatory Variable[a]	Parameter	t-Value
NDUM	32.0264	0.605
LNPCINC	138.6310	2.947
OFFFARMR	−321.4556	−1.664
TRCPYRT	332.9173	1.445
HHSIZE	−32.5912	−6.115
Constant	1,029.9980	3.491
$\bar{R}^2$	0.39	
F-value	17.30	

NOTE: The dependent variable is calorie consumption.
[a]Definitions of variables:
NDUM = new adopter = 1, otherwise = 0.
LNPCINC = expected income proxy; logarithm of per capita income.
OFFFARMR = ratio of off-farm income to total income.
TRCPYRT = ratio of tree crop income to total income.
HHSIZE = household size.

there is clear evidence that calorie intake is significantly higher in smaller households. An increased share of tree crop income in total income has no significant impact on calorie intake.

The second model estimates the impact of tree crop production on the nutritional status of children. The analysis is limited to the weight-for-age indicator. It is assumed that weight-for-age is a function of the following parameters: per capita expenditure, income share of tree crops and nonfarm activities, children's age, percentage of children in the household, and the household size.

The explanatory power of the model is low, presumably because it is not capturing health- and care-related determinants of child nutrition. The model was estimated for the entire sample, with special regard for the new adopters and old plantation farmers (table 21.6). Model results suggest that children in the group of new adopters and old plantation farmers are—other things held constant—comparatively worse off than the children of subsistence farmers in the sample. Off-farm income showed a positive impact, while the tree crop income share negatively influenced the weight-for-age of children. A 10.0 percent increase in off-farm income would increase the percentage of weight-for-age by 2.1 percent, while a 10.0 percent increase in tree crop income would reduce the percentage of weight-for-age by 1.2 percent. The income elasticity[5] of nutritional improvement at the mean was small but significant (0.35

---

5. Income elasticity is calculated with total expenditure figures, since expenditures are expected to provide more reliable information on permanent income.

TABLE 21.6 Determinants of weight-for-age for children

Explanatory Variable[a]	Parameter	t-Value
TRCPYRT	−12.2891	−2.040
SEXDUM	−1.3079	−0.934
PCEXP	0.0032	2.056
PCEXPSQ	−2.45871E-07	−2.223
AGESQ	0.0398	5.244
OFFFARMR	20.5249	3.398
CHILD5	0.2494	0.489
NDUM	−4.3484	−2.441
CHILD514	0.7825	1.496
HHSIZE	−0.2218	−0.886
AGE	−1.5875	−5.801
Constant	86.3841	14.185
$\bar{R}^2$	0.18	
F-value	5.90	

NOTE: The dependent variable is weight-for-age.
[a]Definitions of variables:
TRCPYRT = ratio of tree crop income to total income.
SEXDUM = male = 1, female = 0.
PCEXP = per capita expenditure.
PCEXPSQ = (per capita expenditure)^2.
AGESQ = (age)^2.
OFFFARMR = ratio of off-farm income to total income.
CHILD5 = children under five years of age.
NDUM = new adopter = 1, otherwise = 0.
CHILD514 = children between five and fourteen years over household size.
HHSIZE = household size.
AGE = age in months.

percent for a 10.0 percent increase in income). The significant negative squared term of income indicates nonlinearity. Thus the income elasticity of nutritional improvement was somewhat higher in the lowest expenditure tercile. Nutritional status here would improve by 0.45 percent for a 10.00 percent increase in income.

Conclusions

When farm households adopt tree crops, a significant reallocation of land and labor resources occurs in both the short and long runs. These changes impinge on consumption and nutrition in this study area, where income is generally very low. Resource commitments to perennial crops —the trees—require investment of time, land, and cash, but income flows start only after a considerable time has elapsed. Thus, there is a temporary reduction in the flow of income. New adopter farmers were found to borrow significantly more than subsistence farmers for the investment as well as to bridge the gap. Still, at the time of adoption, average income was found to be lower in the new adopter group than in the subsistence farmers group from which they emerged. As income elasticities of calorie consumption are found to be very high in this poor setting, these (temporary) income reductions may have an adverse effect on household food security. However, the new adopters do not stand out with a lower level of average calorie consumption.

More striking are the findings regarding children's undernutrition. Stunting is significantly higher among the poor in the two groups of old tree crop farmers and new adopters than among subsistence farmers, and the weight-for-age indicator is worse in these two groups than among subsistence farmers. Multivariate analysis also suggests adverse nutritional-status effects (weight-for-age) of new adoption in this poor setting. This may be related to increased time constraints of women in food crops, where they have partly taken over from men. Time-saving technology for women and improved access to credit in the early adoption phase are key elements of program activities to mitigate potential adverse effects of tree crop adoption on child nutrition. Furthermore, the potential food availability constraint arising in the short run from reduced subsistence food area needs to be addressed by yield-increasing technological change in these crops to facilitate a food-secure adoption of three crops.

22 Nutritional Effects of Commercialization of a Woman's Crop: Irrigated Rice in The Gambia

JOACHIM VON BRAUN
KEN B. JOHM
DETLEV PUETZ

Introduction

The Gambia (West Africa) had a population of approximately 0.75 million and a total agricultural cultivated area of only 1,850 square kilometers in 1987-88. By virtue of its small size, The Gambia is prone to an open market economy, and commercialization and specialization of agriculture have to be elements of any viable growth strategy. Besides its small size, The Gambia is almost completely surrounded by Senegal, except for the Atlantic Ocean on the west. Consequently, there is vigorous cross-border trade in agricultural commodities and inputs, in addition to basic nonfood commodities. In 1985, reexport of goods accounted for 69 percent of total domestic exports.

Introduction of technological change and commercialization in The Gambia's agricultural sector usually proceeds along two lines. The first instance is through formal state intervention in large-scale, mostly external donor-financed projects such as the Cotton Development Project, the Irrigated Rice Development Project, or the recent Horticultural Development Project. In the second instance, change comes slowly through farmer-initiated informal networks and mostly originates from neighboring countries. Examples of this second instance include the adoption of improved weeders and other agricultural implements, as well as simple grain processing machines. This analysis focuses more on the first type of project-centered technological change and commercialization process.

Scope of Analysis

This chapter deals with the effects of commercialization and new technology in food crop production on household-level food security and nutrition in The Gambia. The complexity of linkages between technological change, commercialization, food security, and nutrition increases with the complexity of household organization. The study setting has a complex production-consumption system characterized by communal and individual farming and by separate, though interacting, production by men and women farmers. It was into this setting that modern irrigation for rice was introduced. The system and the nature of the technological change are briefly described in the following section.

Two specific research questions are addressed in this chapter:

- Through what mechanisms and to what extent does rapid commercialization and technological change in food crop production improve household-level food security?
- How, and to what extent, does this change translate into measurable nutritional improvement?

Food security is understood in this context as the ability of all members of a household to acquire sufficient amounts of food continuously over time for a healthy life.

The Production System in the Study Area and Nature of the Technological Change

The empirical research results presented in this chapter are based on a detailed farm-household survey of 200 compounds, conducted during 1985–86 and 1987–88 in The Gambia by IFPRI in collaboration with The Gambia's Programming, Planning, and Monitoring Unit for the Agricultural Sector (PPMU), now the Department of Planning.

The study area is located in central Gambia, 300 kilometers east of Banjul (the capital). Four different ethnic groups (Fula, Wolof, Mandinka, and Serahuli) are affected by the program of technological change. Farming in the study area is organized into two distinct parts: a communal farm under the control of the compound head, and a set of private farms that comprise groups of fields allocated to individuals for growing (cash) crops (mainly groundnuts, cotton, and rice) under their personal control. The communal crops (mainly millet, sorghum, maize, and rice) are produced by the combined labor of all compound members—all men and women have a customary obligation to provide labor to the communal fields.[1] Very little of the communally grown food crops, which are grown to meet domestic food consumption needs for the whole compound, is sold. Households, commonly referred to as compounds in The Gambia, are frequently large (17 persons on average) and

---

1. A detailed descriptive account of the system is given by Webb (1987) and Dey (1982).

income-earning and consumption activities are institutionalized through complex consumption and production subunits.[2]

The technological change under consideration in this setting is the upgrading of rice production technologies through the introduction of a large-scale rice irrigation project. Implemented in 1983, this project covers over 1,500 hectares and is designed to raise rice yields and output through the centralized control of water pumping and draining around plots of 0.5 hectare in size. The project is owned and managed by the state, with ploughing services and a package of fertilizers and improved seeds provided on a credit basis. All other farming activities are carried out by the smallholders who are registered as tenants of the plots. This is quite a contrast of management styles with the rice scheme in west Kenya (chapter 17), where the scheme is also centrally managed and farmers' produce is centrally purchased. This project attempted to maintain the traditional use rights of female farmers to rice land by giving priority to women during the official registration of plots.

The rice production technology in the new scheme is not uniform, and neither are the technological conditions outside the scheme. Four distinct rice production technologies can be distinguished, of which some are further subdivided at the location but remained aggregated for the purpose of this chapter.

1. Production in traditional swamps irrigated in the rainy season by the flooding of the River Gambia (partly deep-water rice): yields average 1.3 tons per hectare per crop.
2. Production in small pump-irrigation schemes established in the 1970s: some plots produce two crops per year, but most produce only one. Yields range between 2.5 and 4.5 tons per hectare per crop.
3. Production under partly water-controlled conditions in leveled perimeters: irrigation is achieved either through tidal fluctuations of the river or simply through precipitation. One crop per year is produced from these plots in the rainy season. Yields range between 2.5 and 3.5 tons per hectare per crop.
4. Production under fully water-controlled conditions in perimeters that are centrally irrigated and drained: two crops per year are produced. Yields are between 5.5 and 6.5 tons per hectare per crop.[3]

---

2. Instead of the term *household*, the term *compound* is used here, since it is not only the locally used word for the unit but it also expresses neatly the "compounded" nature of the production-consumption unit at the location that consists of various subsystems.

3. The yield figures quoted here refer to averages for 1984 and 1985.

The study area is well known for its high prevalence of child undernutrition, especially in the rainy season. The survey found that 35 percent of children under five years of age were below 80 percent of the weight-for-age standard in the 1985 rainy season. During the rainy season, workloads (energy expenditures) as well as food energy prices are at their peak, leading to a "hungry season" syndrome. It is hypothesized that improved crop technology, that is, a second rice crop, reduces this imbalance, at least from the energy supply side.

Table 22.1 shows that calorie consumption levels differ substantially by income (total expenditure) levels, and that differences are particularly wide between the bottom and top income quartiles in the hungry (wet) season. In 1985–86 17.1 percent of children aged 6–59 months and 6.9 percent of children aged 60–120 months were stunted. In the wet season of 1985–86, 38.5 percent of children aged 6–59 months were seriously underweight, but in the dry season, this proportion dropped to 28.8 percent. Differences in the prevalence of malnutrition in terms of weight-for-age between income groups were more pronounced among older children (aged 60–120 months) than among the younger ones (aged 6–59 months). During the wet season, 32.8 percent of the older children (60–120 months) in the bottom income quartile were malnourished (below 80 percent weight-for-age), while only 15.2 percent of the children belonging to the top income quartile were found to be so. Apparently, seasonality affects food consumption and nutritional status of all households, but the seasonality effect is more pronounced for the lower-income households.

Conceptual Framework

The analysis concentrates on four linkages that are crucial for relationships between technological change and household-level food security but are particularly poorly understood, especially in the West African setting:

1. the adoption of new technology in a farm production system comprising both individual and communal farming, where the technological change applies to one, mostly individually grown, crop—rice (traditionally rice is a "woman's crop" in the area);
2. the reallocation of resources (labor) between crops as a consequence of adoption of new technology and changes in crop control (with its related effects on labor productivity differentials by gender within the household);
3. the food consumption effects of increased income deriving from technical change and related changes in sources of income (in kind, cash, and so forth);

TABLE 22.1 Food energy consumption and nutritional status of children, by household per capita income quartile and season, The Gambia, 1985–86

Income Quartile[a]	Calorie Consumption per Capita per Day		Prevalence of Malnutrition in Children					
			<90 Percent of Height-for-Age Standard[b]		<80 Percent of Weight-for-Age Standard[b]			
					6–59 Months		60–120 Months	
	Wet Season	Dry Season	6–59 Months	60–120 Months	Wet Season	Dry Season	Wet Season	Dry Season
			(percent of children below stated percentage of standard)					
First (lowest)	1,893	2,176	30.1	11.0	40.5	28.3	32.8	16.7
Second	2,222	2,226	13.8	5.6	37.9	30.1	29.2	19.0
Third	2,622	2,811	13.1	7.7	36.9	36.3	24.3	20.6
Fourth	2,917	2,972	14.3	3.0	39.3	20.8	15.2	10.5
Average	2,380	2,522	17.1	6.9	38.5	28.8	25.8	16.7

SOURCE: Data from the household survey made by the International Food Policy Research Institute and the Programming, Planning, and Monitoring Unit for the Agricultural Sector, The Gambia, 1985–86.

[a]Total expenditures (including value of own-produced food) used as income proxy.
[b]NCHS/WHO standards are used for the identification of the nutritional status.

4. the effects on nutritional status, especially of children and women, as a result of changes in food consumption interacting with the health and sanitation situation.

Adoption of New Technology: Complications When Communal and Individual Farming Coexist

Adoption of new rice technology by farmers in the study area took the form of a rationed distribution of access to the improved irrigated land. Demand for the improved irrigated land greatly exceeded supply. The adoption was mediated through managerial decisions at the program level and through village- and compound-level decisions on "who gets what."

With regard to traditional cereal production, there is a fairly clear distinction between men's and women's crops. Men traditionally organize upland cereals production; more than 90 percent of the upland fields are under male responsibility. Nearly all of the upland cereals production is communal farming. Women, on the other hand, organize swamp rice production. More than 90 percent of traditional rice fields, but hardly any of the upland cereals fields, are under female responsibility. Most of the traditional rice production is considered as individual farming activities. Both men and women are in charge of organizing noncereal cash crop production in their individual upland farm enterprises. Women, for instance, are responsible for 28 percent of the area under groundnuts and 27 percent of the area under cotton.

Distribution of the new rice land to farmers in the irrigation scheme poses problems of equity. Yet, the distributional issue here does not relate well to the criterion of "large or small" farmers frequently used in describing adoption-of-technology issues in Asian or Latin American agriculture. Here, the distributional issue is an issue of differentiation between villages, between compounds of the same village, and—most important—between male and female farmers of various socioeconomic characteristics.

The project's land distribution policy specifically aimed at assuring women access to the project rice land in order to increase women's income directly. Women were given land titles (long-term leases) in a formal registration process. The survey assessed the extent to which this process was translated into women's de facto control over land and its independent cultivation. At the study location, it turns out that women's control over rice fields is inversely correlated with technology adoption (yield levels). The higher the yield level, which coincides with the increase in production inputs and cash needs to finance production, the lower is women's participation versus men's (table 22.2). The distribu-

TABLE 22.2 Yields, distribution of rice fields, labor allocation, and marketed surplus, by type of rice technology, The Gambia, 1984–85

Type of Rice Technology	Average Yield Wet Season (tons per hectare)	Fields Under Women's Responsibility (percent)	Fields Under Communal Production (percent)	Women's Labor Share in Family Labor (percent)	Sales Share of Production	
					Women[a] (percent)	Men[a] (percent)
Fully water-controlled fields in new rice scheme	6.6	14	99	29	9	7
Pump-irrigated fields in old rice scheme	2.9	54	67	68	0	5
Partly water-controlled fields in new rice scheme	2.2	70	87	60	7	7
Traditional swamp rice	1.3	95	17	77	21	15

SOURCE: Data from the household survey made by the International Food Policy Research Institute and the Programming, Planning, and Monitoring Unit for the Agricultural Sector, The Gambia, 1985–86.

[a]Refers to women's and men's fields, respectively. Men's fields include the compound head's controlled fields.

tion of land titles to women, who formally hold more than 80 percent of the irrigated land in the scheme, has apparently not altered this general relationship between technology adoption and control at this study location.

While the new rice technology increased yields substantially, it did not have a big impact on marketed surplus of rice. In fact, in relative terms, more traditional swamp rice grown by women was sold than rice grown under the new technology controlled by the compound head largely as a communal crop (table 22.2). Traditional swamp rice was and still is, to a large extent, a source of cash income for women and not just a source of subsistence food provision.

It is hypothesized that women's status within the compound and compound sociodemographic characteristics are important factors that shape the outcome of the adoption process. A PROBIT model was designed that addressed the "adoption" issue and assessed the factors that enhanced or reduced the probability of women's access to a new project rice field. The estimation results of the model are reported in table 22.3 and show that a disaggregated view is necessary even within the group of women farmers—not just men versus women—to comprehensively judge the access to technology issue. There are three major findings:

- Women of higher status (that is, women who are older and women who are heads of a cooking unit) had a significantly higher probability of obtaining a field.
- Women who were traditional rice growers had a significantly increased probability of obtaining a field.
- In large compound subunits, the individual woman had a lower probability of obtaining her own field because the new rice fields became "communal."

While the model gives some insights on what determines differential access to new technology, it does not address the strong negative correlation found between technological change and women's access to new technology (table 22.2). This relates to women's ability versus that of men to organize the operation of production and to finance the new technology in the household production system. These issues are now addressed.

Reallocation of Labor, Labor Productivity, and Instability in Rice Production

The pattern of female labor inputs into rice production follows the pattern of control over fields. Women provide the highest share of labor

TABLE 22.3 Determinants of access of individual women to fields in the new rice scheme (PROBIT estimate)

Explanatory Variable	Parameter	t-Value	Significance Level	Change in Probability of Access to Field[a]
Age of woman[b]	0.5122	3.47	0.001	+0.21
Woman is head of a cooking unit	0.6697	3.59	0.000	+0.26
Woman is married to compound head	−0.0240	0.18	0.989	—
Woman was traditional rice grower[c]	0.5522	3.14	0.002	+0.23
Number of women of working age in cooking unit	−0.1157	−3.26	0.001	−0.04
Compound has more than one cooking unit	0.3341	2.05	0.041	+0.13
Size of land in rice scheme for whole cooking unit[d]	1.3588	4.70	0.000	+0.54
Constant	−4.3331	−5.44	0.000	—
Chi-square	86.55			

NOTE: The dependent variable is a dummy variable; if woman had her own field in the scheme or shared one with other women = 1, else = 0.
[a] The probability is approximated from the estimated parameter by multiplying with a factor of 0.4 following an approximation approach described by Amemiya (1981).
[b] In age groups of 15–19, 20–39, or 40–59 years).
[c] Dummy variable = 1, if woman had her own rice field in the area before the scheme and a personal rice store; else = 0.
[d] In irrigated-land equivalent.

input into those rice fields they control, as do men. However, considerable mutual assistance between the sexes also exists, as indicated by the more even distribution of labor time allocation than distribution of control (table 22.2, column 4).

While upland fields, in general, are not a scarce resource in the study area, labor in the wet season certainly is scarce. Any increase in labor demand in one production activity in that season induces substitution effects. Increased labor productivity in rice due to new technology is, therefore, expected to pull labor away from other crops, at least in the medium run, as long as the local labor supply situation does not adjust with in-migration. The labor substitution effects induced by the new rice fields on the various crops were found highly significant. Two noteworthy quantitative findings are reported here:

- Evaluated at the respective sample means, access to an average piece of project land reduces household labor inputs to traditional rice by

21 percent; to upland cereals by 19 percent; and to groundnuts by 22 percent because less of such cropland is farmed and less labor is allocated per unit of land under these crops (von Braun, Puetz, and Webb 1989).

- Tracing the substitution effects in labor allocation to production by means of production functions estimated for the same crop groups shows that, for an additional ton of rice produced in the new scheme, 380 kilograms of cereals (millet, sorghum, and traditional rice) and 340 kilograms of groundnuts are forgone (von Braun, Puetz, and Webb 1989).[4] In monetary terms, these substitution effects mean that for one dollar incrementally earned from the new rice, 64 cents of net income from other wet-season crop production is given up.

One might suspect that fully water-controlled rice production provides for increased, steady, and reliable income streams and thereby improves household food security. While this is the case for a well-managed and sustainable project situation, instability is a new risk factor for participant farmers who commit their resources to the new irrigation technology. Yields in this scheme, when traced over four years in the IFPRI/PPMU surveys, were far from stable. Year-to-year yield fluctuations in the fully water-controlled fields were 30.5 percent of 1984–87 mean yields and were nearly twice as high in the partly water-controlled fields (table 22.4). Surprisingly, both these values exceed the yield fluctuations in traditional rice and groundnuts. Even if the year 1987, which was particularly bad in the scheme because of flooding and other factors, is excluded, the general picture does not change. However, in these comparisons of wet-season crops it should not be overlooked that the fully water-controlled rice offers incremental dry-season crops at a time when employment and food stocks are down. The favorable effects of this are traced below.

Average labor productivity in upland crops over the growing season exceeded labor productivity in partly water-controlled rice production in 1985 when the scheme was well functioning, but it did not exceed labor productivity in fully water-controlled conditions (which benefit from a substantial subsidy on the irrigation infrastructure) (table 22.5). Yet, marginal returns to labor in upland crops (both millet and groundnuts) were higher than in the new rice fields in the wet season of 1985. The low marginal labor productivity, despite high average labor productivity, in

---

4. The production function estimates are presented in von Braun, Puetz, and Webb (1989).

TABLE 22.4 Yield fluctuations in irrigated and upland crops, Jahally-Pacharr project area and neighboring villages, The Gambia, 1984–87

Year	Irrigated Rice				Upland Crops	
	Fully Water-Controlled Rice in Scheme		Partly Water-Controlled Rice in Scheme	Traditional Rice	Millet	Groundnuts
	Wet Season	Dry Season	Wet Season	Wet Season	Wet Season	Wet Season
	(yields per hectare in tons)					
1984	6.55	7.20[a]	2.19	1.26	0.57	1.01
1985	5.17	6.49	2.86	1.11	0.80	1.09
1986	4.07	5.07	1.89	1.18	1.03	1.51
1987	2.06	3.05	0.94	0.79	0.64	1.29
Mean yields, 1984–87	5.01	6.12	1.58	1.14	0.76	1.23
Average year-to-year change in mean yields (percent)	30.5	22.7	58.4	17.8	37.3	19.5

SOURCE: Data from the household survey made by the International Food Policy Research Institute and the Programming, Planning, and Monitoring Unit for the Agricultural Sector, The Gambia, 1985–86, 1986–87, and 1987–88.

[a]Source of this figure: Jahally-Pacharr project management.

TABLE 22.5 Marginal and average labor productivity of major crops in the wet season, The Gambia, 1985–87

Crop	Marginal Labor Productivity Derived from Production Functions[a] (1985) (dalasi per person per day)[d]	Average Labor Productivity in Gross Margin per Labor Day[b]	
		1985	1987[c]
Upland cereals (millet, sorghum, and maize)	7.50	8.88	6.76
Groundnuts	4.61	8.70	8.99
Traditional rice	3.20[e]	5.70	1.39
Partly water-controlled rice in scheme	3.20[e]	7.40	3.24
Fully water-controlled rice in scheme	1.42[f]	14.70	4.33

SOURCE: Data from the household survey made by the International Food Policy Research Institute and the Programming, Planning, and Monitoring Unit for the Agricultural Sector, The Gambia, 1984–85, 1985–86, and 1987–88.

[a]The marginal labor productivity is evaluated at respective mean values of labor input and average sales prices of output. Estimation results are presented in von Braun, Puetz, and Webb (1989).
[b]Gross margin (net return) is calculated as the difference between value of (marketable) output and variable costs (including hired labor).
[c]In 1985 prices.
[d]In 1985, 1 dalasi equalled about US $0.20 on the parallel market.
[e]Traditional rice and partly water-controlled rice in the scheme are dealt with in one model.
[f]The parameter on which this value is based is statistically significant at the 15 percent level only.

the fully water-controlled rice in 1985 does not come as a surprise. Farmers participating in the scheme are under considerable pressure to perform work tasks defined by the project's management at a given schedule, which is geared toward maximizing yields rather than equating marginal returns with marginal (opportunity) costs of labor. Farmers not following the schedule and achieving inappropriate yield levels run the risk of being excluded from the scheme. In 1987, average labor productivity in the scheme dropped drastically, by 71 percent in the fully water-controlled rice technology, and by 58 percent in the partly water-controlled rice technology, because of the yield problem just mentioned. Also, because of relative price changes in cereals versus groundnuts in the context of economic adjustment programs, the profitability of the latter increased while those of the former declined (table 22.5).

In sum, labor productivity is highest in the fields of the technology type that was largely taken into communal food production under the supervision of the male heads of households. Thus, women, in general, have benefited less than men from the new technology, at least in a direct way, on the production and income-earning sides. For households as a whole, however, favorable income and food supply effects were identified, along with major substitution effects in labor allocation. How did these changes in agricultural resource allocation and income transmit to food consumption and nutrition? This question is addressed in the remainder of this chapter.

Effects on Food Consumption

In the wet (hungry) season, per capita daily calorie consumption in the lowest income quartile was, on average, 13 percent below the dry season's level (table 22.1); 49 percent of the households in that quartile were consuming less than 80 percent of the recommended calorie requirement, whereas the corresponding proportion in the highest quartile was only 2 percent. The main rural consumption policy issue is to identify a strategy to effectively increase the calorie consumption of low-income people in the wet season. This consumption analysis, therefore, focuses on calories and on the wet season.

While a high correlation is found between per capita income (or total expenditure) and calorie consumption, availability of cereals from own production and calorie consumption are only weakly correlated. This becomes further evident when compound cooking units—the unit of observation for this consumption analysis—are arranged by quartiles of per capita expenditure (including value of own-produced food) versus quartiles of per capita cereals production in table 22.6. From the table, it

TABLE 22.6 Calorie consumption by quartile of cereal availability from own production and by quartile of total expenditure, The Gambia, wet season 1985–86

Quartile	Calorie Consumption per Adult-Equivalent per Day	
	Cereal Availability Quartile[a]	Expenditure Quartile
First (lowest)	2,734	2,293
Second	2,780	2,672
Third	3,018	3,162
Fourth	3,133	3,428

SOURCE: Data from the household survey made by the International Food Policy Research Institute and the Programming, Planning, and Monitoring Unit for the Agricultural Sector, The Gambia, 1985–86.

[a]From own production. Derived by subtracting seed, waste after harvest, and about 10 percent donation to the mosque from cereal production per capita.

can be seen that it is income that matters for consumption and food security, rather than cereal production in a physical sense. The following analysis of the determinants of calorie consumption sheds further light on this.

Consumption consists of food prepared for joint meals in the common cooking units (*sinkiro*) and of individually acquired snack food. Technological change in rice may impinge on food consumption via its effect on income, local prices, and determinants mediated via the control over and the pattern of the income stream in the compound. We develop four hypotheses:

1. Following conventional demand theory, calorie consumption is determined by income level—with a decreasing effect as income rises—and price (total expenditure including the value of own-produced food is used as an income proxy, and the price of rice, the major traded staple, is used as a price indicator for calories).
2. A higher share of cereals produced under women's control has a positive consumption effect over and above the income effect of cereal production because women are assumed to have a higher propensity to provide for household calories.
3. Actual storage levels, in the context of expected and current income, increase offtake and, thus, calorie consumption, especially in the wet season, when markets are occasionally disrupted (wet-season storage levels are largely determined by access to fully water-controlled rice in the scheme, thus, a result of technological change).
4. An increased share of income in the form of cash reduces the consumption of calories when income level is controlled (readily fungible cash may end up in nonfood expenditure or more luxury purchased foods to a larger extent than does subsistence food income as speculated in numerous studies).[5]

A calorie-consumption model is specified and estimated for the critical wet season. The model results are presented in table 22.7. Three noteworthy findings are highlighted:

1. Calorie consumption per adult-equivalent is significantly income elastic, but with increased income the increase in calorie consumption is reduced, as indicated by the negative parameter estimated for the income squared term. A 10 percent increase in income (total expenditure)—at sample means—leads to a 4.8 percent increase in wet season (hungry season) calorie consumption.

---

[5]. A discussion of mechanisms and of various research results that address this issue is provided in von Braun and Kennedy (1986).

TABLE 22.7 Calorie consumption model for the wet season, 1985

Explanatory Variable[a]	Parameter	t-Value
INCOME	1.59329	7.218
INCOMESQ	−2.067E-04	−5.316
RICEPR	−572.59043	−1.590
STORAGE	4.26204	1.702
WOMCER	321.83780	2.088
CASHIN	67.74398	0.444
LOCATION	−360.16405	−3.321
SIZE	−16.89481	−1.949
WOMSH	817.08670	1.823
CHLD5SH	624.70833	1.038
CHLD10SH	834.17860	1.812
CHLD14SH	981.11329	1.867
Constant	1,616.86638	2.937
R^2	0.37	
F-value	9.65	
Degrees of freedom	197	

NOTE: The dependent variable is calorie consumption per adult-equivalent person per day.

[a]Definitions of variables:

INCOME = Total expenditure (as income proxy) per adult-equivalent person for year 1985–86 (in dalasi).
INCOMESQ = Income squared.
RICEPRE = Rice price in dalasi per kilogram actually paid for purchases by household (if no purchase, village mean price of season).
STORAGE = Storage of all grains (in rice equivalents) per adult-equivalent person in respective season.
WOMCER = Share of cereals produced in compound under women's control.
CASHIN = Share of cash income over total income earned by sinkiro in respective season.
LOCATION = Lowland village = 1 (else = 0).
SIZE = Size of cooking unit *sinkiro* in adult-equivalent persons.
WOMSH = Share of women in adult-equivalent persons.
CHLD5SH = Share of children under 5 years of age in adult-equivalent persons.
CHLD10SH = Share of children between 5 and 10 years of age in adult-equivalent persons.
CHLD14SH = Share of children between 10 and 14 years of age in adult-equivalent persons.

2. A reduced share of cereals from women's production significantly reduces calorie consumption. Thus, it can be expected that as a consequence of the new rice technology, a shift in control of the traditional woman's crop to the control of the compound head leads to reduced calorie consumption, if income is held constant. Income growth from technological change in rice must, therefore, be higher to compensate for this adverse intrahousehold effect if beneficial effects for calorie consumption are to be achieved. This compensa-

tion, by higher income, would require an increase of 4.6 percent of income at sample mean, which is only a small fraction of the income increase deriving from the technological change in rice. The relative diversion of income from calories after women partly lost control is, thus, less than the favorable effect for calorie consumption from total household income. Nevertheless, in some households of the scheme, the trade-off may adversely affect calorie consumption. The gainer-loser patterns at the micro- and intra-household levels are extremely complex.
3. An increased share of cash income does not, as hypothesized, reduce calorie consumption once household income is controlled for. Increased monetization of the economy in this West African setting appears not to have adverse consequences for food consumption.

Effects on Nutritional Status

When technological change raises income and rising income in turn raises food consumption, especially in the hungry season, as found above, these positive effects of technological change should be identifiable in an improved nutritional status. The relationships are not straightforward, however. Multifaceted relationships of interacting determinants establish the nutritional status situation and changes therein. A high prevalence of malnutrition and high mortality rates among children are related to low birth weights, especially for those children born late in the wet season. These low birth weights are partly the result of stress to the mothers due to high-energy expenditures and low-energy intakes during the wet season. This energy stress also reduces breast-milk production, which in turn reduces weight gains of children approaching weaning age (Prentice 1980). Older children are affected by availability and quality of food and by frequency of servings in the *sinkiros*. Household-level calorie availability may be of more direct relevance for the nutritional status of older children.

Separating the actual effects of unsatisfactory food consumption from health problems such as infectious diseases and diarrhea remains difficult. It is the interaction between them, rather than a clearly separable cause-and-effect chain, that leads to the alarming nutritional deterioration in many children by the age of 12 to 24 months, as indicated by anthropometric indicators of weight and height. During the wet season, 35 percent of children under five years of age were found to be below 80 percent of the weight-for-age standard (NCHS median). During the dry season, this proportion was 26 percent.

Nutritional Improvement of Women

A woman's body weight fluctuates, on average, by 2.2 kilograms between the wet season and the dry season in The Gambia. Detailed research at Keneba Station (The Gambia) shows that this fluctuation in weight is a result of seasonal low-energy intakes (because of shortages in the dry season), and high-energy expenditures (because of labor peaks in field preparation and weeding) (Lawrence et al. 1989). Reduction of this seasonal stress on mothers in the wet season, by means of increased food consumption or reduced work load, or both, could be expected to have a favorable effect on both mothers and children. How does the new rice production technology have an impact on seasonality? Women's weight fluctuations are reduced in households with greater access to land in the new rice scheme. While the weight-for-height of women from compounds with the least access to the new rice technology fluctuated by 5.4 percent (2.9 kilograms), this indicator fluctuated by only 1.8 percent (1.1 kilograms) in women from households with the most access to new rice technologies (that is, those in the top quartile). Women's reduced weight stress over the seasons may be seen as a substantial benefit of the new rice technology in that it levels out seasonal imbalances.

Nutritional Improvement of Children

A model is specified to explain differences in nutritional status of children in the context of food consumption, health, and the sanitation environment of the compounds in which they live. The link to technological change in agriculture is established with the above analysis in which the income and food consumption effects of technological change were assessed.

The nutritional status variable explained in the model is formed of Z-score values[6] of weight-for-age. A set of hypotheses forms the basis for this model:

- Increased per capita calorie consumption in the household also increases food consumption of children, who capture a part of the greater food availability. This increased consumption reduces their nutrition problem, but decreasingly so at the margin.
- The nutritional status (weight) of a child who has had diarrhea for an extended time preceding the survey is expected to be impaired.

---

6. The Z-score is computed as follows for each child's observation:

$$\frac{\text{(Actual measurement} - 50 \text{ percentile standard)}}{\text{Standard deviation of 50 percentile standard}}$$

- While short-term nutritional status is affected by diarrhea and related diseases, bad water quality is an adverse factor in the nutrition and health environment in which the child lives, and negatively affects growth.
- In large compounds with many small children, we expect a minimal adverse effect on child welfare.
- Children of the compound head are expected to be better taken care of than others.
- Children born in the "bad" season (late in the wet season or early in the dry season) with low birth weight may have a persistent nutritional status problem.
- Increased access to the rice project is expected to have favorable income and consumption effects and, thus, nutritional effects, but we diagnose adverse effects from the scheme for women's control over income. Spending income on nutritional improvement may suffer if women have a different marginal propensity to spend on child welfare than do men. This issue (over and above the income-consumption effect) is expected to be addressed by the variable on size of lands in the new rice scheme.

It must be stressed at the outset that this type of model that attempts to explain differences in anthropometric measures of children faces a lot of random variance in the dependent variable. Children in high-income countries develop in very different patterns as well. A low R-square is thus to be expected and does not pose a problem as long as the high variance in the dependent variable is actually random "noise." Significance and stability of parameter estimates are of greater importance.

From the estimation results in table 22.8, several findings can be highlighted:

- Increased food energy consumption at the compound level (CAL) reduces children's undernutrition problem significantly. A 10 percent increase in per capita calorie consumption significantly increases the weight-for-age Z-score at the sample mean by 0.27 (2.3 percent). Increased calorie availability clearly had a favorable effect on children's nutritional improvement.
- Weight-for-age of children is significantly reduced by infectious diseases (DIAR), as expected. An additional day of diarrhea attack during the recall period (four weeks) reduced weight-for-age by 2.2 percent at the mean.
- There appears to be a strong intrahousehold bias: children of the compound head (CHCHD) are found to be significantly better off than other children in the same compound unit. Distribution of investment in children's health and nutrition is obviously not equal

TABLE 22.8 Determinants of children's nutritional status, 1984–85

Explanatory Variable[a]	Parameter	t-Value	Mean Value of Sample
CAL	0.0091559	2.713	2,832.00
DIAR	−2.50525	−2.690	0.99
WATER	−0.01006	−2.248	1,459.85
SIZE	−0.71560	−1.826	12.75
NCHILD	−0.10040	−0.547	0.44
SEX	−8.18773	−1.501	.49
AGE	1.67620	4.308	56.63
AGESQ	−7.69439E-03	−2.426	4,126.00
SBORN	−13.64637	−2.480	0.54
HEIGHTM	1.85207	4.487	160.25
CHCHD	17.84216	3.074	0.52
VERDAT	−19.99746	−3.132	0.52
RICPROJ	−109.19153	−0.740	0.03
DRYS	19.29197	3.507	0.48
Constant	−467.02280	−6.746	—
$\hat{R}^2$	0.164		
F-value	17.28		
Degree of freedom	1,233		

NOTE: The dependent variable is weight-for-age Z-scores of children aged between 7 and 120 months (mean value of weight-for-age Z-score is −118.28 [Z-scores multiplied by 100]).

[a]Definitions of variables:

CAL = calories per adult-equivalent person per day from all sources, including snacks.
DIAR = number of days with diarrhea in preceding four weeks.
WATER = bacteria count of most-used well in village used for drinking water.
SIZE = number of adult-equivalent persons in *sinkiro*.
NCHILD = ratio of number of children under 10 years of age over number of persons in *sinkiro*.
SEX = sex of child (boys = 1; girls = 0).
AGE = age of child in months.
AGESQ = (age of child in months)².
SBORN = if child born in "bad" season = 1 (else = 0), born during March–August.
HEIGHTM = height of mothers in centimeters.
CHCHD = if child is son or daughter of compound head = 1, else = 0.
VERDAT = if date of birth is verified by record = 1, else = 0.
RICPROJ = size of lands in new rice scheme in hectare irrigated-land equivalent per adult-equivalent person.
DRYS = if observation in dry season = 1, else = 0.

within the compound. A shift of control over resources to the compound heads, identified as occurring with the new rice technology, may further strengthen such intrahousehold inequality effects.

- There are no indications that the degree of access to the new irrigated rice scheme has an effect (negative or positive) on nutrition over and above any of the factors already included in the model.

This means that the income-consumption relationships, including the income control changes, seem to transmit their impact on nutrition via those factors actually captured in the above-specified nutrition model.

Conclusions

Technological change in African agriculture is a necessary condition for achieving sustained increases in food production.[7] The special nutrition problems in Africa relate not only to low levels of food availability but also to seasonality and high year-to-year variability in food intakes. Policymakers and program planners must be aware of the complex linkages between technological change in agriculture, commercialization, and nutritional improvement in order to arrive at nutritional benefits in an effective and sustainable way.

The more complex the household structure and production organization in agriculture are, the less straightforward are the predictions on how technological change may affect nutritional improvement. In this study setting, new rice production technology ended up contributing primarily to expanded communal farming under the control of male household heads. The commercialization effects largely occurred on the input side. Not much incremental marketed surplus was achieved. Women farmers lost out relatively in terms of reduced opportunities for expanded individual farming. Holding household income constant, women's reduced crop control had an adverse effect on calorie consumption. The overall household income effect was, however, positive and much stronger and, despite the shift in control over the crop from women to men, this led to increased levels of, and more stability in, food consumption, which in turn improved the nutritional status of children and women. Incremental household calories improved nutritional status of children, especially when the children were malnourished.

---

7. The theme is extensively dealt with in Mellor, Delgado, and Blackie (1986).

PART VI
Conclusions

23 Conclusions for Agricultural Commercialization Policy

JOACHIM VON BRAUN
EILEEN KENNEDY

Agricultural commercialization, economic development, and nutrition are linked with one another. Policies influence the strength and direction of these linkages and welfare outcomes. Ignoring the linkages may be to the disadvantage of the nutritional welfare of the poor; opportunities to improve the well-being of the poor may be lost. In this concluding chapter, we attempt to draw some generalized lessons from the preceding chapters.

Integration of traditional smallholder agriculture into the exchange economy is part of a successful development strategy. Specialization and commercialization of farming households within a more diversified agricultural and rural economy are part of the development process. Specialization and development of markets and trade, which are characteristic of commercialization, are fundamental to economic growth.

Agricultural commercialization in low-income countries will generally grow over the coming decades due to urbanization and growing incentives for regional and farm-specific specialization in the context of diversifying rural economies. An optimistic scenario of a smooth transition from subsistence-oriented smallholder production systems to commercialized agricultural systems, cannot, however, be assumed. The commercialization of agriculture for economic development and nutritional improvement is not a matter of isolated projects but of a range of policies. The policies needed for a smooth transition to overcome the disadvantages of subsistence agriculture are discussed below. They include macro policy reform, infrastructure policy, agricultural technology development and dissemination, land tenure, rural finance policies, and complementary measures in education and health.

A pessimistic scenario where commercialization of agriculture would hamper economic development can be easily painted, too. This could happen in areas where rural infrastructure is deteriorating and the policy and security environment is such that rural households are forced into subsistence orientation. Large parts of Sub-Saharan Africa's rural areas may be threatened by such antidevelopmental trends. A related set of market failure factors may accelerate the shift into subsistence agriculture in several former centrally planned economies, such as in Central Asia. The mix of earlier excessive specialization in agriculture, disruption of interregional exchange, and the absence of social safety nets may further stimulate subsistence orientation. Thus, we may expect that the commercialization of agriculture will not progress smoothly and that there will be backlashes resulting from past or current policy failures. This topic will probably stay with the development research community for longer than we hope.

However, the developing world cannot afford the inefficiencies in resource allocation, especially of human and land resources, that subsistence agriculture entails under a long-term perspective. Of course, given current infrastructure, technology, education, and social security systems—or rather the lack of all these—subsistence agriculture is often the only feasible and most efficient mode of economic activity in rural areas of low-income countries. To overcome the subsistence orientation, however, in such a way that the poor are not adversely affected even in the short run remains the challenge of policy on commercialization of agriculture for economic development and nutritional improvements.

Scope for Public Policy

The foreign exchange constraints and heavy debt burdens of many developing countries provide further impetus for greater export orientation of agriculture. An expanded and more efficient agricultural export sector is a cornerstone of many economic reform programs in low-income countries.

Obviously, successful development in the staple food sector, through technological change and appropriate sectoral policies, and growth in the cash crop sector are not mutually exclusive. Appropriate policies for input supply, output marketing, and rural infrastructure development benefit both sectors and are crucial for their growth (chapter 6).

While commercialization of agriculture might essentially be considered a matter of stimulated private sector activity, it is also true that public action is essential for facilitation of the power of its "driving forces." As principal driving forces of commercialization, we identified (in Part IV) macro and trade policies, market reform, rural infrastructure improvement, and the development of legal and contractual (institutional) environments in which farmers and processors may operate. Policies related to these driving forces will strongly influence the nature and speed of the agricultural commercialization process, which, in turn, will determine to which extent and how soon the risks of subsistence agriculture for farmers and nonfarmers, that is, the risks of thin and volatile markets, can be reduced.

The trade and exchange rate policies of countries are of critical importance for the profitability of crops. The picture of protection and taxation in the study settings is mixed: while sugar in the Philippines and Kenya and rice in The Gambia were protected, export vegetables in Guatemala were implicitly taxed (chapters 13, 16, 22, and 12). An open trade environment, both domestically and internationally, is a prerequisite for success in capturing the long-run gains from specialization (chapters 6 and 7).

It should be stressed that not only at the (macro) level of the driving forces does public and state action play a key role in shaping the commercialization process. It also applies to the program level. In virtually every one of the 11 cases studied, central or local government action or donor policies impacted on stimulation of commercialization and its outcomes.

High risks to poor farm households and high transaction costs are the basic reasons for high prevalence of subsistence farming. Subsistence farming must be phased out in low-income countries via developmental progress in the driving forces of commercialization mentioned above. Policy must facilitate a transition that does not unduly replace (old) subsistence-related production risks with (new) market and policy failure risks, which poor smallholders may not be in a position to estimate. Avoidance of trade shocks and concern for appropriate scheduling of input and output market reforms are important considerations in this respect.

Even with well-functioning factor and product markets, it is easy to construct scenarios in which some poor producers would lose from commercialization. Such scenarios discussed below include the "agricultural treadmill," late access to new commercialization and technical options, and a host of "bad policies."

Increased market supply facing highly inelastic demand is one such scenario in which some producers lose. The resulting agricultural treadmill—increased supply leading to lower prices—is a reality with important regional and international dimensions. However, its potentially serious damage is often diluted by inbuilt compensating effects. In particular, the favorable effects for consumers should be taken into account when one is weighing the disadvantages of the agricultural treadmill for small nonadopters. Assessing the effects of commercialization and technical change from the perspective of producers only is misleading. Once the consumption effects and other general equilibrium effects are included in the assessment, the treadmill effects are usually seen to be diffused (Binswanger and von Braun 1991). Commercialization and specialization are usually introduced for commodities whose demand is elastic—often as a means of bypassing the problem of inelastic demand faced by traditional commodities. It is, therefore, difficult to construct scenarios in which commercialization by itself—unaided by failures of institutions, policies, or markets—has adverse consequences.

The relative seriousness for the poor of the various scenarios differs; the worst outcomes arise when several scenarios or effects coincide. Late adoption of new technology or commercialization options is a case in point. The risks associated with new technology or new crops discourage poor farmers from adopting them early; when combined with treadmill effects, late adoption, then, is likely to injure the profits of poor farmers or close their doors to adoption, or both. Many of the case studies in this book show that late adoption is not a general problem, and that policy and program designs have a key role to play. Potential adverse effects can also be mitigated by government action. Credit policies and extension services are often biased against the poor. Government policies can facilitate market or capacity expansion where doors have been closed and can help the poor to seize opportunities related to commercialization and technical change and thereby derive the benefits (Binswanger and von Braun 1991).

Many adverse circumstances arise not because of the inherent nature of the commercialization opportunity but because of bad policy. Constraints on trade (chapter 20), coercion in production, and ill-advised tenancy laws are government actions that may turn a promising opportunity into a disadvantage for the poor. The answer to many of these issues, then, is policy reform rather than reversal or deceleration of technological advance and commercialization (Nerlove 1988).

The growth potentials of commercialization undoubtedly offer opportunities to "extract surplus" or to steal and exploit where not much was available to rob in the first place under subsistence production conditions. This, however, can hardly be a convincing argument to avoid commercialization for such reasons. Obviously, exploitative policies need to be corrected where they exist. The conclusion that policy changes can either avert or mitigate adverse effects of commercialization of subsistence agriculture is based on the assumption that the policy and institutional responses are exogenous and independent of the expanded commercialization. In some cases, however, institutional changes and policy responses are not exogenous but reflect existing conflict among social groups. The perverse responses, then, are a logical outcome of these conflicts and cannot easily be altered by a benevolent policy. Some cases of tenant eviction fall in this category (chapter 13) (von Braun, de

Haen, and Blanken 1991). Where institutional and policy responses are endogenous in this way, more pessimistic conclusions are warranted about the benefits of commercialization for politically weak and poor groups. An important issue for further empirical research, therefore, is the extent to which these responses are endogenous (de Janvry and Sadoulet 1989).

One of the generalized conclusions from this study of actual commercialization experiences is that the effects for the poor are specific to location, implementation, and policy environment. Still, some general policy conclusions regarding food security, employment, income, women's role, and child nutrition can be derived from the experiences studied.

Subsistence Production and Food Security

There is a conscious effort by smallholder producers in all study settings to maintain subsistence food production along with new commercial production, despite apparently higher returns to land and labor from the cash crops. While cultural and taste factors may play a role, this reliance on food from own production under household control is a response to high transaction costs and risks related to market, employment, and production. It can largely be viewed as an insurance policy of farm households in response to a risky income environment. The higher the transactions costs are in food markets and the closer households are to food insecurity, such as in the extremely poor study environments of Rwanda and Malawi, the stronger is their preference for high shares of subsistence production (chapters 18 and 20). The poor are the ones most forced into adopting this strategy. Theoretically, this strategy of farm households may be viewed as a second-best solution, compared to full market integration as related economic benefits of specialization are forgone. However, given risky economic environments and missing insurance markets, maintenance of own food supplies can be economically a first-best strategy under existing circumstances. Agricultural policy can effectively assist in its transformation by promoting technological change in staple (subsistence) foods. This also provides further room for specialization at the farm level, and, thereby, permits further gains from commercialization and market integration of smallholders to be captured. Improved technology thus helps subsistence farmers to commercialize in low-risk ways. Development of financial and insurance markets would be complementary and could count on payoffs in terms of gains from commercialization.

The positive effects of commercialization for household food security are greatest when incremental income and employment from commercialization are most concentrated among the malnourished poor. Generally, the smallest farm households in the study settings participated less than proportionally in their respective commercialization schemes, but when they did participate, they tended to be the more radical adopters of the new commercial crops. Efforts to integrate the smallest farms into the schemes can be enhanced through legal arrangements, as shown in The Gambia, where opening up of access had favorable effects on food security at the household level (chapter 22).

Research and extension policies as well as supplies of inputs such as seeds and fertilizer for subsistence crops are critical for a viable commercialization strategy that meets smallholders' demands. Extension services in commercialization schemes with new crops or livestock (for example, dairy in India) have to assist farmers to avoid management mistakes. Crop management failures in the more input-intensive cash crops may pose a risk of greatly increased losses. Export vegetable production in Guatemala, which has input costs per hectare that are four times higher than those of traditional vegetables and twelve times higher than those of maize, is a case in point (chapter 12).

Employment On and Off the Farm

The employment effects for the poor that result from commercialization are very crop specific and are a function of the local labor market and the technologies introduced. Choice of crop and technology, therefore, has a major influence on the actual outcome of the employment effects. Program and policy design in this field can go a long way to maximize the income benefits for the poor through agricultural development. This applies not only to creation of on-field employment, as exemplified by the substantial employment increases in vegetable production in Guatemala (chapter 12) and in potato production in Rwanda (chapter 18), but also to employment in processing and trading that results from commercialization (chapters 9, 11, and 14).

Commercialization of agriculture entails a substantial expansion in the demand for hired labor in virtually all study environments, but particularly so when much processing is involved. To the extent that hired labor households rank among the malnourished poor, this employment effect is expected to be of particular benefit (chapters 8, 11, 12, and 14).

Off-farm nonagricultural employment and income already play a significant role in all study settings, with their share of total household income ranging between 20 percent and 60 percent. Much of this nonagricultural employment is in supply of local goods and services, which, in the study settings characterized by densely populated environments (for example, Guatemala, the Philippines, Kenya, and Rwanda), may suggest favorable indirect multiplier effects for income and employment resulting from agricultural commercialization. Infrastructure improvement plays an important role in fostering these multiplier effects, especially in high-potential, densely populated areas (chapter 8). The hidden demand for labor-intensive commercialization of agriculture is particularly high where alternative employment opportunities are lacking and where the trade infrastructure for the commodity in question can be put in place at low cost.

In one study setting, the Philippines, the rapid expansion of commercialization—in this case, sugarcane—contributed to the creation of a landless class of households that used to be tenants growing corn on rented land (chapter 13). An important contributing factor to the consolidation of landholdings was a long-run decline in corn productivity on land brought into cultivation in recent decades, which discouraged smallholders, tenants, and landowners from continuing to produce corn, and which resulted in declining incomes of the poor before the introduction of sugarcane. More appropriate policy responses to help the poor would have been to encourage smallholder sugar production (for example, by awarding tenants sugar contracts with the mill), and to raise smallholder corn productivity through technology extension programs.

Income and Consumption Improvement

Some of the case study settings had been selected because there were suspicions or preliminary hypotheses suggesting adverse effects of schemes on income and nutrition. This was especially so for the cases of sugarcane (Kenya, the Philippines), export vegetables (Guatemala), dairy development (India), and tobacco (Malawi).

While income effects of commercialization programs and projects were positive in general in most of the study settings, they were not necessarily sustained for all households and for all components of the commercialization process. For example, the income streams from cardamom production in Papua New Guinea were not sustained (chapter 14) and tree crop promotion in Sierra Leone was met by deteriorating terms of trade (chapter 21).

Although substantial, the net income gains, in general, were much less than the gross income gains from the commercialized crops because of large substitution effects within agricultural production and between agriculture and off-farm employment. The latter was particularly the case in Guatemala; off-farm income earning was reduced when labor-intensive export vegetable production drew family labor and hired labor back into agriculture (chapter 12). In The Gambia, double-cropped irrigated rice production gained, to a large extent, at the cost of upland crops such as groundnuts and millets (chapter 22).

Some households lost income, at least in the short run, due to the commercialization schemes. This group is rather small and heterogeneous across the study settings. For instance, in Rwanda, farmers who were relocated because of an extended industrial tea cultivation program lost out, and, in Kenya, farm households who lost their land to the factory because of the sugar scheme were found to be worse off in terms of food consumption but not in terms of nutritional status (chapters 22 and 16). Also, the introduction of tree crops, which yield revenues only after long time periods, may pose difficulties in poor environments, as in Sierra Leone (chapter 21). Careful ex ante assessment of possible creation of absolute losers is required. General employment expansion cannot be relied upon to reach out to these groups in the short run.

Judging the distribution of benefits only from the production and labor sides may be misleading, as becomes clear when spending patterns of income in the study settings are reviewed.

The critical issues relating commercialization of agriculture and household-level food security and nutrition are not just whether incremental income is earned by the poor and whether such incremental income is sustainable, but also how such incremental income is spent by the poor. A much debated issue is whether incremental cash income controlled by male heads of households is disproportionately spent on nonfoods and on items that do not improve the welfare of the households in general, and of women and children in particular. In all study settings, however, it is observed that with rising income from commercial crops, absolute spending for food consumption increases (chapter 4). In some settings, the substitution between food items from own production and from purchases is quite complex. Focusing on individual items rather than on aggregate nutrients and diet diversity is particularly misleading in this case—as shown, for example, in the Indian dairy case study (chapter 15).

In some study settings, adverse effects of increased commercialization for household spending on food were found at the margin. It should be stressed that these effects at the margin are not the net effects of increased commercialization. In Rwanda, increased shares of cash income (which was controlled by men) led to a less than expected increase in household calorie consumption; in other words, holding income constant, increased cash income compared to subsistence food income led to a smaller increase in calorie consumption in that setting (chapter 18). In Guatemala, the income elasticity of calorie consumption among export vegetable producers was less than that among other farm households in the same income range (chapter 12). In both settings, however, the overall income increase due to commercialization was much larger than the deviation effects of commercialization for spending on food at the margin. Thus, the income effects more than compensated for the marginal deviation effects. For cases of "commercialization without growth" such as in the Sierra Leone tree crops case (chapter 21), this, however, was not the case.

Women and Commercialization

Income and employment benefits of commercialization are not spread equally within households. Generally, women's work in agriculture is reduced not only relatively but also absolutely with rising income, which correlates with increased farm size in most of the study locations (chapters 13, 14, 20, and 21). From this angle, there is little support for the hypothesis that commercialization of agriculture leads to increased work loads for women, with potential adverse effects on child care and nutrition.

Sociocultural situations determine quite different effects of commercialization on women's work in agriculture. In Guatemala, for instance, the dramatic increase in labor demand associated with export vegetable production led to an absolute increase in labor input by both men and women (chapter 12). However, with rising farm size, women's labor was relatively reduced, while the shares of hired labor and child labor increased. This was not, however, the case for men's family labor. In The Gambia, on the other hand, the increased labor demand from double-cropped irrigated rice was, to a large extent, fulfilled by a shift of male labor from upland crops into rice, but, in the final assessment, it turns out that the overall work load of men remained more or less constant, while women's labor input increased somewhat (chapter 22).

A common feature in most study settings was that women's work in commercialized crops and women's direct control over income from these crops was much less than that of men, and, frequently, even disproportional to the labor input into the crops. In none of the schemes studied did women play a significant role as decision makers and operators of the more commercialized crop production lines. Policy and program design has thus far shown little explicit concern for this despite potentially far-reaching welfare implications (chapter 5). Women's constraints and potentials in the commercialization of agriculture need to be explicitly taken into account at the planning stage of programs. In participatory program design, special attention is to be given to legal security (where women's land rights are affected), credit schemes, and extension systems for women farmers and for women in the processing activities that are often so important. The general problem of women's bad health status found in many of the case study settings requires due attention to overcome this fundamental constraint to productive participation in commercialization programs.

Children's Nutrition and Commercialization

The effects of commercialization on children's welfare are mediated, in part, through the income-consumption link, which is found to have favorable effects on children's nutritional status. Potential linkages between income, children's education, and demand for children's labor that may result from new labor-intensive commercialized crops may also be relevant. At the aggregate level, it was found that in the poorest households (at a per capita income level of US$100 per year), a 10 percent increase in income led to an improvement in children's nutritional status (weight-for-age) 1.1 percent in Guatemala and the Philippines, 1.9 percent in The Gambia, and 2.5 percent in Rwanda. Income was not significant only in the Kenyan example, which had a particularly bad health situation (chapter 3). The observed deviations in expenditures at the margin from food to nonfood in some of the study areas did not translate into measurable adverse effects on nutrition (chapter 5), with the exception of Sierra Leone (chapter 21).

In general, no strong relationships were found between income and children's health in the period of the study. Poor household and community health and sanitation environments overshadow potential positive effects of income for health improvement. Community health services have to move in tandem with the agricultural development process, and increased income and wealth of communities fostered by agricultural commercialization may provide the resources required at the community level to sustain local health services. Certainly, the effective demand for health care, both curative and preventive, can increase when higher incomes are combined with better knowledge of how to eradicate the sources of disease. Better quality water and more of it at community and household levels is one such factor, and it was found in The Gambia study setting that it can contribute substantially to children's nutritional improvement. Local initiatives for community development can be effectively stimulated when the resource base of the communities expands, as occurred in the Guatemalan setting with the strengthening of the local health services (chapter 12).

The net nutritional effect (in terms of anthropometric indicators) of incremental income is modest, except in very poor households under acute food consumption constraints (for example, Rwanda), since the increased income does not decrease morbidity. Increased income and increased food availability contribute to solving the hunger problem but not the problem of preschool children's malnutrition, which results from a complex interaction of lack of food and morbidity (chapter 5).

Concluding Remarks

Public policy has to protect farms' choices to facilitate access to commercialization options at low risks. Policy responsibility arises where commercialization generates new food security risks with which small farmers are not able to cope. Sometimes, however, the problem is not absence of "good" policies, but presence of "bad" ones, as discussed above. Any development from a state of low public action and policy intensity—typically found in the situation of widespread subsistence farming—toward commercialization with active program promotion induces risks for policy failures. Prevention of these through appropriate institutional arrangements (for safety of contractual arrangements, assurance of competition, and so forth) must be called for. To call, alternatively, for prevention of commercialization—as is sometimes done—is a misleading conclusion and, as shown in the great majority of in-depth studies, would bar the poor from access to a basic force of rural modernization and employment growth.

There are two areas of policy that require particular attention in order to both foster the agricultural commercialization process with its developmental effects and reduce the risks of commercialization for food security and nutrition. These two areas are financial systems and social security systems, both of which should be accessible by the poor. Improved understanding of existing (including indigenous) systems is a precondition to enhance both systems from the bottom up rather than from the top down. Improved social security and related insurance systems reduce the pressure on households to self-insure through subsistence, and permit therefore an opening up toward commercialization options. The institutional options for social security are manifold and can build on a range of local and international experience (Ahmad et al. 1991).

Financial systems development can go a long way toward risk reduction too. If the poor are effectively included, again the pressure to rely on subsistence is reduced (Zeller et al. 1993). In order to maximize the development potentials of increased income, policies and programs parallel to the commercialization of agriculture have to accommodate the increased ability of households to save and build productive asset bases in order to avoid savings in the form of nonproductive assets. A rapid development of rural financial markets in the commercialization process is, therefore, also important from a growth perspective. It is particularly called for in environments where commercialization of agriculture leads to large lumpy payments of cash a few times a year. Larger development schemes for commercialization can provide the critical mass required for efficient rural banking with low overhead costs. Such banking facilities are to be expressly open to all individuals and not just to male heads of households that are enrolled in the commercialization schemes. Through access to rural financial institutions, the benefits of commercialization can be spread much more widely across the community and be less limited to the actual direct participants in commercialization schemes.

In summary, there are five policy and program design issues that are important for maximizing the potential benefits from agricultural commercialization and for minimizing damage:

- promotion of technological change in subsistence food crops along with commercial crop production for household food security in areas with risky food markets;
- improvement of infrastructure, especially in remote areas, when a change in production towards nonfoods may lead to the switch to a net food import balance and thereby drastic price changes;
- openness for effective integration of women farmers and of the smallest farms' households into schemes for commercialization and attention to land tenure and resulting land allocation problems when net returns to land increase substantially;
- development of effective rural financial systems to generate savings and make credit available not only to scheme participants but also to the community as a whole;
- development and promotion of community health and sanitation services in order to maximize the health and nutrition returns of increased income.

References and Bibliography

Abbott, J. C. 1988. *Agricultural Processing for Development.* Aldershot, England: Gower.

Adelhelm, R., and Kotschi, J. 1985. "Development and Introduction of Self-Sustaining Agricultural Practices in Tropical Smallholder Farms." *Entwicklung und Laendlicher Raum* 19(4):17–20.

Ahmad, Ehtisham, and Hussain, Athar. 1991. "Social Security in China: A Historical Perspective." In *Social Security in Developing Countries,* ed. Ehtisham Ahmad, Jean Drèze, John Hills, and Amartya Sen. Oxford: Clarendon Press.

Ahmad, Ehtisham; Drèze, Jean; Mills, John; and Sen, Amartya, eds. 1991. *Social Security in Developing Countries.* Oxford, U.K.: Oxford University Press.

Ahmed, Raisuddin. 1981. *Agricultural Price Policies under Complex Socioeconomic and Natural Constraints: The Case of Bangladesh.* Research Report 27. Washington, D.C.: International Food Policy Research Institute.

Ahmed, Raisuddin, and Bernard, Andrew. 1989. *Rice Price Fluctuations and an Approach to Price Stabilization in Bangladesh.* Research Report 72. Washington, D.C.: International Food Policy Research Institute.

Ahmed, Raisuddin, and Donovan, Cynthia. 1989. "Evaluating Infrastructural Development: A Review of the Literature." Washington, D.C.: International Food Policy Research Institute. Mimeo.

———. 1992. *Issues of Infrastructural Development: A Synthesis of the Literature.* Washington, D.C.: International Food Policy Research Institute.

Ahmed, Raisuddin, and Hossain, Mahabub. 1987. "Infrastructure and Development of Rural Economy." Washington, D.C.: International Food Policy Research Institute and Bangladesh Institute of Development Studies.

———. 1990. *Developmental Impact of Rural Infrastructure in Bangladesh.* Research Report 83. Washington, D.C.: International Food Policy Research Institute and Bangladesh Institute of Development Studies.

Alderman, Harold. 1986. *The Effect of Food Price and Income Changes on the Acquisition of Food by Low-Income Households.* Washington, D.C.: International Food Policy Research Institute.

———. 1987. *Cooperative Dairy Development in Karnataka, India: An Assessment.* Research Report 64. Washington, D.C.: International Food Policy Research Institute.

Alderman, Harold, and Mergos, George. 1987. "The Food and Nutrient Consumption Effects of Operation Flood in Karnataka and Madhya Pradesh."

Paper presented at International Food Policy Research Institute Workshop on the Economics of Dairy Development in Selected Countries. Copenhagen, January 6-8.

Alderman, Harold; Mergos, George; and Slade, Roger. 1987. *Cooperative Dairy Development in India: Evolution, Debate, and Evidence*. Working Paper on Commercialization of Agriculture and Nutrition 2. Washington, D.C.: International Food Policy Research Institute.

Alvarez, Claude. 1983. "Operation Flood: The White Lie," *The Illustrated Weekly of India* (October):8-13.

Amemiya, Taveshi, 1981. "Qualitative Response Models: A Survey." *Journal of Economic Literature* 10(December):1483-1536.

An, Xiji. 1989. "The Development and Improvement of Agricultural Marketing in China." In *China's Rural Development Miracle*, ed. John W. Longworth. St. Lucia, Australia: University of Queensland Press.

Asian Productivity Organization. 1978. *Asian Food Processing Industries*. Tokyo: APO.

Baldeston, J. B.; Wilson, A. B.; Freire, M. E.; and Simonen, M. S. 1981. *Malnourished Children of the Rural Poor*. Boston: Auburn House Publishing Company.

Bank of Sierra Leone. 1988a. *Economic Review July-December 1986 and January-June 1987*. Freetown: Bank of Sierra Leone.

———. 1988b. *Economic Trends: April to June 1988*. Freetown: Bank of Sierra Leone.

———. 1988c. *Sierra Leone in Figures*. Freetown: Bank of Sierra Leone.

Barclay, Albert. 1977. "The Mumias Sugar Project: A Study of Rural Development in Western Kenya." Ph.D. diss., Columbia University.

———. 1980. *Agro-Industrial Production and Socioeconomic Development: Case Studies of Sugar Production on Muhoroni and Mumias*. Roskilde, Denmark: Rotschild University Centre.

Bautista, Romeo M. 1987. *Production Incentives in Philippine Agriculture: Effects of Trade and Exchange Rate Policies*. Research Report 59. Washington, D.C.: International Food Policy Research Institute.

Becker, Gary S. 1965. "A Theory of the Allocation of Time." *The Economic Journal* 75(299):493-517.

Beckford, G. 1972. *Persistent Poverty: Under-Development in the Plantation Economies of the Third World*. New York: Oxford University Press.

Bellin, Friederike. 1991. "Auswirkungen des Anbaus von Kaffee, Kakao und Ölpalmen auf Einkommen und Ernahrung der Kleinbäuerlichen Haushalfe in Süd-Sierra Leone." Ph.D. diss., Justus Liebig-Universität, Giessen.

Berry, Eileen; Berry, L.; Campbell, D. J.; and Major, David C. 1982. "Regional Reconnaissance of Rwanda, Burundi, Kivu, Province of Zaire." Final report, vol. 1, to United States Agency for International Development. Vancouver: International Development Program, Clark University.

Binswanger, Hans, and von Braun, Joachim. 1991. "Technological Change and Commercialization in Agriculture: The Effect on the Poor." *The World Bank Research Observer* 66(January):57-80.

Binswanger, Hans, and Rosenzweig, Mark. 1986. "Behavioral and Material

Determinants for Production Relations in Agriculture." *Journal of Development Studies* 22(April):503–39.
Binswanger, Hans; Khandakar, Shadhidur R.; and Rosenzweig, Mark. 1988. "The Effect of Infrastructure on Agricultural Production." Washington, D.C., World Bank. Mimeo.
———. 1989. *How Infrastructure and Financial Institutions Affect Agricultural Output and Investment in India.* Working Paper 163. Washington, D.C.: World Bank.
Bo-Pujehun Rural Development Project. 1986. *Farming Systems Survey.* Bo, Sierra Leone: Ministry of Finance, Development, and Economic Planning.
———. 1989. "Oil Palm Study: A Review of B-P RDP Oil Palm Package." Bo, Sierra Leone, Bo-Pujehun Rural Development Project. Mimeo.
Bossen, L. H. 1984. *The Redivision of Labor: Women and Economic Choice in Four Guatemalan Communities.* Albany, N.Y.: New York Press.
Bouis, Howarth E., and Haddad, Lawrence J. 1989. "Estimating the Relationship Between Calories and Income: Does Choice of Survey Variable Matter?" Washington, D.C.: International Food Policy Research Institute. Mimeo.
———. 1990. *Effects of Agricultural Commercialization on Land Tenure, Household Resource Allocation, and Nutrition in the Philippines.* Research Report 79. Washington, D.C.: International Food Policy Research Institute.
———. 1992. "Are Estimates of Calorie-Income Elasticities Too High?: A Recalibration of the Plausible Range." *Journal of Development Economics* 39:333–64.
Bouis, Howarth E., and Kennedy, Eileen. 1989. "Traditional Cash Crop Schemes' Effects on Production, Consumption, and Nutrition: Sugarcane in the Philippines and Kenya." Paper presented at the International Food Policy Research Institute/Institute of Nutrition of Central America and Panama Workshop on Commercialization of Agriculture and Household Food Security: Lessons for Policies and Programmes. Antigua, Guatemala, March 9–11.
von Braun, Joachim. 1988a. "Commercialization and Food Security in Guatemala." In *IFPRI Policy Brief No. 3.* Washington, D.C.: International Food Policy Research Institute.
———. 1988b. "Effects of Technological Change on Food Consumption and Nutrition: Rice in a West African Setting." *World Development* 16(9):1083–98.
von Braun, Joachim, and Kennedy, Eileen. 1986. *Commercialization of Subsistence Agriculture: Income and Nutrition Effects in Developing Countries.* Working Paper on Commercialization of Agriculture and Nutrition 1. Washington, D.C.: International Food Policy Research Institute.
von Braun, Joachim, and Paulino, Leonardo. 1990. "Food in Sub-Saharan Africa: Trends and Policy Challenges for the 1990s." *Food Policy* 15(6):505–17.
von Braun, Joachim, and Webb, Patrick. 1989. "The Impact of New Crop Technology on the Agricultural Division of Labor in a West African Setting." *Economic Development and Cultural Change* 37(3):513–34.

von Braun, Joachim; de Haen, Hartwig; and Blanken, Jürgen. 1991. *Commercialization of Agriculture under Population Pressure: Effects on Production, Consumption, and Nutrition in Rwanda*. Research Report 85. Washington, D.C.: International Food Policy Research Institute.

von Braun, Joachim; Hotchkiss, David; and Immink, Maarten D. C. 1989. *Nontraditional Export Crops in Guatemala: Effects on Production, Income, and Nutrition*. Research Report 73. Washington, D.C.: International Food Policy Research Institute.

von Braun, Joachim; Puetz, Detlev; and Webb, Patrick. 1989. *Technological Change in Rice and Commercialization of Agriculture in a West African Setting: Effects on Production, Consumption, and Nutrition*. Research Report 75. Washington, D.C.: International Food Policy Research Institute.

von Braun, Joachim; Bouis, Howarth; Kumar, Shubh; and Pandya-Lorch, Rajul. 1992. *Improving Food Security of the Poor: Concept, Policy, and Programs*. Washington, D.C.: International Food Policy Research Institute.

Bryan, William Jennings. 1967. "Cross of Gold Speech." In *W. J. Bryan, Selections*, ed. R. Ginger. New York: Bobbs-Merrill Company.

Buch-Hansen, Mogens. 1980a. *Agro-Industrial Production and Socioeconomic Development: A Case Study of KTDA*. Roskilde, Denmark: Rotschild University Centre.

———. 1980b. *Agro-Industrial Production and Socioeconomic Development: Case Studies of Sugar Production on Muhoroni and Mumias*. Roskilde, Denmark: Rotschild University Centre.

Business International Corporation. 1985. "Mhlume Sugar Company, Swaziland." Study prepared for the United States Agency for International Development. Swaziland: Business International Corporation.

Byrd, William. 1983. *China's Financial System: The Changing Role of Banks*. Boulder, Colo.: Westview.

Celis, Rafael; Milimo, John T.; and Wanmali, Sudhir. 1991. *Adopting Improved Farm Technology: A Study of Smallholder Farmers in Eastern Province, Zambia*. Washington, D.C.: International Food Policy Research Institute.

Chavas, Jean-Paul, and Holt, Matthew T. 1990. "Acreage Decisions under Risk: The Case of Corn and Soybeans." *American Journal of Agricultural Economics* 72(3):529–38.

Chen, Dongsheng, and Chen, Jiyuan. 1988. *A Study of the Economic Structures in China's Different Areas*. Beijing: Shanxi People's Publishing House and the China Social Sciences Press.

Chen, Jiyuan. 1990. "Commercialization of the Rural Economy for the Elimination of Poverty: China." Paper prepared for the International Food Policy Research Institute/Institute of Nutrition of Central America and Panama Workshop on Commercialization of Agriculture and Household Food Security: Lessons for Policies and Programs. Antigua, Guatemala, March 9–11.

Chen, Junsheng. 1989. "Helping the Poor Areas to Develop Their Economic Potential Remains a Long-Term Task." Summary of a talk at the seventh plenary meeting of the State Council's leading group for economic development in poor areas. People's Daily (February 11).

Chen, Lincoln; Huq, E.; and d'Souza, S. 1981. "Sex Bias in the Family Allocation of Food and Health Care in Rural Bangladesh." *Population and Development Review* 7(1):55.

Chowdhury, Omar H., and Hossain, Mahabub. 1985. *Roads and Development?: A Case Study of Return on Investment in Rural Roads in Bangladesh.* Dhaka: Center for Development Science.

Christiansen, R. E., and Southworth, V. R. 1988. "Agricultural Pricing and Marketing Policy in Malawi: Implications for a Development Strategy." Paper prepared for the Symposium of Agricultural Policies for Growth and Development. Mangochi, Malawi.

Corpus, Velona A.; Ledesma, Antonio J.; Limbo, Azucena B.; and Bouis, Howarth E. 1987. "The Commercialization of Agriculture in the Southern Philippines: The Income, Consumption, and Nutrition Effects of a Switch from Corn to Sugar Production for Ten Case Study Households in Bukidnon." Washington, D.C.: International Food Policy Research Institute. Mimeo.

Crotty, Raymond. 1983. "Operation Flood and the EEC." *Economic and Political Weekly* (April):552.

Dehua, Jiang. 1988. "Research on the Classification and Development of the Poor Areas of China." Washington, D.C.: International Food Policy Research Institute. Mimeo.

Delepierre. 1985. *Les Tendances Évolutives de l'Agriculture Vivrière (Analyse des Données Statistiques, 1966–1983).* Kigali: Ministère du Plan, République Rwandaise.

Devres International. 1980. *Socioeconomic and Environmental Impacts of Low Volume Rural Roads: A Review of the Literature.* Washington, D.C.: United States Agency for International Development.

Dewey, Kathryn. 1979. "Commentary—Agricultural Development, Diet, and Nutrition." *Ecology of Food and Nutrition* 8.

Dey, Jennie. 1982. "Development Planning in The Gambia: The Gap Between Planners' and Farmers' Perceptions, Expectations, and Objectives." *World Development* 10(May):377–96.

Dinham, Barbara, and Hines. Colin. 1984. *Agribusiness in Africa.* Trenton, N.J.: Africa World Press.

Dixon, John. 1981. *The Chinese Welfare System, 1949–1979.* New York: Praeger.

DPI (Department of Primary Industry). 1984. *Agriculture, Forestry and Fisheries Sector: Medium-Term Development Strategy.* Volume 2: General objective and situation report. Konedobu, Papua New Guinea: Department of Primary Industry.

Eastern Africa Economic Review. August 1989. Special Issue on Contract Farming and Smallholder Outgrower Schemes in Eastern and Southern Africa.

Economic Daily, February 11, 1989.

Ellis, Frank. 1977. *A Study of Employment in the Banana Export Industry of Panama and Central America.* Geneva: International Labour Organization.

Evenson, Robert E. 1978. "Time Allocation in Rural Philippine Households." *American Journal of Agricultural Economics* 60(2):322–30.

———. 1986. "Infrastructure, Output Supply, and Input Demand in Philippine

Agriculture: Provisional Estimates. *Journal of Philippine Development* 13(1 and 2):62–76.

Fafchamps, Marcel. 1992. "Cash Crop Production, Food Price Volatility, and Rural Market Integration in the Third World." *American Journal of Agricultural Economics* 74(1):90–99.

Finkelshtain, Israel, and Chalfant, James A. 1991. "Marketed Surplus Under Risk: Do Peasants Agree with Sandmo?" *American Journal of Agricultural Economics* 73(3):557–67.

Finlayson, M. P.; McComb, J. R.; Hardaker, J. B.; and Heywood, P. F. 1991. "Commercialization of Agriculture at Karimui, Papua New Guinea: Effects on Household Income, Consumption, and the Growth of Children." Report of a joint project of the Papua New Guinea Institute of Medical Research, Madang, and the Department of Agricultural Economics and Business Management, University of New England, Armidale, Australia.

Fleuret, P., and Fleuret, A. 1980. "Nutrition, Consumption, and Agricultural Change." *Human Organization* 39:250–60.

FAO (Food and Agriculture Organization of the United Nations). 1983. *Production Yearbook 1982*. Rome.

———. 1984. *Integrating Nutrition into Agricultural and Rural Development Projects: A Manual*. Rome.

———. 1987. *Agricultural Price Policy Issues and Policies*. Rome.

———. Various years. *Data Tapes*, various issues. Rome.

FAO/WHO (Food and Agriculture Organization of the United Nations/World Health Organization). 1974. *Handbook on Human Nutritional Requirements*. Rome: FAO.

Frederickson, P. C. 1985. "Electrification and Regional Economic Development in the Philippines." *The Journal of Philippine Development* (September)22:12.

Freiwalds, J. 1983. "Jamaica Broilers." *Agribusiness Worldwide* (October/November).

Garfield, Elsie B. 1979. *The Impact of Technical Change on the Rural Kenyan Household: Evidence from the Integrated Agricultural Development Program*. Institute for Development Studies Working Paper 358. Nairobi: University of Nairobi.

George, Shanti. 1983. "Cooperatives and Indian Dairy Policy: More Anand than Pattern." In *Cooperatives and Rural Development*, ed. Donald Attwood and B. S. Baviskar. New Delhi: Oxford University Press.

———. 1984. "Diffusing Anand: Implications of Establishing a Dairy Cooperative in a Village in Central Kerala." *Economic and Political Weekly* (December):2161–69.

———. 1986. *Operation Flood: An Appraisal of Current Indian Dairy Policy*. New Delhi: Oxford University Press.

George, Susan. 1977. *How the Other Half Dies: The Real Reasons for World Hunger*. Montclair, N.J.: Allanheld, Osmun.

Glover, David J. 1984. "Contract Farming and Smallholder Outgrower Schemes in Less Developed Countries." *World Development* 12(3):1143–57.

———. 1986. "Agrarian Reform and Agro-Industry in Honduras." *Canadian Journal of Development Studies* 7(1).

Glover, David, and Kusterer, Ken. 1990. *Small Farmers, Big Business: Contract Farming and Rural Development*. London: Macmillan.

Glover, David, and Ghee Lim, Teok, eds. Forthcoming. *Contract Farming and Outgrower Schemes in Southeast Asia*. Kuala Lumpur: University of Malaysia Press.

Greer, Joel, and Thorbecke, Erik. 1986. *Food, Poverty, and Consumption Patterns in Kenya*. Geneva: International Labour Organization.

Griffin, Keith, ed. 1984. *Institutional Reform and Economic Development in the Chinese Countryside*. Hong Kong: Macmillan.

Grilli, Enzo R., and Yang, Maw Cheng. 1988. "Primary Commodity Prices, Manufactured Goods Prices, and the Terms of Trade of Developing Countries: What the Long Run Shows." *The World Bank Economic Review* 2(1):1–47.

Groos, A. (ed.). 1988. "Nutrition Surveys of Karimui District, South Simbu, 1985/86." Papua New Guinea Institute of Medical Research.

Groos, A., and Hide, R. L. 1989. "Nutrition Surveys of Karimui and Gumine Districts, Simbu Province, 1987/1988." Final Report to the South Simbu Rural Development Project. Papua New Guinea Institute of Medical Research.

Gross, D., and Underwood, B. 1971. "Technological Change and Calorie Costs: Sisal Agriculture in Northern Brazil." *American Anthropologist* 73:725–40.

Guyer, Jane. 1980. *Household Budget and Women's Income*. African Studies Center Working Paper 28. Boston: Boston University.

Haaga, J.; Mason, J.; Omoro, F. Z.; Quinn, V.; Rafferty, A.; Test K.; and Wasonga, L. 1986. "Child Malnutrition in Rural Kenya: A Geographic and Agricultural Classification." *Ecology of Food and Nutrition* 18:297–307.

Haddad, Lawrence J. 1987. "Agricultural Household Modelling with Intrahousehold Food Distribution: A Case Study of Commercialization in a Southern Philippine Province." Ph.D. diss., Stanford University.

Haggblade, Steve, and Minot, Nicholas. 1987. "Opportunities for Enhancing Performance in Rwanda's Alcoholic Beverage Subsector." Mimeo.

Harper, M., and Kavura, R. 1985. *Private Enterprise and Rural Development*. Rome: Food and Agriculture Organization of the United Nations.

Harvey, P. W. J., and Heywood, P. F. 1983. "Nutrition and Growth in Simbu." Research report of the Simbu Land Use Project. Vol. IV. Papua New Guinea Institute of Applied Social and Economic Research.

Haugerùd, Angelique. 1985. "Farmer's Use of Potato Cultivars in Eastern Africa." Paper presented at the 1985 Annual Review Meeting of The International Potato Center. Lima, Peru.

Hausman, J. A. 1978. "Specification Tests in Econometrics." *Econometrica* 46(6, November):1251–71.

Hayami, Y.; Kawagoe, T.; Morooka, Y.; and Siregar, M. 1987. *Agricultural Marketing and Processing in Upland Java: A Perspective from a Sunda Village*. ESCAP CGPRT Centre 8. Bogor, Indonesia: CGPRT Centre.

———. 1988a. "Income and Employment Generation from Agricultural Processing and Marketing: The Case of Soybean in Indonesia." *Agricutural Economics* 1:327–329.
———. 1988b. *Agricultural Marketing in Transmigration Areas of Sumatra*. Bogor, Indonesia: UN/ESCAP and CGPRT.
Hazell, Peter. 1983. *Rural Growth Linkages: Household Expenditure Patterns in Malaysia and Nigeria*. Research Report 41. Washington, D.C.: International Food Policy Research Institute.
Hazell, Peter B. R., and Ramasamy, C. 1991. *The Green Revolution Reconsidered: The Impact of High-Yielding Rice Varieties in South India*. Baltimore: Johns Hopkins University Press for the International Food Policy Research Institute.
Hernandez, Mercedes et al. 1974. "Effect of Economic Growth on Nutrition in a Typical Community." *Ecology of Food and Nutrition* 4(March):283–91.
Heywood, P. F. 1983. "Growth and Nutrition in Papua New Guinea." *Journal of Human Evolution* 12:133–43.
Heywood, P.; Singleton, N.; and Ross, J. 1988. "Nutritional Status of Young Children—the 1982/83 National Nutrition Survey." *Papua New Guinea Medical Journal* 31:91–101.
Hide, R. L. 1980. "Cash Crop and Food Production in Chimbu." In *History of Agriculture Working Paper No. 44*. Port Moresby: University of Papua New Guinea/Department of Primary Industry.
Hill, Polly. 1986. *Development Economics on Trial*. Cambridge: Cambridge University Press.
Hinderink, J., and Sterkenburg, J. J. 1987. *Agricultural Commercialization and Government Policy in Africa*. New York: Methuen.
Hintermeister, Alberto. 1986. "Report on 'Cuatro Pinos'." Report submitted to Institute of Nutrition of Central America and Panama by International Food Policy Research Institute. Mimeo.
Hirschman, Albert O. 1958. *The Strategy of Economic Development*. New Haven, Conn.: Yale University Press.
Hitchings, Jon. 1982. "Agricultural Determinants of Nutritional Status Among Kenyan Children with Model of Anthropometric and Growth Indicators." Ph.D. diss., Stanford University.
Hiwa, S. 1988. "Agricultural Development Policy, Food Production, and Consumption." Paper prepared for the Workshop on Household Food Security and Nutrition. Zomba, Malawi.
Holtham, G., and Hazelwood, A. 1976. *The Mumias Sugar Company: Aid and Inequality in Kenya*. London: Croom Helm.
Hoorweg, J., and Niemeijer, R. 1982. *The Effects of Nutrition Rehabilitation at Three Family Life Training Centres in Central Province, Kenya*. Research Report 14. Leiden, The Netherlands African Studies Centre.
———. 1989. "*Intervention in Child Nutrition: Evaluation Studies in Kenya*." London: Kegal Paul International.
Hoorweg, J.; Niemeijer, R.; and Steenbergen, W. V. 1983. *Nutrition Survey in Murang'a District, Kenya. Part 1: Relations between Ecology, Economic and*

Social Conditions, and Nutritional State of Preschool Children. Research Report 19. Leiden, The Netherlands African Studies Centre.

Hoque, Nazrul. 1987. "Rural Electrification and Its Impact on Fertility: Evidence from Bangladesh. Ph.D. diss., Pennsylvania State University.

Horton, S. 1985. "The Determinants of Nutrient Intake." *Journal of Development Economics* 19:147–162.

Houtman, C. B. 1981. *Report of Agro-Economic Studies in the Ahero and West Kano Pilot Irrigation Schemes.* Report 34. Nairobi: Integrated Rural Development Program.

IDS Bulletin. 1988. 19(2).

ILO (International Labour Organization). 1972. *Employment, Incomes, and Equality: A Strategy for Increasing Productive Employment in Kenya.* Geneva.

———. 1974. *Sharing in Development: A Programme of Employment, Equity, and Growth for Philippines.* Geneva.

Immink, Maarten; Kennedy, Eileen; Sibrian, Ricardo; Pinto-Paiz; and de Valverde, Christa. 1992. "Nontraditional Export Crops on Smallholder Farms and Production, Income, Nutrition, and Quality of Life Effects. A Comparative Analysis, 1985–1991." International Food Policy Research Institute, Washington, D.C. Mimeo.

India, Government of. 1985. *National Sample Survey, 38th Round,* No. 319. New Delhi: Department of Statistics.

Ishikawa, Shigeru. 1968. "Agrarian Reform and Its Productivity Effect: Implications of the Chinese Pattern." In *The Structure and Development in Asian Economics.* Tokyo: Japan Economic Research Center.

———. 1986. "Patterns and Processes of Intersectoral Resource Flows: Comparison of Some Cases in Asia." Paper prepared for the Twenty-Fifth Anniversary Meeting of the Economic Growth Center, Yale University, on The Current State of Development Economics. New Haven, Conn., April.

Islam, Nurul. 1990. *Horticultural Exports of Developing Countries: Past Performances, Future Prospects, and Policy Issues.* Research Report 80. Washington, D.C.: International Food Policy Research Institute.

Jabbar, S. 1972. "A Case Study of Production Orientation: Malayan Canned Pineapple." *European Journal of Marketing* 6(3).

Jaetzold, R., and Schmidt, H. 1982. *Farm Management Handbook of Kenya, Vol. II/A, West Kenya.* Nairobi: Ministry of Agriculture.

de Janvry, Alain. 1987. "Dilemmas and Options in the Formulation of Agricultural Policies in Africa." In *Food Policy,* ed. James Price Gittinger, Joanne Leslie, and Caroline Hoisington. Baltimore: Johns Hopkins University Press for the World Bank.

de Janvry, Alain, and Sadoulet, Elisabeth. 1989. "Alternative Approaches to the Political Economy of Agricultural Policies: Convergence of Analytics, Divergence of Implications." In *Agriculture and Government in an Interdependent World,* ed. A. Maunder and A. Valdés. Proceedings of the Twentieth International Conference of Agricultural Economists, Buenos Aires, August 1988. Aldershot, U.K.: Dartmouth.

Jha, Dayanatha; Hojjati, Behjat; and Vosti, Stephen. 1991. "The Use of Improved Agricultural Technology in Eastern Province." In *Adopting Improved Farm Technology: A Study of Smallholder Farmers in Eastern Province, Zambia*, ed. Rafael Celis, John T. Milimo, and Sudhir Wanmali. Washington, D.C.: Rural Development Studies Bureau (University of Zambia)/National Food and Nutrition Commission/Eastern Province Agricultural Development Project (Government of the Republic of Zambia)/International Food Policy Research Institute.

Jiang, Dehua, Yaoguang Zhang, Liu Yong, and Shao Fan Hou. 1989. *Zhongguo de Pinkum Diqu Leixing ji Kaifa* (The Types and Development of China's Poor Areas). Beijing: Tourism Education Publishing House.

Johnny, M. 1985. "Informal Credit for Integrated Rural Development in Sierra Leone." Ph.D. diss., Verlag Weltarchiv, Hamburg, Germany.

Johnson, D. Gayle. 1989. "Economic and Noneconomic Factors in Chinese Rural Development." Paper prepared for the American Agricultural Economics Association Annual Meeting. Baton Rouge, La.

Jones, William E., and Egli, Roberto. 1984. *Farming Systems in Africa: The Great Lakes Highlands of Zaire, Rwanda, and Burundi*. Washington, D.C.: World Bank.

Kaiser, L. L., and Dewey, K. G. 1991. "Migration, Cash Cropping, and Subsistence Agriculture: Relationships to Household Food Expenditures in Rural Mexico." *Social Science and Medicine* 33(10): 1113–26.

Keller, B., and Mbewe, D. 1988. "Impact of the Present Agricultural Policies on the Role of Women, Extension and Household Food Security." Paper presented at Workshop on Strategies for Implementing National Policies in Extension and Food Security, 8–10 December 1988, Itezhi-tezhi, Zambia. Ministry of Agriculture and Cooperatives, Planning Division, Lusaka, Zambia.

Kennedy, Eileen. 1989. *The Effects of Sugarcane Production on Food Security, Health, and Nutrition in Kenya: A Longitudinal Analysis*. Research Report 78. Washington, D.C.: International Food Policy Research Institute.

Kennedy, Eileen, and Bouis, Howarth E. 1993. *Linkages Between Agriculture and Nutrition: Implications for Policy and Research*. Washington, D.C.: International Food Policy Research Institute.

Kennedy, Eileen, and Cogill, Bruce. 1987. *Income and Nutritional Effects of the Commercialization of Agriculture in Southwestern Kenya*. Research Report 63. Washington, D.C.: International Food Policy Research Institute.

Kennedy, Eileen, and Oniang'o, Ruth. 1993. "Determinants of Household and Preschooler Vitamin A Consumption in Southwestern Kenya." *Journal of Nutrition* 123:841–846.

Kennedy, Eileen, and Peters, Pauline. 1992. "Influence of Gender of Head of Household on Food Security, Health, and Nutrition." *World Development* 20:1077–1085.

Kenya, Central Bureau of Statistics. 1979. *Child Nutrition in Rural Kenya*, Nairobi.

———. 1981. *Kenya Population Census, 1979*, Vol. 1. Nairobi.

———. 1988. *Economic Survey, 1988*. Nairobi.

Kirchback, F. 1986. *Structural Change and Export Marketing Channels in Developing Countries.* Geneva: International Trade Center.

Kliest, T. 1984. *The Agricultural Structure of the Lower Kano Plain.* Food and Nutrition Studies Programme Report 4. Nairobi/Leiden, The Netherlands: Ministry of Planning and National Development, African Studies Centre.

———. 1985. Regional and Seasonal Food Problems in Kenya. Food and Nutrition Studies Programme Report 10. Nairobi/Leiden, The Netherlands: Ministry of Planning and National Development, African Studies Centre.

Koester, Ulrich. 1985. "Regional Cooperation in the Food Sector Among Developing Countries to Improve Food Security." Paper presented at the 19th International Conference of Agricultural Economists. Málaga, Spain, August 26–September 24.

Korte, R. 1969. "The Nutritional and Health Status of the People Living on the Mwea-Tebere Irrigation Settlement." In *Investigations into Health and Nutrition in East Africa,* ed. H. Kraut and H. D. Cremer. Munich: IFO Institut for Wirtschaftsforschung.

Krueger, Anne O.; Schiff, Maurice; and Valdés, Alberto. 1988. "Agricultural Incentives in Developing Countries: Measuring the Effect of Sectoral and Economywide Policies." *World Bank Economic Review* 2(3):255–71.

Kumar, Shubh. 1986. "The Nutrition Situation and Its Food Policy Links." In *Accelerating Food Production in Sub-Saharan Africa,* ed. John W. Mellor, Christopher Delgado, and Malcolm J. Blackie. Baltimore: Johns Hopkins University Press for the International Food Policy Research Institute.

Kusterer, Ken. 1981. *The Social Impact of Agribusiness: A Case Study of ALCOSA in Guatemala.* Washington, D.C.: United States Agency for International Development.

———. 1982. *The Social Impact of Agribusiness: A Case Study of Asparagus Canning in Peru.* Washington, D.C.: United States Agency for International Development.

Kydd, J., and Christiansen, R. E. 1982. "Structural Change in Malawi Since Independence: Consequences of a Development Strategy Based on Large-Scale Agriculture." *World Development* 10(5):355–75.

Lamb, Geoffrey, and Muller, Linda. 1982. *Control, Accountability, and Incentives in a Successful Development Institution: The Kenya Tea Development Authority.* Staff Working Paper 550. Washington, D.C.: World Bank.

Lambert, J. N. 1979. "The Relationship Between Cash Crop Production and Nutritional Status in Papua New Guinea." In *History of Agriculture Working Paper 33.* Port Moresby: University of Papua New Guinea/Department of Primary Industry.

Lancaster, Kelvin J. 1966. "A New Approach to Consumer Theory." *Journal of Political Economy* 74(2):132–57.

Lappé, F. M., and Collins, J. 1977. *Food First: Beyond the Myth of Scarcity.* Boston: Houghton Mifflin Co.

Lardy, Nicholas R. 1978. *Economic Growth and Distribution in China.* Cambridge: Cambridge University Press.

———. 1983. *Agriculture in China's Modern Economic Development.* Cambridge: Cambridge University Press.

Laure, Joseph. 1981. *Nutrition et Population en Vue de la Planification Alimentaire*. Kigali, Rwanda: IAMSEA/ORSTOM/Food and Agriculture Organization of the United Nations.

———. 1982. *Des Vivres ou du Thé? Ou l'Alimentation et les Conditions de Vie de Familles Rwandaises (Dépouillement de l'Enquête Faite à Kigali en Avril 1979)*. Kigali, Rwanda: OCAM/IAMSEA/ORSTOM.

Lawrence, Mark; Lawrence, F.; Cole, T. J.; Coward, W. A.; Singh, J.; and Whitehead. R. G. 1989. "Seasonal Pattern of Activity and its Nutritional Consequences in Gambia." In *Causes and Implications of Seasonal Variability in Household Food Security*, ed. David Sahn. Baltimore: Johns Hopkins University Press for the International Food Policy Research Institute.

Leach, M. 1990. "Images of Propriety: The Reciprocal Constitution of Gender and Resource Use in the Life of a Sierra Leonean Forest Village." Ph.D. diss., University of London.

Lee, C. Y. 1976. "Pineapple Canning, Siam Food Products." In *Marketing—An Accelerator for Small Farmer Development*. Bangkok: Food and Agriculture Organization of the United Nations.

Lehmann, D.; Howard, P.; and Heywood, P. 1988. "Nutrition and Morbidity: Acute Lower Respiratory Tract Infection, Diarrhea, and Malaria." *Papua New Guinea Medical Journal* 31(2):109–17.

Lele, Uma. 1975. *The Design of Rural Development: Lessons from Africa*. Baltimore: Johns Hopkins University Press.

———. 1989a. *Agricultural Growth, Domestic Policies, the External Environment, and Assistance to Africa: Lessons of a Quarter Century*. MADIA Discussion Paper 1. Washington, D.C.: World Bank.

———. 1989b. *Structural Adjustment, Agricultural Development, and the Poor: Some Lessons from the Malawian Experience*. MADIA study. Washington, D.C.: World Bank.

Lele, Uma; van de Walle, Nicolas; and Gbetibouo, Mathurin. 1989. *Cotton in Africa: An Analysis of Differences in Performance*. MADIA Discussion Paper 7. Washington, D.C.: World Bank.

Lim, Antonio. 1987. "Export Crops in the Philippines with Special Emphasis on Mindanao: A Political Economy Study." Washington, D.C.: International Food Policy Research Institute. Mimeo.

Lipton, Michael. 1985. "Indian Agricultural Development and African Food Strategies: A Role for the EC." In *India and the EC*, ed. W. Caliewaert. Brussels: Centre for European Policy Studies.

Lipton, Michael, and Longhurst, Richard. 1989. *New Seeds and Poor People*. London: Century Hutchinson.

Longhurst, Richard. 1988. "Cash Crops, Household Food Security, and Nutrition." *IDS Bulletin* 19(2):28–36.

Low, Allan. 1986. *Agricultural Development in Southern Africa: Farm Household Economics and the Food Crisis*. London: James Currey.

Lunven, Paul. 1982. "The Nutritional Consequences of Agricultural and Rural Development Projects." *Food and Nutrition Bulletin* 4:17–22.

Maddala, G. S. 1983. *Limited Dependent and Qualitative Variables in Econometrics*. Cambridge: Cambridge University Press.

Malawi, Centre for Social Research. 1988. *The Characteristics of Nutritionally Vulnerable Subgroups Within the Smallholder Sector of Malawi: A Report from the 1980/1 NSSA.* Zomba.

Marter, Alan. 1978. *Cassava or Maize: A Comparative Study of the Economics of Production and Market Potential of Cassava and Maize in Zambia.* Lusaka, Zambia: RDSB-UNZA.

Martorell, Reynaldo, and Habicht, Jean-Pierre. 1986. "Growth in Early Childhood in Developing Countries." In *Human Growth: A Comprehensive Treatise*, ed. F. Falkner and J. M. Tanner, 241–262. New York: Plenum Publishing Corporation.

Massell, B. 1969. "Consistent Estimation of Expenditure Elasticities from Cross-Section Data on Households Producing Partly for Subsistence." *Review of Economics and Statistics* 51(May):136–42.

Mathema, P. R. 1976. "Ratna Feed Industries, Nepal." In *Marketing—An Accelerator for Small Farmer Development.* Bangkok: Food and Agriculture Organization of the United Nations.

Maxwell, Simon. 1988. "Cash Crops in Developing Countries." *IDS Bulletin* 19 (2):1–4.

Maxwell, Simon, and Fernando, A. 1989. "Cash Crops in Developing Countries: The Issues, the Facts, the Policies." Paper presented at Research Seminar on Rural Development Studies. Institute of Social Studies, The Hague.

McGuirk, Anya, and Mundlak, Yair. 1991. *Incentives and Constraints in the Transformation of Punjab Agriculture.* Research Report 87. Washington, D.C.: International Food Policy Research Institute.

Mellor, John W. 1986. "Agriculture on the Road to Industrialization." In *Development Strategies Reconsidered*, ed. John P. Lewis and Valeriana Kallab. New Brunswick, N.J. Transaction Books.

Mellor, John W., and Ahmed, Raisuddin (eds.). 1988. *Agricultural Price Policy for Developing Countries.* Baltimore and London: Johns Hopkins University Press for the International Food Policy Research Institute.

Mellor, John W., and Johnston, Bruce F. 1984. "The World Food Equation: Interrelations Among Development, Employment, and Food Consumption." *Journal of Economic Literature* 22(June):531–74.

Mellor, John W., and Lele, Uma. 1973. *Growth Linkages of the New Foodgrain Technologies.* Occasional Paper 50. Ithaca, N.Y.: Cornell University, Department of Agricultural Economics.

Mellor, John W.; Delgado, Christopher; and Blackie, Malcolm J. (eds.). 1986. *Accelerating Food Production in Sub-Saharan Africa.* Baltimore: Johns Hopkins University Press for the International Food Policy Research Institute.

Meneguay, M. R., and Huang, K. R. 1976. *Farmer Factory Contract System for Processing Tomato.* Taiwan: Asian Vegetable Research and Development Centre.

Mergos, George, and Slade, Roger. 1987. *Dairy Development and Milk Cooperatives: The Effects of a Dairy Project in India.* World Bank Discussion Paper 15. Washington, D.C.: World Bank.

Mettrick, H. No date. "Botswana Meat Corporation." Prepared for the Interna-

tional Course for Development Oriented Agriculture. Wageningen, Netherlands.
Michler, W. 1989. "Wem der Kaffee Bitter Schmeckt: Die Verheerenden Auswirkungen des Kaffeepreisverfalls." In *Welternährung, Zeitung für Freunde und Förderer der Welthungerhilfe*, 18. Jahrgang, 4. Quartal.
Milimo, J. et al. 1988. "Agricultural Policy and Its Impact on Food Security: The Zambian Case Study." Paper presented at the 4th Annual Conference on Food Security Research in Southern Africa. Harare, Zimbabwe.
Ministère du Plan de Rwanda. 1983. *Stratégie Alimentaire de Rwanda: Objectifs et Programmes d'Action*. Kigali.
Minot, Nicholas. 1986. *Contract Farming and Its Impact on Small Farmers in LDCs*. East Lansing: Michigan State University and United States Agency for International Development.
Mittendorf, H.J. 1968. "Marketing Aspects in Planning Agricultural Processing Enterprises in Developing Countries." *FAO Monthly Bulletin of Agricultural Economics and Statistics* 17(4).
Morooka, Y., and Mayrowani, H. 1988. "Soybean-Based Farming System in Upland Java: Palawija and a Village Economy." Bogor, ESCAP CGPRT Centre. Mimeo.
Nakajima, Chihiro. 1970. "Subsistence and Commercial Farms: Some Theoretical Model of Subjective Equilibrium." In *Subsistence Agriculture and Economic Development*, ed. C. R. Wharton. London: Cass.
———. 1986. *Subjective Equilibrium Theory of the Farm Household*. New York: Elsevier Science Publishers.
Narain, Dharm. 1981. *Distribution of Marketed Surplus of Agricultural Produce by Size of Holdings in India*. Delhi, India: Institute of Economic Growth.
National Economic and Development Authority. 1974. *1971 Census of Agriculture*. Manila: National Economic and Development Authority, National Census and Statistics Office.
———. 1985. *1980 Census of Agriculture*. Manila: National Economic and Development Authority, National Census and Statistics Office.
NCDP. 1983. *Agricultural Baseline Data for Planning*. Edited by P. Ncube. Lusaka: NCDP/University of Zambia.
Nerlove, Marc. 1988. *Modernizing Traditional Agriculture*. Occasional Paper 16. Panama: International Centre for Economic Growth.
Niemeijer, R.; Geuns, M.; Kliest, T.; Ogonda, V.; and Hoorweg, J. 1985. *Nutritional Aspects of Rice Cultivation in Nyanza Province, Kenya*. Food and Nutrition Studies Programme Report 14. Nairobi/Leiden, The Netherlands: Ministry of Planning and National Development, African Studies Centre.
———. 1988. "Nutrition in Agricultural Development: The Case of Irrigated Rice Cultivation in West Kenya." *Ecology of Food and Nutrition* 22:65–81.
Nieves, I. 1987. "Exploratory Study of Intrahousehold Resource Allocation in a Cash-Cropping Scheme in Highland Guatemala." Washington, D.C.: International Center for Research on Women. Mimeo.
Noy, F., and Niemeijer, R. 1988. *Resident Tenants at Ahero Irrigation Scheme: Household Economics and Nutrition*. Food and Nutrition Studies Pro-

gramme Report 10. Nairobi/Leiden, The Netherlands: Ministry of Planning and National Development, African Studies Centre.
Nyongesa, Dismas. 1987. "The Sugar Industry in Kenya." Paper presented at the International Food Policy Research Institute workshop. Nairobi, Kenya, June 29–July 1.
Oyejide, Ademola. 1986. *The Effects of Trade and Exchange Rate Policies on Agriculture in Nigeria*. Research Report 55. Washington, D.C.: International Food Policy Research Institute.
Oyugi, L. A.; Mukhebi, A. W.; and Mwangi, W. M. 1987. "The Impact of Cash Cropping on Food Production: A Case Study of Tobacco and Maize in Mogori Division, South Myanza District of Kenya." *Eastern Africa Economic Review* 3:43–50.
Pacey, Arnold, and Payne, Philip (eds.). 1985. *Agricultural Development and Nutrition*. London: Hutchinson.
Pakistan, Government of. 1986. *Household Expenditure Study, 1984–85*. Karachi: Federal Bureau of Statistics.
Payne, Philip. 1987. "Undernutrition: Measurement and Implications." Paper prepared for the World Institute of Development Economics Research Conference on Poverty, Undernutrition, and Living Standards. Helsinki.
People's Daily, February 8, 1989.
Perkins, Dwight. 1988. "Reforming China's Economic System." *Journal of Economic Literature* 26(June):601–49.
Peters, P., and Herrera, G. 1989. "Cash Cropping, Food Security and Nutrition: The Effects of Agricultural Commercialization Among Smallholders in Malawi." Harvard Institute for International Development, Cambridge, Mass. Mimeo.
Pinckney, Thomas C. 1989. *The Demand for Public Storage of Wheat in Pakistan*. Research Report 77. Washington, D.C.: International Food Policy Research Institute.
Pinstrup-Andersen, Per. 1983. *Export Crop Production and Malnutrition*. Occasional Paper 10. Chapel Hill: University of North Carolina.
———. 1985. "The Impact of Export Crop Production on Human Nutrition." In *Nutrition and Development*, ed. Margaret Biswas and Per Pinstrup-Andersen. London: Oxford University Press.
Pinstrup-Andersen, Per; de Londoño, Norha Ruiz; and Hoover, Edward. 1976. "The Impact of Increasing Food Supply on Human Nutrition: Implications for Commodity Priorities in Agricultural Research and Policy." *American Journal of Agricultural Economics* 86(May):131–142.
Pitt, Mark M., and Rosenzweig, Mark R. 1985. "Health and Nutrient Consumption Across and Within Farm Households." *The Review of Economics and Statistics* 67(May):212–223.
Poapongsakorn, N., et al. 1985. *Food Processing and Marketing in Thailand*. Geneva: United Nations Conference on Trade and Development.
Poats, Susan. 1981. "La Pomme de Terre au Rwanda: Résultats Préliminaires d'une Enquête de Consommation." *Bulletin Agricole du Rwanda* (April):82–91.

Popkin, Barry M. 1980. "Time Allocation of the Mother and Child Nutrition." *Ecology of Food and Nutrition* 9:1-14.

―――. 1983. "Rural Women, Work, and Child Welfare in the Philippines." In *Women and Poverty in the Third World*, ed. Mayra Buvinic, Margaret Lycette, and William P. McGreevey. Baltimore: Johns Hopkins University Press.

Prentice, A. M. 1980. "Variations in Maternal Dietary Intake, Birth Weight, and Breast Milk Output in The Gambia." In *Maternal Nutrition During Pregnancy and Lactation*, ed. H. Aebi and R. G. Whitehead. Bern, Switzerland: Hans Huber Publishers.

Programme National pour l'Amélioration de la Pomme de Terre au Rwanda. 1984. "Développement de la Culture de la Pomme de Terre." Séminaire interdisciplinaire tenu à Ruhengeri dans le cadre du 5ᵉ anniversaire de PNAP, July 24-26.

Ranis, Gustav, and Stewart, Frances. 1987. "Rural Linkages in the Philippines and Taiwan." In *Macropolicies for Appropriate Technology in Developing Countries*, ed. F. Stewart. Boulder, Colo.: Westview Press.

Reda, K. 1970. *The Broiler Boom in Lebanon*. Rome: Food and Agriculture Organization of the United Nations.

Roll, C. Robert. 1975. "Incentives and Motivation in China: The 'Reality' of Rural Inequality." Paper presented at the Annual Meeting of the American Economic Association. Dallas, Texas, December 28.

Rozelle, Scott. 1991. "The Economic Behavior of Village Leaders in China's Reform Economy." Ph.D. diss., Cornell University.

Ruigu, G. M. 1988. *Large-Scale Irrigation Development in Kenya: Past Performance and Future Prospects*. Food and Nutrition Studies Programme Report 23. Nairobi/Leiden, The Netherlands: Ministry of Planning and National Development, African Studies Centre.

Rwanda, Republic of. 1985. *Résultats de l'Enquête Nationale Agricole, 1984*, Vols. 1, 2, 3. Kigali: Ministère de l'Agriculture, de l'Élevage et des Forêts.

Sahn, David, and Shively, Gerald. 1991. "Household Welfare, Nutrition, and Agricultural Earnings in Southern Malawi." Washington, D.C.: Cornell Food and Nutrition Policy Program. Mimeo.

Schultz, Theodore W. 1983. *Transforming Traditional Agriculture*. Chicago and London: The University of Chicago Press.

Scott, Gregory J. 1986. *La Pomme de Terre en Afrique Centrale: Une Étude sur le Burundi, le Rwanda, et le Zaire*. Lima, Peru: International Potato Center.

Sen, Amartya K. 1985. "Women, Technology, and Sexual Divisions." *Trade and Development* 6:195-223.

Shand, R. T. (ed.). 1986. *Off-farm Employment in the Development of Rural Asia*, Vol. 2. Australia: National Center for Development Studies, Australian National University.

Sharif, Mohammed. 1986. "The Concept and Measurement of Subsistence: A Survey of the Literature." *World Development* 14(5):555-577.

Sharpley, Jennifer. 1986. *Economic Policies and Agricultural Performance: The Case of Kenya*. Paris: OECD Development Centre.

Sierra Leone, Government of. 1978. *Sierra Leone National Nutrition Survey.* Freetown.
Singh, Inderjit; Squire, Lyn; and Strauss, John (eds.). 1986. *Agricultural Household Models: Extensions, Applications, and Policy.* Baltimore: Johns Hopkins University Press for the World Bank.
Singh, S. P., and Kelly, Paul. 1981. *AMUL: An Experiment in Rural Economic Development.* New Delhi: Macmillan.
Smith, Adam. 1937. *The Wealth of Nations.* New York: Random House.
Spencer, Dunstan, and Byerlee, Derek. 1976. "Technical Change, Labor Use, and Small Farmer Development: Evidence from Sierra Leone." *American Journal of Agricultural Economics* (December):874–80.
SSB (State Statistical Bureau of China [Zhongguo Guojia Tongjiju]). Various years. *Zhongguo Tongji Nianjian* (Statistical Yearbook of China). Beijing: Statistical Publishing House of China.
———. Various years. *Zhongguo Nongcun Tongji Nianjian* (Rural Statistical Yearbook of China). Beijing: Statistical Publishing House of China.
———. Various years. *Zhongguo Fenxian Nongcun Jingji Tongji Gaiyao* (Statistical Abstract of China's Rural Economy by County). Beijing: Statistical Publishing House of China.
State Council, Office of the Leading Group for Economic Development in Poor Areas. 1989. *Outline of Economic Development in China's Poor Areas.* Beijing: Agricultural Publishing House.
State Council, Research Center for Rural Development and Central Committee of the Communist Party of China, Office for the Study of Rural Policies, Office for Rural Socioeconomic Investigations. 1988. *Shifting Toward the Orbit of a Commodity Economy—Investigation Report Based on Surveys at Regular Observation Points in Rural Areas Throughout China.* Beijing: Publishing House for Rural Readers.
Stauffer, F. C. 1986. "A White Elephant Project Made Good." Paper presented to the 7th International Agriculture Forum. Geneva, May.
Sterkenburg, J. J.; Brandt, J.; and von Beinum, G. G. 1982. *Rural Housing Conditions in Kisumu District, Kenya.* Utrecht: Department of Geography of Developing Countries, University of Utrecht/Housing Research and Development Unit, University of Nairobi.
Stigler, George, and Becker, Gary. 1977. "De Gustibus Non Est Disputandum." *American Economic Review* 67 (March):76–90.
Stone, Bruce. 1988. "Relative Prices in the People's Republic of China: Rural Taxation Through Public Monopsony." In *Agricultural Price Policy for Developing Countries,* ed. John W. Mellor and Raisuddin Ahmed. Baltimore: Johns Hopkins University Press for the International Food Policy Research Institute.
———. 1989. "Fertilizer's Greener Pastures." *The China Business Review* (September-October).
———. 1990. "Evolution and Diffusion of Agricultural Technology in China." In *Sharing Innovation: Global Perspectives on Food, Agriculture, and Rural Development,* ed. Neil G. Kotler. Washington, D.C.: Smithsonian Institution Press.

Streeten, Paul. 1988. *What Price Food? Agricultural Price Policies in Developing Countries.* Ithaca, N.Y.: Cornell University Press.

Strohl, R. J. 1985. "Frozen Orange Juice Concentrate." In *Successful Agribusiness Management,* ed. J. Freiwalds. Brookfield, Vt.: Gower.

Tavis, L. A. (ed.). 1982. *Multinational Managers and Poverty in the Third World.* Notre Dame, Ind.: University of Notre Dame Press.

Taylor, Jeff. 1988. "Rural Employment Trends and the Legacy of Surplus Labor, 1978-86." *The China Quarterly* 116:736-66.

Terhal, Piet, and Doornbus, Martin. 1983. "Operation Flood: Development and Commercialization." *Food Policy* 8(August):235-39.

Thomas, Duncan. 1989. "Intrahousehold Resource Allocation: An Inferential Approach." New Haven, Conn.: Yale University. Mimeo.

Tinker, Irene. 1979. *New Technologies for Food Chain Activities: The Imperative of Equity for Women.* Washington, D.C.: U.S. Agency for International Development, Office of Women in Development.

Tosh, J. 1980. "The Cash Crop Revolution in Tropical Africa: An Agricultural Appraisal." *African Affairs* 79:79-94.

Travers, Lee. 1984. "Post-1978 Rural Economic Policy and Peasant Income in China." *The China Quarterly* (June):241-259.

Travers, Lee, and Stone, Bruce. 1991. "Poverty in China: Status and Dynamics." International Food Policy Research Institute, Washington, D.C. Mimeo.

de Treville, Diana. 1986. *An Annotated and Comprehensive Bibliography on Contract Farming with a Focus on Africa.* Binghampton, NY: Institute for Development Anthropology.

Truitt, G. A. 1980. "Peppers for Tabasco." Case study prepared for Harvard Business School.

Tschajanow, A. 1923. *Die Lehre von der bäuerlichen Wirtschaft. Versuch einer Theorie der Familienwirtschaft im Landbau.* Berlin.

Tshibaka, Tshikala B. 1986. *The Effects of Trade and Exchange Rate Policies on Agriculture in Zaire.* Research Report 56. Washington, D.C.: International Food Policy Research Institute.

Tuinenburg, K. 1987. "Experience with Food Strategies in Four African Countries." In *Food Policy,* ed. James Price Gittinger, Joanne Leslie, and Caroline Hoisington. Baltimore: Johns Hopkins University Press for the World Bank.

UNCTAD (United Nations Conference on Trade and Development). 1987. *Commodity Yearbook.* Geneva.

Vaughan, M. 1987. *The Story of an African Famine: Gender and Famine in Twentieth-Century Malawi.* Cambridge: Cambridge University Press.

Veenstra, Y. 1987. "Luo vrouwen en geld: Economische Afhankelijkheidsrelaties in een Afrikaanse Samenleving." M.A. thesis, University of Groningen.

Vis, H. L.; Yourassowsky, C.; and van der Borght, H. 1975. *A Nutritional Survey in the Republic of Rwanda.* Tervuren, Belgium: Musée Royal de l'Afrique Centrale.

Webb, Patrick. 1987. "Changes in Intrahousehold Decisionmaking and Resource Allocation: The Impact of New Rice Production Technology and

Increased Commercialization in a Rural West African Setting." International Food Policy Research Institute, Washington, D.C. Mimeo.

———. 1989. "When Projects Collapse: Irrigation Failure in The Gambia from a Household Perspective." *Journal of International Development* 3(4): 339–53.

White, L. 1987. *Magomero: Portrait of an African Village.* Cambridge: Cambridge University Press.

Wiebe, K. D. 1992. "Household Food Security and Cautious Commercialization in Kenya." Ph.D. diss., University of Wisconsin at Madison.

Winarno, F. G.; Haryadi, Y.; and Satiawiharja, B. 1985. "Special Traditional Foods of Indonesia." In *Proceedings of the IPB-JICA International Symposium on Agricultural Product Processing and Technology,* ed. F. Srikandi, A. Matsuyama, and A. Kamaruddin. Bogor, Indonesia: IPB and JICA.

World Bank. 1981. *Accelerated Development in Sub-Saharan Africa: An Agenda for Action.* Washington, D.C.

———. 1985. *China to the Year 2000.* Washington, D.C.

———. 1988a. *Commodity Trade and Price Trends.* Washington, D.C.

———. 1988b. *Papua New Guinea: Policies and Prospects for Sustained and Broad-based Growth.* Vol. 1. The Main Report. A World Bank Country Study. Washington, D.C.

———. 1990. *China: Between Plan and Market.* Washington, D.C.

———. Various years. *World Development Report.* New York: Oxford University Press.

Ye, Qiaolun. 1991. "Analyzing the Effects of Market and Nonmarket Factors on Fertilizer Use in China's Reform Economy." Ph.D. diss., Stanford University.

Yearwood, J.; Quinn, J.; and Kai-Kai, F. 1983. *Agricultural Production in the Bo and Pujehun Districts of Sierra Leone.* Bo, Sierra Leone: Planning and Coordinating Office.

Zeller, Manfred; Schrieder, Gertrud; von Braun, Joachim; and Heidhues, Franz. 1993. "Rural Finance for Food Security of the Poor: Concept, Review and Implications for Research and Policy." International Food Policy Research Institute, Washington, D.C. Mimeo.

Contributors

John C. Abbott is an independent consultant. He was formerly chief of Marketing Service at the Food and Agriculture Organization of the United Nations.

Raisuddin Ahmed is director of the Markets and Structural Studies Division at the International Food Policy Research Institute. Before coming to IFPRI he was deputy chief of the Agriculture and Water Resources Division of the Planning Commission in Bangladesh and chief agricultural economist of the Ministry of Agriculture.

Harold Alderman is a senior economist in the Policy Research Department, Poverty and Human Resources Division of the World Bank. He was formerly a research fellow at the International Food Policy Research Institute.

Friederike Bellin is an assistant lecturer on nutrition in developing countries at the University of Giessen, Germany. She serves as a consultant to the German Technical Cooperation for Development.

Jürgen Blanken is a senior agricultural economics adviser at the Agricultural Research Station at Rubona, Rwanda.

Howarth Bouis is a research fellow in the Food Consumption and Nutrition Division at the International Food Policy Research Institute.

Chen Jiyuan is a professor and director of the Rural Development Research Institute at the Chinese Academy of Social Sciences, Beijing.

Hartwig de Haen is assistant director-general, Agriculture Department, Food and Agriculture Organization of the United Nations. At the time this volume was written he was a professor of agricultural economics at the University of Goettingen, Germany.

M. P. Finlayson is a consultant with Hassal & Associates, Canberra, Australia. He was formerly a research officer with the Papua New Guinea Institute of Medical Research, Madang, Papua New Guinea.

David Glover is director of the Economy and Environment Program for Southeast Asia, an activity managed by the International Development Research Centre (IDRC), Singapore. He was formerly director of the Economic Policy Program of IDRC in Ottawa.

Lawrence J. Haddad is a research fellow in the Food Consumption and Nutrition Division of the International Food Policy Research Institute. He was formerly a lecturer in quantitative development economics at the University of Warwick, England.

J. Brian Hardaker is a professor of agricultural economics at the University of New England, Armidale, New South Wales, Australia.

M. Guillermo Herrera is a faculty lecturer in nutrition at the Harvard School of Public Health and research associate at the Harvard Institute for International Development.

Peter F. Heywood is a professor of community nutrition at the University of Queensland, Australia. He was formerly deputy director of the Papua New Guinea Institute of Medical Research, Madang, Papua New Guinea.

Jan Hoorweg is head of the Department of Social and Economic Studies at the African Studies Center, Leiden, The Netherlands, and a professor of food security and nutritional intervention at the Department of Human Nutrition of the Agricultural University, Wageningen. He was formerly director of the Food and Nutrition Studies Programme in Kenya.

Maarten D. C. Immink is a research fellow in the Food Consumption and Nutrition Division of the International Food Policy Research Institute. He was previously chief of the Division of Food Planning and Nutrition, Institute of Nutrition of Central America and Panama.

Nurul Islam is the senior policy adviser at the International Food Policy Research Institute. Before coming to IFPRI he was the assistant director general of the Food and Agriculture Organization of the United Nations, Economic and Social Policy Department. Earlier he was a fellow at St. Antony's College, Oxford, and Queen Elizabeth House, Oxford University.

Jiang Dehua is a professor of rural and agricultural geography at the Institute of Geography, Chinese Academy of Sciences, Beijing.

Ken B. Johm is director of the Department of Planning, Ministry of Agriculture, The Gambia.

Toshihiko Kawagoe is a professor of economics at Seikei University, Tokyo. He was previously an agricultural economist at the Regional Coordination Centre for Research and Development of Coarse Grains, Pulses, Roots, and Tuber Crops in the Humid Tropics of Asia and the Pacific in Bogor, Indonesia.

Eileen Kennedy is administrator of the Nutrition Research and Education Agency in the U.S. Department of Agriculture. She was formerly a research fellow in the Food Consumption and Nutrition Division of the International Food Policy Research Institute and a visiting professor of nutrition at Tufts University.

Shubh K. Kumar is a research fellow in the Food Consumption and Nutrition Division of the International Food Policy Research Institute. She was formerly a member of the nutrition faculty at Punjab Agricultural University, Ludhiana, India.

John R. McComb is a Ph.D. candidate at the University of Queensland, Australia. He was formerly a research officer with the Papua New Guinea Institute of Medical Research, Madang, Papua New Guinea.

Rudo Niemeijer is a consultant on household and nutrition surveys, anthropological field studies, and data analysis. At the time this volume was written, he was head of the Research Methods and Data Analysis Division with the African Studies Center, Leiden, The Netherlands.

Pauline E. Peters is a research associate at the Harvard Institute for International Development and lecturer in the Department of Anthropology, Harvard University.

Detlev Puetz is an independent consultant. He was formerly a postdoctoral fellow in the Food Consumption and Nutrition Division of the International Food Policy Research Institute.

Scott Rozelle is an assistant professor of agricultural economics at the Food Research Institute, Stanford University.

Catherine Siandwazi is coordinator of the Food and Nutrition Programme at the Commonwealth Regional Health Community Secretariat for East, Central, and Southern Africa. She was formerly deputy director at the National Food and Nutrition Commission in Zambia.

Bruce Stone is an analyst of Chinese agricultural development. He was formerly a research fellow at the International Food Policy Research Institute.

Tong Zhong is a visiting researcher in the Outreach Division of the International Food Policy Research Institute.

Joachim von Braun is a professor of food economics and food policy at the University of Kiel, Germany. He was formerly director of the Food Consumption and Nutrition Division at the International Food Policy Research Institute.

Xu Zhikang is professor emeritus of economic geography at the Institute of Geography, Chinese Academy of Sciences, Beijing.

Index

Abbott, John C., 153, 155, 158, 161, 162, 163
Africa: anti-development trends in Sub-Saharan, 366; cash crops productivity, 117; discrimination against agriculture, 109; food budget share of income, 71, 74; food consumption, 73; food crops production, 76, 104; gender differences in income expenditures, 93; nonfood expenditures, 70
Agricultural commercialization
  Described, 3-5, 11, 103
  Determinants, 12-13; demographic changes, 13-14; technological change, 14-15
  Driving forces to stimulate, 5; contract farming, 166, 173-75; crop development strategy, 103-06; infrastructure development, 141-44; market reforms, 109, 111-14; processing enterprises, 153-54; proposed policies for, 365, 375-76; terms of trade, 106-08
  Farm size and, 47, 52
  Forms, 11-12, 42, 47
  Potential role in agricultural growth, 5-6
  Project failures, 154-55, 218
  Scheme participation: employment effect, 54, 62-63; gainer-loser situation, 59-61; income effect, 50, 59, 63; market surplus, 55-56; and profitability, 52-54; price risks, 57-58; smallest farm, 47-48, 49, 62; staple foods production, 48-52, 55
  Wage rates and, 55, 58
Agricultural Development and Marketing Corporation, Malawi, 309, 316, 318
Agricultural fundamentalism, 239, 240
Agricultural production
  Cooperatives for, 87, 190-96, 198, 200-01
  Effects on consumption, 249-51
  Fertilizer and 145, 146
  Food availability from: community, 26; household, 27-29; national, 25-26
  High-yield varieties, 128, 131, 133
  Infrastructure development and, 141-44
  Irrigation and, 128, 131, 145, 146, 264, 266-69, 345, 358, 359
  Plantations for, 219, 220, 233, 237, 238, 371
Ahmad, Ehtisham, 123, 375
Ahmed, Raisuddin, 5, 113, 141, 143, 144, 145, 146, 147
Alderman, Harold, 239, 240, 243n, 244, 245, 249n
Alvarez, Claude, 239
An, Xiji, 120
Asia: cash crop productivity, 117; discrimination against agriculture, 109; food crops, 104

402 Index

Baldeston, J. B., 200
Bananas: Honduras production cooperative, 170; Papua New Guinea, 218; price variability, 112
Bangladesh
  Benefit/cost measure of roads, 152
  Infrastructure effects: consumption, 149; employment, 148; household income, 147–48; social attitudes, 149
  Infrastructure role in commercialization, 7, 144–46, 152
Bautista, Romeo M., 62
Beans, 193
Becker, Gary S., 17, 242
Beckford, G., 154
Beinum, G. G. von, 269, 270n
Bellin, Friederike, 328, 331
Bernard, Andrew, 113
Binswanger, Hans, 146, 243, 368
Blackie, Malcolm J., 362n
Blanken, Jürgen, 12n, 155, 276, 293n, 369
Bolivia, dairy processing enterprise, 158
Bossen, L. H., 193
Bouis, Howarth, 11, 25, 65, 66, 73n, 204, 213
Brandt, J., 269, 270n
Braun, Joachim von, 3n, 4, 5, 11, 12, 14n, 25, 28, 37, 52, 92, 98, 105, 113, 146, 155, 189, 190, 200, 276, 293n, 343, 352, 356n, 365, 368
Brazil, orange juice processing enterprises, 161, 164
Bukidnon Sugar Company, Philippines, 204, 217
Byrd, William, 122

Calorie consumption, 29
  Household, 66; adequacy, 76–77; cost, 72; female-controlled, 93–94; food marketing costs and, 74–76; income and, 72, 74, 99; recommended daily allowance, 77
  Preschoolers: determinant of weight, 87, 91; income and, 91; morbidity and, 98
Cardamom production
  Guatemala, 220n
  India, 220n
  Papua New Guinea: plantation produced, 219–20, 222, 238, 371; smallholders, 219, 237
  Prices, 220
Cassava
  Venezuela processing plants, 155
  Village-level farming and processing, Java, 177, 178, 179; marketing margin, 184; prices, 182; value added from home processing, 181
Celis, Rafael, 296
Central America: contract farming, 168, 171; road density effect on agriculture, 146
Cereal crops: distinction between men's and women's, 348; export subsidies by developed countries, 108; nominal rate of protection for, 109; price trend, 107–08; price variability, 112; yields, 51
Chalfant, James A., 22
Chavas, Jean-Paul, 23
Chen, Jiyuan, 119, 131n, 133n
Children
  Commercialization effect on nutrition, 15, 29–30, 43, 47, 374
  Preschoolers: calorie consumption, 87, 91, 98; morbidity, 79, 82–83, 85, 87; in male- versus female-headed households, 93–94, 96, 99; nutritional status, 83–85, 87, 91
Chilies, 219, 223
China
  Agricultural commercialization, 7, 119; changing production and crop policies, 121–22; constraints on, 139; elements involved in, 120; labor and credit market changes, 122–23; poor counties, 135–36; success with, 139–40
  Economic rural reforms, 119; effect on poorer regions, 124–25
  Foodgrains, 121, 134, 135–36
  Gross value of agricultural output, 122n
  Poor counties: agricultural diversification, 133–34; described, 127, 129–30; economic structure, 131, 133; income growth and distribution, 136, 138–39; irrigation, 128, 131; special government assistance for, 125–26
  Poverty, 123–24; reduction in, 126–27
  Successful poor areas: described, 127; grain-producing, 127, 128; irrigation facilities, 128

Chowdhury, Omar H., 152
Christiansen, R. E., 318
CIMMYT. *See* International Maize and
  Wheat Improvement Center
Cocoa. *See also* Tree crops
  Malaysia, 115
  Price variability, 112
  Sierra Leone, 328, 329
Coffee. *See also* Tree crops
  Papua New Guinea, 219, 223, 226
  Sierra Leone, 328; exports 329, 330
Cogill, Bruce, 253
Collins, J., 4, 154
Commonwealth Development
  Corporation, 153, 167
Contract farming
  Defined, 167
  Effects: employment, 168, 171;
    extension service, 172;
    infrastructure development,
    171-72; market development, 172
  Food security and: crop requirements
    for, 170; income generation for,
    168-69; recommended
    improvements in, 173-75
  Types, 167-68
Cooperatives: Guatemalan export
  crop-producing, 87, 190-96, 198,
  200-01; Indian dairy development,
  243-47, 248, 249
Corn production. *See* Philippines
Cotton: China, 121; The Gambia, 343,
  344
Crotty, Raymond, 244

Dairy development
  Commercialization: effect on supply
    of by-products, 244-45; and
    incomes of poor, 247-48;
    nutritional implications, 240, 242,
    248, 249; price effects, 245, 248
  Evaluation, 240
  Karnataka Dairy Development
    Project cooperative, 243-47, 249
  Madhya Pradesh cooperative, 244,
    248, 249n
  Operation Flood, 239-40
de Haen, Hartwig, 12n, 155, 276, 293n,
  369
Delepierre, 276
Delgado, Christopher, 362n
de Londoño, Norha Ruiz, 243n
Demographic change, and
  commercialization, 13-14

Developing countries
  Agricultural commercialization,
    61-62
  Contract farming, 166; employment
    effects, 168; food security and,
    168-73; measures to improve,
    173-75; multipartite arrangement,
    167
  Discrimination against agriculture, 109
  Domestic versus foreign market
    relative prices, 109
  Export crop production: national
    food security and, 25; taxation, 62
  Food crops' share of total crop
    output, 104
  Processing enterprises: by commodity
    group, 160-62; contract
    negotiations, 164; factors leading to
    success of, 158; problems, 153-55;
    raw material suppliers and,
    158-60; technology transmission
    and, 163-64
  Subsistence production, 4
Devres International, 146
Dey, Jennie, 344n
Dinham, Barbara, 6
Dixon, John, 123
Donovan, Cynthia, 144
Doornbus, Martin, 244

EC. *See* European Community
Economic growth, agricultural
  commercialization and, 3, 365
Ecuador, processing enterprise, 163
Egli, William E., 276
Ellis, Frank, 171
Employment: agricultural
  commercialization effect on,
  26-27, 54-55, 58, 370-71; cash
  versus export crop production, 118;
  contract farming and, 168, 171;
  from crop processing, 184;
  infrastructure development and,
  144, 148
European Community (EC), 108, 240
Evenson, Robert E., 17, 146
Exchange rate: equilibrium versus
  nominal, 109; and food import
  price changes, 114; overvaluation
  effect on export crops, 110; and
  return from agriculture, 109
Ezemenari, Kene, 22n

Fafchamps, Marcel, 22
FAO. *See* Food and Agriculture Organization
Fernando, A., 266
Fertilizers, 145, 146
Finkelshtain, Israel, 22
Finlayson, M. P., 218, 222n, 223, 228
Fleuret, P. and A., 266
Food and Agriculture Organization (FAO), 105, 154, 155
Food crops
  Basic: cash crops versus, 11; percent of total crop output, 104–05; productivity growth, 116; proposed priority for production, 106–07; shift to cash crops from, 115–16
  Cash: characteristics, 12; described, 11, 103; effect on morbidity, 79, 82–83; income from, 28, 71–72; nutritional effects, 79, 82, 92, 240, 252; production strategy for, 115; productivity growth, 116–17
  Choice between nonfood crops and, 103–06
  Contract farming for, 169–70
  Discrimination against, 109
  Export: choice between food crops and, 117–18; discrimination against, 110–11; domestic market prices, 109, 111; protection for, 109; specialization in, 106; terms of trade and, 106–07; trade restrictions on, 108; world market prices, 106–07; 111–14
  Research on, 115–16
Food expenditures, 65; budget shares, 71–72; and calorie consumption, 72–74, 76–77; market dependency and, 74–76; nonfood expenditures versus, 65, 70; by women, 72
Foodgrains, China, 121, 122; declining production, 134; marketing, 135–36; purchases on per capita basis, 136
Food security
  Contract farming relevant to, 166; by crop displacement, 169–70; by income generation, 168–69; recommended improvements in, 173–75; by risk sharing and reduction, 170
  Food supply and: access to food imports and, 106; nonfood crop production and per capita, 105–06; world price instability and, 113
  Household level: cash crops implication for, 12; commercialization effect on, 12, 37, 118, 369–70; market dependency and, 74–76
  Subsistence production and, 369–70
Frederickson, P. C., 147
Freiwalds, J., 160, 163
Fruit, processing enterprises, 161–62

Gambia, The
  Agricultural commercialization, 6, 37, 343; double-crop rice production and, 42, 371–72, 373; employment effects, 54, 58; food expenditures effect, 66, 71, 72; and household calorie intake, 66, 73, 77; morbidity effects, 79, 81; nutritional effects, 66, 73, 77, 85, 93, 98, 344, 374; and role of women, 93–94, 95; technological change and, 343, 362
  Rice production technological change, 54, 343; effect on food consumption, 355–58; and household food security, 344, 346, 348, 370; impact on yield, 350–52; irrigation scheme, 267, 345, 348, 350; labor inputs and productivity, 350–52, 354–55; land distribution policy, 348; nutritional effects, 358–59
George, Shanti, 244
George, Susan, 154
Ghana: lime processing enterprise, 161; raw material supplies for processing plants, 155
Glover, David, 166, 168, 171, 172
Grilli, Enzo R., 107
Groos, A., 229, 230
Guatemala, 6, 37
  Agricultural commercialization: for cash versus subsistence crops, 49; degree of, 42; employment effects, 54, 58; gainer-loser situation, 61; and household calorie intake, 66, 73, 74; for new crops, 50; profitability, 53, 54; scheme participation, 48; staple foods production, 50, 56
  Cardamom production, 220n
  Contract farming, 168, 171, 172
  Export vegetables, 189, 367, 370

Health services, 200, 374
Preschool children: commercialization effect on, 85; compared with older children, 98; illness, 82, 87; nutritional status, 85
Smallholder cooperatives for export crops, 7, 187, 189; described, 190; food availability from, 198; household income and food expenditures, 195–96, 201; income growth from crop diversification, 203; nutritional effects, 198, 200–01; results, 201–02; vegetable production, 192–95

Haaga, J., 266
Haddad, Lawrence J., 66, 73n, 204
Hardaker, J. Brian, 218
Harper, M., 158, 160
Harvey, P. W. J., 219, 229
Haryadi, Y., 178n, 184
Haugerùd, Angelique, 277n
Hausman, J. A., 249
Hayami, Yujiro, 147, 176, 183
Hazell, Peter B. R., 5
Health, 15; and children's growth, 262; commercialization impact on, 79, 96–97; expenditures, 70; sanitation and, 289, 292, 294; wealth and, 82. *See also* Morbidity; Nutritional status
Hernandez, Mercedes, 5
Herrera, M. Guillermo, 309, 310n
Heywood, Peter F., 218, 219, 229, 230
Hide, R. L., 219, 230
High-yielding varieties (HYVs)
China: grain production, 128; improved fertilizer and, 133; poor counties use of, 131
Developed infrastructure and use of: Bangladesh, 145; India, 146
Hinderink, J., 274
Hines, Colin, 6
Hintermeister, Alberto, 190
Hirschman, Albert, 141
Hiwa, S., 310
Holt, Matthew T., 23
Honduras, cooperative for banana production, 170
Hoorweg, Jan, 264, 266
Hoover, Edward, 243n
Horton, S., 249
Hossain, Mahabub, 143, 144, 145, 147, 152

Hotchkiss, David, 98, 190, 200
Households
Commercialization at level of, 11–12; consumption effects, 15, 22, 27–29; gainer-loser situations from, 59–61; participation or nonparticipation in, 47
Consumption versus production decisions, 240; nutrition and, 242, 243; prices, 241; welfare impact on, 241, 243
Expenditures: nonfarm goods and services, 144; nonfood, 70; per capita total, 65, 66
Food expenditures, 66, 70; and calorie consumption, 72–74, 76–77; income and, 71–72; by male- versus female-headed, 93–94; own-produced food versus, 74–76
Income: calorie intake and, 72, 74, 99; effect on hunger versus malnutrition, 99
Infrastructure development effect on, 144, 147–48
Models: allocation decisions approach, 24; role of market risks in, 20–23; subjective equilibrium, 16–17; time allocation, 18–20; two-person, 20, 30–33
Houtman, C. B., 268n, 270n
Howard, P., 230
Huang, K. R., 159, 161, 164
Hussain, Athar, 123

ILO. *See* International Labour Organisation
IMF. *See* International Monetary Fund
Immink, Maarten D. C., 98, 189, 190, 200
Income
Cash crop, 71–72, 79; Guatemala vegetables, 195–96, 201; Kenya sugarcane, 253, 255; Malawi diversified, 310, 312–13; West Kenya diversified, 270–71
Commercialization effect on household, 12, 23, 371–73; consumption effects, 27–29; and demand for food, 26; nutritional improvement from, 43, 47, 79, 84, 97–99
Commercialization effect on total, 58–59
Crop processing, 184–85

Income (cont'd)
  Household expenditures from, 65-66; diet changes and, 72-74; food, 66, 71; marginal propensities, 71-72; nonfood, 65, 70
  India dairy development and, 247-49
  Papua New Guinea subsistence, 224, 228, 255
India, 6
  Cardamom production, 220
  Dairy processing enterprises, 158, 163
  Infrastructure development, 146
  Karnataka Dairy Development Project, 243; and household expenditures, 248; marketed share of milk, 249; price and income effects on nutrition, 245-47; producer prices, 244
  Madhya Pradesh dairy cooperative, 244, 248, 249n
Indonesia
  Contract farming, 168
  Java village-level farming and processing, 177-78; cottage industries, 178-79; income, 184-86; marketing by middlemen, 179-80; market price, 182-84; production structure, 180-81; value added from home processing, 181-82, 186
  Palm oil production, 115
  Rural growth benefits from commercialization, 176
Infrastructure development, rural
  Commercialization role in strengthening, 141, 142, 144
  Constraints on, 151-52
  Contract farming and, 171-72
  Defined 141
  Effects of, 144, 152; on agricultural production, 145-46; on employment, 148-49; on household income, 147-48; on social attitudes, 149
International Labour Organisation (ILO), 176
International Maize and Wheat Improvement Center (CIMMYT), 297
International Monetary Fund (IMF), Cereal Financing Facility, 114
Irrigation: China, 128, 131; developed infrastructure and, 145, 146; The Gambia, 345, 348, 350; India, 146; West Kenya, 264, 266-67, 268-69

Ishikawa, Shigeru, 122, 123
Islam, Nurul, 103, 108

Jabbar, S., 162
Jaetzold, R., 270n
Jamaica, poultry processing enterprises, 161, 163
de Janvry, Alain, 369
Jiang, Dehua, 119, 129
Java. *See* Indonesia
Johm, Ken B., 343
Johnson, D. Gayle, 119
Jones, Roberto, 276

Karimui Cardamom Growers Association, 220
Karimui Spice Company: changing economic status, 222; failure, 220, 233, 238; wage earnings, 226
Kavura, R., 158, 160
Kawagoe, Toshihiko, 176
Keller, B., 297
Kelly, Paul, 248
Kennedy, Eileen, 3n, 4, 11, 14n, 28, 29, 52, 79, 84, 92, 96, 105, 213, 252, 253, 260, 356n, 365
Kenya. *See also* West Kenya
  Agricultural commercialization: cash crops, 264; income effect on nutritional status, 43, 265-67
  Contract farming, 168, 169, 171
  Food expenditures, 66, 71, 72
  Household calorie intake, 66, 72n, 73n, 77, 256-57
  Preschool children, 28; calorie consumption, 87, 91, 99, 257, 262; commercialization effect on, 85, 97; illness, 82, 87, 94, 258, 260, 262; nutritional status, 84, 85, 258, 260
  Processing enterprises: pineapple, 161-62; tea, 158, 162, 165
  Sugarcane commercialization; and household expenditures, 260; income effects, 253, 255, 262; land use, 49; and nutritional status, 266; profitability, 53, 255; smallholders, 6-7, 252; and subsistence crop production, 256
  Women: household heads, 94, 96; illness, 92, 96; income control by, 92, 94
Khandakar, Shadhidur R., 146

Kliest, T., 264, 268, 269n
Korte, R., 266
Krueger, Anne O., 5, 62, 109, 110
Kumar, Shubh, 25, 295
Kusterer, Ken, 168, 171, 172
Kydd, J., 318

Lamb, Geoffrey, 162, 169
Lambert, J. N., 5, 219
Lancaster, Kelvin J., 17
Lappé, F. M., 4, 154
Lardy, Nicholas R., 123, 127
Latin America: cash crop productivity, 117; discrimination against agriculture, 109; food crops' share of total crop output, 104; nonfood cash crops, 106
Lawrence, Mark, 359
Leach, M., 337n
Lebanon, poultry processing enterprises, 160, 163, 164
Lee, C. Y., 158, 159, 162
Lehmann, D., 230
Lesotho, processing enterprise, 163
Li Jianguang, 136n
Lim, Ghee, 166
Lipton, Michael, 47n, 240
Longhurst, Richard, 4, 11

McComb, John R., 218
McGuirk, Anya, 146
Maize
  Guatemala export crop, 193, 194, 198
  Kenya: consumption, 271, 273; production, 49; profitability, 255
  Malawi: household produced and used, 316; income and landholding effects on supply, 316–18; land and labor input, 314; liberalization of market, 327n; price control, 318
  Price variability, 112
  Rwanda land use for, 278, 279
  Zambia hybrid, 295; adoption, 297, 299–300; child nutrition effects, 302, 304–06; food consumption effects, 300–02; women's involvement with, 307
Malawi
  Agricultural commercialization, 6, 309, 314; effect on preschool children, 85, 87, 91, 97; and food expenditures, 66, 71; road density and, 146
  Contract farming, 168
  Food insecurity, 369
  Household economy: diversified income, 310, 312–13; food acquisition and expenditures, 320–21; role of women, 312–13; subsistence, 314
  Maize, 312; household produced and used, 316; income and landholding effects on supply, 316–18; land and labor input for, 314; liberalization of market, 327n; price control, 318
  Nutritional status, 309, 321; extent of cultivated tobacco and, 318, 325–26; female role in, 93, 96; food crop production and, 319–20, 326–27; morbidity and, 324–25; seasonal pattern for, 324
  Tobacco cultivation, 310; food crops production versus, 76, 319–20; household land for, 320; impact on calorie consumption, 325–26; income variability from, 318–19, 326
Malaysia: cocoa and palm oil production, 115; fish meal processing enterprise, 158; pineapple processing enterprise, 162
Malnutrition. *See also* Nutritional status
  Hunger versus, 99
  Multifaceted causes, 29
  Preschoolers and, 85
Market(s)
  Development, 3, 172
  Domestic, 109, 111
  Household food security and, 74–76
  Infrastructure development and, 142–43
  Outlets for processing enterprises, 155
  World: price instability, 112–14; price trend of export crops, 106–08; production costs effect on supplies in, 114–15
Massell, B., 249
Mathema, P. R., 163
Maxwell, Simon, 266
Mbewe, D., 297
Mellor, John W., 5, 362
Meneguay, M. R., 159, 161, 164
Mergos, George, 240, 244, 248, 249n
Milimo, John T., 296
Milk. *See also* Dairy development
  Consumption: cooperatives effect on, 246; producer effect on, 247, 250–51

Milk (cont'd)
Prices: marketing cooperatives and, 244; retail versus farmgate, 250, 251
Minot, Nicholas, 166n
Mittendorf, H. J., 154, 164
Morbidity, preschoolers, 79; age differences, 83; commercialization and, 82; determinants, 85, 87; wealth and, 82
Morooka, Yoshinori, 176n
Muller, Linda, 162, 169
Mundlak, Yair, 146

Nakajima, Chihiro, 16, 18
Narain, Dharm, 146
National Irrigation Board, West Kenya, 264, 265, 268
Near East, cash crop productivity, 117
Nepal, poultry feed processing enterprise, 160, 163
Nerlove, Marc, 58
Niemeijer, Rudo, 264, 265, 266, 267, 269n, 270n, 271, 273
Nieves, I., 196
Nonfood expenditures, 65, 70
Noy, Frank, 264n, 265, 266, 270n, 271, 273
Nutritional status
Agricultural commercialization effect on, 4, 12, 15, 37; elasticity of, 47; income-related improvement in, 43, 47, 79, 84; for preschoolers, 29–30, 79, 83–85, 96–97; for women, 29–30, 79
Cash crops production and, 4–5, 12, 79, 98
Education for, 263
Income-based aggregate model to explain, 43, 84, 85
Preschoolers: calorie consumption and, 87; gender of household head and, 93–94, 96, 99; household income and, 87
Sanitation and, 292

Oniang'o, Ruth, 29
Operation Flood, 239; criticisms of, 244; nutritional effects, 240
Oshima, , 176
Oyejide, Ademola, 62

Pacey, Arnold, 266
Palm oil. *See also* Tree crops
Malaysia, 115

Sierra Leone, 328, 329, 330
Pandya-Lorch, Rajul, 25
Papua New Guinea, 6
Agricultural commercialization: income effect, 43, 59; nutritional status and, 43, 237; role of women in, 237
Agricultural production: cash cropping, 218–19, 223; subsistence, 218, 226, 227, 228
Cardamom plantation production, 219; income effects, 228–29, 371; paid employment for, 220; reasons for failure, 220, 233, 238; smallholder production versus, 219, 237
Contract farming, 168
Households in Karamui division: children's nutritional status, 229–33; expenditures, 226–27; female role in, 226, 228; food energy consumption, 227–29, 232; material possessions, 223–24; morbidity, 230, 238; wage-earning and non-wage earning, 222–26, 237
Malnutrition, 219

Payne, Philip, 266, 289
Peanuts, 177
Peas, Rwanda: land use for, 278, 279; marketed share, 285
Perkins, Dwight, 120
Peru, contract farming, 171
Peters, Pauline E., 96, 309, 310n
Philippines
Agricultural commercialization, 6; employment effect, 54; and food expenditures, 66, 71; and household calorie intake, 66, 73, 74, 77; household losers from, 216–17; income-related nutritional status, 47; and nonfarm expenditures, 70; profitability, 54; smallholder farming and, 7, 47, 48, 204–06
Corn production: household, 75, 205–06; nutrition effects, 209, 213; profitability, 216; recommendations for improving, 217; yield, 213–14
Infrastructure development, 146, 148
Pineapple production, 162
Preschool children: calorie consumption, 87, 91, 209, 213, 216; commercialization effect on,

85, 97; illness, 82, 87; nutritional status, 85, 209, 213
Sugarcane production, 6–7, 204; nutrition effects, 206, 209, 213; profitability, 206, 216; protection, 53, 367; smallholder, 205–06, 217; source of commercialization, 42, 371
Pinckney, Thomas C., 113
Pineapples, processing enterprises, 159, 161–62
Pinstrup-Andersen, Per, 243, 266
Poapongsakorn, N., 160, 163
Potatoes. *See* Rwanda, Potato production
Prentice, A. M., 358
Preschoolers. *See* Children
Prices, agricultural
  Comparative costs of production and, 114–16
  Domestic: protection for, 109, 110–11; relative, 109
  Household endogenous and exogenous, 240–42
  World market: instability, 112–14; long-run trends in relative, 106–08
Processing enterprises
  Developing countries, 153; market outlets for, 155; problem of raw material supply, 154–55
  Factors leading to success, 158–60
  Farmer's contract problems, 164
  Relationships with growers, 172
  Technology transfer and, 163–64
  Transnational versus national, 162–63
Puetz, Detlev, 343, 352

Ramasamy, C., 5
Raw materials, agricultural
  Price: instability, 112; long-run trend, 107–08
  Suppliers for processing plants, 154–55
  Terms of trade, 107
Reda, K., 160, 163
Rice production. *See also* Gambia, The; West Kenya
  Irrigation schemes: The Gambia, 267, 345, 348, 350; West Kenya, 264, 265, 266–67, 268–69
  Land distribution policy, 348
  Technological change in, 54, 343
  Yield, 350, 352

Rosenzweig, Mark, 146, 243
Ross, J., 219
Rozelle, Scott, 119, 122
Ruigu, G. M., 264, 265
Rwanda, 6, 37
  Agricultural commercialization, 42, 56; and calorie consumption, 66, 73, 287–89, 293, 372; and food expenditures, 66, 71, 72; and health and sanitation, 289, 292; income effects, 47; nutritional effects, 53, 289, 292, 294; staple food production, 48
  Agricultural production: land use shares, by crop, 278–79; national strategy for, 276; subsistence oriented, 284–85, 293
  Potato production: and calorie production, 284, 285; cash requirements, 282; and commercialization, 42, 48, 278, 285; employment effects, 282, 293, 370; expansion, 276, 277; income effects, 286–87, 293; labor input, 282–83, 287; land and labor productivity, 280, 282; major regions for, 276–77; smallholder, 278; for subsistence consumption, 285; technology for, 277, 278, 279, 293
  Preschool children: calorie consumption, 87; commercialization effect on, 85, 97; illness, 82, 87; nutritional status, 84, 85
  Tea cultivation program: farms displaced by, 61, 372; household producers, 61; processing plant, 155
  Women: income effect on household food consumption, 92–93; labor input, 282–83, 287; and nutritional improvement, 288–89

Sadoulet, Elisabeth, 369
Sahn, David, 113, 326
Sanitation: and nutritional status, 15, 289, 292, 294; in study sites, 97, 98
Satiawiharja, B., 178n, 184
Schiff, Maurice, 5, 62, 109, 110
Schmidt, H., 270n
Schultz, Theodore W., 3
Scott, Gregory J., 277n
Shand, R. T., 148

Shively, Gerald, 326
Siandwazi, Catherine, 295
Sierra Leone
　Agricultural production: Bo-Pujehun Rural Development Project, 329, 330, 331; commercialization, 59, 336; role in national economy, 328–29; women's role in food crop, 333
　Contract farming, 168
　Tree crop production: and food expenditures, 336–37; impact on nutrition, 330, 337, 340–42; revenues from, 372; technology for, 330; terms of trade, 371; women's role in, 336
Singh, Inderjit, 22, 23, 24, 240n, 241, 243n, 248
Singleton, N., 219
Siregar, Masdjidin, 176n
Slade, Roger, 244, 248, 249n
Smallholder agriculture
　Cardamom production, 219–20
　Commercialization, 6–7, 189
　Contract farming and, 168–70
　Cooperatives for export crops, 87, 190; food availability, 198; household income and food expenditures, 195–96, 201; income growth, 203; results, 201–02; study of, 191–92; vegetable production, 192–95
　Crop mixes, 242
　Processing enterprises, 153–54
　Subsistence crop production, 62
　Sugarcane production, 204–06, 252–53
Sorghum, Rwanda: land use for, 278, 279; marketed share, 285
Southeast Asia, nutrition improvement from commercialization, 8
Soybeans, Java
　Village-level farming and processing, 176, 177, 178, 179; marketing margin, 184; prices, 182
Specialization: for crops with elastic demand, 368; and economic growth, 3; export crops, 106; gains from, 52; self-sufficiency versus, 5
Squire, Lyn, 22, 23, 24, 240n, 241, 243n
Staple foods production
　Agricultural commercialization and, 48–52; market surplus from, 55–57; profitability and productivity, 52–54

　Cash crop acreage and, 105
　HYV coverage of China's, 131
　Percent of total crop output, 104
　Production of household used, 316
Steenbergen, W. V., 266
Sterkenburg, J. J., 269, 270n, 274
Stigler, George, 242
Stone, Bruce, 119, 122, 124, 126n, 128, 131
Strauss, John, 22, 23, 24, 240n, 241, 243
Strohl, R. J., 161, 164
Subsistence agriculture: causes of high prevalence, 367; constant flow of income from, 28; food market risks and, 20; and food security, 369–70; for home consumption, 3–4; income from, 224, 228, 255; labor productivity, 53; percent of total agricultural output, 218; policies to overcome disadvantages, 365; semi-, 28; tree crops production versus, 49, 332–33
Sudan, sugar processing plant, 155
Sugarcane
　Price variability, 56–57, 112
　Processing, 162
　Production: Kenya, 6, 49, 53, 252–53, 255–56; Philippines, 204–06, 209, 213, 216
Swaziland: contract farming, 170; sugar processing enterprise, 162
Sweet potatoes, Rwanda: land use for, 278, 279; marketing share, 285

Taiwan, contract terms in processing enterprises, 159, 161, 164
Tanzania, contract farming, 171
Taylor, Jeff, 122
Tea
　Processing enterprises, 158, 162, 165
　Production: and commercialization, 42; household, 61, 372; smallholders, 162, 165
Tea Development Authority, Kenya, 158, 162, 165
Technological change
　Agricultural commercialization and, 14–15, 118
　Costs of production response to, 114
　Crop production and; maize hybrid adoption, 295, 297, 299–300; potatoes, 277, 278, 279, 293; rice, 54, 343–46, 348–50; tree crops, 330
　Infrastructure development and, 143, 145

Irrigation schemes, 267, 345, 348, 359
  Risks associated with, 368
Terhal, Piet, 244
Thailand: contract farming, 168; pineapple processing enterprise, 159, 162; poultry contracts for processing, 163
Thomas, Duncan, 242n
Tinker, Irene, 92
Tobacco
  Malawi cultivation, 76, 309–10, 318–20
  Village-level farming and processing in Java, 177, 178, 179, 181; marketing margin, 184; price, 182, value added for home processing, 186
Tosh, J., 266
Trade policies: constraints on, 108, 368; for profitability of crops, 367; world market prices and, 106–07, 112–14
Travers, Lee, 120, 126n
Tree crops
  Effects: on food expenditures, 336–37; on nutritional status, 330, 337, 340–42; revenue, 372
  Subsistence crops versus, 49, 332–33, 336
  Technology, 330
  Terms of trade, 371
Tschajanow, A., 16
Tshibaka, Tshikala B., 62

United Nations Conference on Trade and Development (UNCTAD), 112, 153
U.S. Agency for International Development (USAID), 167
U.S. National Center for Health Statistics, 231

Valdés, Alberto, 5, 62, 109, 110
Vaughan, M., 309
Veenstra, Y., 270n
Vegetable crops, Guatemala smallholder cooperative: for export, 189, 192, 367, 370; food and nutrition effects, 198–201, 203; production and economic effects, 193–96, 201, 203
Venezuela, cassava processing plant, 155

Wages: agricultural commercialization and rates, 55, 58; household earners versus nonearners, 222–23, 237; plantation employment, 220
Wanmali, Sudhir, 296
Waterworth, J., 296n
Webb, Patrick, 344n, 352
West Kenya, irrigation scheme for rice, 264, 265
  Effect on income and food supply, 266–67
  Resident versus nonresident tenants, 268–69, 270; consumption and nutrition differences, 271, 273–75; income differences, 270–71
White, L., 309
Wiebe, K. D., 23
Winarno, F. G., 178n, 184
Women
  Agricultural commercialization effect on, 92, 373–74; health status, 79; nutritional status 29–30, 79; work, 54, 63–64
  Cash-cropping effect on, 92
  Head of household: and calorie intake, 92–93; income control, 94; and nutritional status, 93–94, 96
  Time allocation, 15–16
World Bank, 120, 128, 151, 153, 167, 218, 219, 277, 309

Xu Zhikang, 119

Yang, Maw Cheng, 107
Ye, Qiaolun, 131

Zambia, 6
  Maize: diet staple, 296; technological change in, 296–97
  Maize hybrid adoption, 297, 299–300; and children's nutritional status, 302, 304; and food and calorie consumption, 300–02; and household morbidity, 304–06; women's involvement with, 307
  Male versus female role in agricultural decision-making, 306–08
  Reassessment of agricultural policies, 295
Zeller, Manfred, 375
Zhong Tong, 119
Zimbabwe, contract farming, 170